# Counteracting Urban Heat Island Effects in a Global Climate Change Scenario

Francesco Musco
Editor

# Counteracting Urban Heat Island Effects in a Global Climate Change Scenario

*Editor*
Francesco Musco
Department of Design and Planning in Complex Environments
IUAV University of Venice
Venice, Venezia, Italy

ISBN 978-3-319-10424-9     ISBN 978-3-319-10425-6   (eBook)
DOI 10.1007/978-3-319-10425-6

Library of Congress Control Number: 2016947952

Printed on acid-free paper

This Springer imprint is published by Springer Nature
The registered company is Springer International Publishing AG Switzerland

# Glossary and Abbreviations

This glossary has been compiled using definitions found on European Commission, Intergovernmental Panel on climate Change and other relevant sources, with brief descriptions provided on main topics.

**Rectal temperature** – Tre (°C).
**Tskm** – mean skin temperature (°C).
**Tskfc** – face skin temperature (°C).
**Mskdot** – sweat production (g/min).
**Shiv** – heat generated by shivering (W).
**wettA** – skin wittedness (%) of body area.
**VblSk** – skin blood flow (%) of basal value.
**DTS** – Dynamic Thermal Sensation

**Physiologically Equivalent Temperature** The Physiologically Equivalent Temperature (PET) is the equivalent temperature at a given place (outdoors or indoors) to the air temperature in a typical indoor setting with core and skin temperatures equal to those under the conditions being assessed. Thereby, the heat balance of the human body with a work metabolism 80 W (light activity, added to basic metabolism) and a heat resistance of clothing 0.9 clo) is maintained (Höppe 1999).

**The Universal Thermal Climate Index UTCI** (Jendritzky et al. 2012) is defined as the air temperature (Ta) of the reference condition causing the same model response as the actual condition. Thus, UTCI represents the air temperature, which would produce, under reference conditions, the same thermal strain as in the actual thermal environment.

Both meteorological and non-meteorological (metabolic rate and thermal resistance of clothing) reference conditions were defined:

- wind speed (v) of 0.5 m/s at 10 m height (approximately 0.3 m/s in 1.1 m),
- mean radiant temperature (Tmrt) equal to air temperature,
- vapor pressure (VP) that represent relative humidity of 50 %, at high air temperatures (>29 °C) the reference air humidity is defined as 20 hPa.

representative activity to be that of a person walking with a speed of 4 km/h (1.1 m/s). This provides a metabolic rate of 2.3 MET (135 W/m$^2$).

**CE**: Central Europe
**DSS**: Decision Support System
**LP**: Lead Partner
**WP**: Work Package
**AF**: Application Form
**UHI**: Urban Heat Island
**ACT**: Action
**DBMS**: Database Management Software
**MBMS**: Model Base Management Software
**DGMS**: Dialogue Generation Management Software
**SMS**: Short Message Service
**TN**: Transnational Network
**M&A**: Mitigation and Adaptation

**Combined heat and power (CHP)** – Also known as cogeneration, this is an efficient, clean and reliable approach to generate electricity (power) and thermal energy from a single fuel source. CHP can greatly increase the facility's operational efficiency and decrease energy costs. At the same time, CHP reduces the emission of greenhouse gases, which contribute to global climate change.

**Conference of the Parties (COP)** – All countries that have ratified the United Nations Framework Convention on Climate Change (UNFCCC) are referred to as the Parties. The COP is responsible for implementing the objectives of the Convention and there have been regular meetings since 1995, these are often referred to as the United Nations Climate Conferences.

**Covenant of Mayors** – A program involving the most pioneering cities joining a permanent network to exchange and apply good practices to improve their energy efficiency and promote low-carbon business and economic development. The development of the Covenant of Mayors was supported by the Directorate Energy (DG Energy) of the European Commission (EC).

**Degression rate** – The degression mechanism was chosen in part as a means for gradually eliminating the premium paid to renewables relative to the so-called market price. It was believed at the time this measure was necessary to circumvent the European Union's prohibition against state aid. This "degression" rate varies with technology.

**District heating** – A system for distributing heat generated in a centralised location for residential and commercial heating requirements such as space heating and water heating. The heat is often obtained from a cogeneration plant burning fossil fuels but increasingly biomass, although heat-only boiler stations, geothermal heating and central solar heating are also used, as well as nuclear power. District heating plants can provide higher efficiencies and better pollution control than localized boilers.

**Electricity from renewable energy sources (RES-E)** – Electricity produced from renewable energy sources shall mean electricity produced by plants using only renewable energy sources, as well as the proportion of electricity produced from renewable energy sources in hybrid plants also using conventional energy sources and including renewable electricity used for filling storage systems, and excluding electricity produced as a result of storage systems.

**Emissions inventory** – An itemised list of emission estimates for sources of air pollution in a given area for a specified time period. It can also include infomation on activities that cause emissions and removals, as well as background on the methods used to make the calculations. Policy makers use greenhouse gas inventories to track emission trends, develop strategies and policies and assess progress. Scientists use greenhouse gas inventories as inputs to atmospheric and economic models.

**Energy efficiency** – Measures undertaken as part of Demand-Side Management to reduce the consumption of electricity for a specific task or function.

**Energy Performance Contracting (EPC)** – An innovative financing technique that uses cost savings from reduced energy consumption to repay the cost of installing energy conservation measures.

**European Union (EU)** – Originally a regional economic integration organisation, known as the EEC (European Economic Community), the European Union has grown into a geographical political and economic entity. Also see Member States.

**Feed-in tariff system** – Renewable energy payment as an incentive structure to encourage the adoption of renewable energy through government legislation, with the government regulating the tariff rate. The price per unit of electricity that a utility or supplier has to pay for renewable electricity from private generators is fixed.

**Fischer–Tropsch (FT) process** – A method for the synthesis of hydrocarbons and other aliphatic compounds. Synthesis gas, a mixture of hydrogen and carbon monoxide, is reacted in the presence of an iron or cobalt catalyst; much heat is evolved, and such products as methane, synthetic gasoline and waxes, and alcohols are made, with water or carbon dioxide produced as a by-product. Combination of biomass gasification and Fischer-Tropsch (FT) synthesis is a possible route to produce renewable transportation fuels.

**Fossil fuels** – Also called mineral fuels, these are finite fuels from fossil carbon deposits such as oil, natural gas and coal. When burned to gain energy, greenhouse gases are released during the combustion processes.

**Gasification** – A thermochemical conversion of a solid fuel to a gaseous fuel.

**Gigawatt (GW)** – A unit of power equal to 1 billion watts; 1 million kilowatts, or 1000 MW

**Global warming** – An increase in the average temperature of the Earth's surface. Global warming is one of the consequences of the enhanced greenhouse effect and will cause worldwide changes to climate patterns.

**Global warming potential (GWP)** – The index used to translate the level of emissions of various gases into a common measure in order to compare the relative radiative forcing of different gases without directly calculating the changes in atmospheric concentrations. The International Panel on Climate Change (IPCC) has presented these GWPs and regularly updates them in new assessments (see http://unfccc.int/ghg_data/items/3825.php)

**Greenhouse effect** – The trapping and build-up of heat in the lower atmosphere near a planet's surface. Some of the heat flowing back towards space from the Earth's surface is absorbed by water vapour, carbon dioxide, methane and other gases in the atmosphere. If the atmospheric concentration of these gases rises, then theory predicts that the average temperature of the lower atmosphere will gradually increase.

**Greenhouse gases (GHGs)** – The atmospheric gases responsible for causing global warming and climate change. The major GHGs are carbon dioxide ($CO_2$), methane ($CH_4$) and nitrous oxide ($N_2O$). Less prevalent – but very powerful – greenhouse gases are hydrofluorocarbons (HFCs), perfluorocarbons (PFCs) and sulphur hexafluoride (SF6).

**Gross domestic product (GDP)** – Defined as the measure of the total output of goods and services for final use occurring within the domestic territory of a given country, regardless of the allocation to domestic and foreign claims.

**Heat pumps** – Heat pumps offer the most energy-efficient way to provide heating and cooling in many applications, as they can use renewable heat sources in our surroundings. A typical electrical heat pump will just need 100 kWh of power to turn 200 kWh of freely available environmental or waste heat into 300 kWh of useful heat.

**Intergovernmental Panel on Climate Change (IPCC)** – A scientific intergovernmental body set up by the World Meteorological Organization (WMO) and by the United Nations Environment Programme (UNEP) to provide the decision-makers and others interested in climate change with an objective source of information about climate change. In accordance with its mandate and as reaffirmed in various decisions by the Panel, the IPCC prepares at regular intervals comprehensive Assessment Reports of scientific, technical and socio-economic information relevant for the understanding of human induced climate change, potential impacts of climate change and options for mitigation and adaptation.

**Kilowatt hour (kWh)** is a unit of energy: is the product of power in kilowatts multiplied by time in hours. Energy delivered by electric utilities is usually expressed and charged for in kWh.

**Light-emitting diodes [LED] lighting** – This is a semiconductor diode that emits light when an electric current is applied in the forward direction of the device, as in the simple LED circuit. The effect is a form of electroluminescence where incoherent and narrow- spectrum light is emitted.

**Local Agenda 21 (LA21)** – Local Agenda 21 is a local-government-led, community-wide, and participatory effort to establish a comprehensive action strategy for environmental protection, economic prosperity and community well-being in the local jurisdiction or area.

**Megawatt hours (MW)** is a unit of energy equal to 1 million watt hours.

**Member states** – The EU-27 countries are split into New Member States (NMS) and Old Member States (OMS), based on their date of their accession into the European Union (EU). The OMS are Austria, Belgium, Denmark, Finland,

France, Germany, Greece, Italy, Ireland, Luxembourg, the Netherlands, Portugal, Spain, Sweden and the United Kingdom. The NMS include the Czech Republic, Estonia, Hungary, Latvia, Lithuania, Poland, Slovakia, Slovenia, Malta and Cyprus, with the most recent expansion including Bulgaria and Romania in 2007.

**Methane** – A hydrocarbon that is a greenhouse gas with a high global warming potential (estimated GWP is 24,5). Methane (CH 4) is produced through anaerobic (without oxygen) decomposition of waste in landfills, animal digestion, decomposition of animal wastes, production and distribution of natural gas and oil, coal production and incomplete fossil fuel combustion.

**Metric tonne carbon dioxide equivalent (Mt CO2 e)** – A metric measure used to compare the emissions from various greenhouse gases based upon their global warming potential (GWP). Carbon dioxide equivalents are commonly expressed as 'million metric tonnes of carbon dioxide equivalents (MMTCDE)'. The carbon dioxide equivalent for a gas is derived by multiplying the tonnes of the gas by the associated GWP.

**Public private partnership (PPP)** – A mechanism to use the private sector to deliver outcomes for the public sector, usually on the basis of a long term funding agreement, in a win-win scenario.

**Renewable energy sources (RES)** – Renewable energy is energy generated from natural resources naturally replenished in a short period of time. The renewable sources used most often are: wind, solar, geothermal heat, wave motion, tidal, hydraulic, biomass, landfill gas, treatment process gas and biogas.

**Renewable heating and cooling (RES-H)** – Heating and cooling are necessary elements of any comprehensive strategy to develop renewables and to achieve sustainability in the energy sector. Renewable heating and cooling can significantly contribute to security of energy supply in the EU and reducing CO2 emissions.

**Stern Review (SR)** – The Stern Review on the Economics of Climate Change, the most comprehensive review ever carried out on the economics of climate change, was published on October 30, 2006 and was lead by Lord Stern. The Review set out to provide the report assessing the nature of the economic challenges of climate change and how they can be met, both in the UK and globally.

**Third party financing (TPF)** – This is an appropriate tool for funding of optimization strategies without financial charge to the final user. This is due to budget savings from increased energy efficiency and more appropriate allocation of financial resources made available.

**Terawatt hours (TWh)** is a unit of energy equal to 1 billion kilowatt-hours

**Parts per million (ppm)** – Commonly used as a measure of small levels of pollutants in air, water, body fluids, etc. This is a way of expressing very dilute concentration of substances. One ppm is equivalent to 1 mg of something per liter of water (mg/l) or 1 mg of something per kilogram soil (mg/kg).

**United Nations Framework Convention on Climate Change (UNFCCC)** – An international treaty signed at the Rio Earth Summit in 1992 in which 150 countries promised stabilisation of greenhouse gas concentrations in the atmosphere at a level that would prevent dangerous anthropogenic interference with the climate system. The UNFCCC supports all institutions involved in the climate change process.

**Urban planning** – Design and regulation of the uses of space that focus on the physical form, economic functions, and social impacts of the urban environment and on the location of different activities within it. Urban planning concerns itself with both the development of open land and the revitalization of existing parts of the city, thereby involving goal setting, data collection and analysis, forecasting, design, strategic thinking, and public consultation. The holistic approach of landscape and urban research was stimulated by the introduction of aerial photography. This proved to be a valuable instrument, not only to make thematic inventories and monitor changes, but also to describe holistic aspects of complex landscapes.

**Urban sprawl** – The expansive growth of an uncontrolled or unplanned extension of urban areas into the countryside. Urban sprawl is commonly used to describe physically expanding urban areas. The European Environment Agency (EEA) has described sprawl as the physical pattern of low-density expansion of large urban areas, under market conditions, mainly into the surrounding agricultural areas. Sprawl is the leading edge of urban growth and implies little planning control of land subdivision. Development is patchy, scattered and strung out, with a tendency for discontinuity. It leap-frogs over areas, leaving agricultural enclaves. Sprawling cities are the opposite of compact cities – full of empty spaces that indicate the inefficiencies in development and highlight the consequences of uncontrolled growth (EU 2008).

Department of Design and Planning                                Francesco Musco
in Complex Environments
IUAV University of Venice, Venice, Italy
Corila, Venice, Italy                                          Alessandro Meggiato
climatechange@iuav.it

# Contents

# Editor Bios and Contributors

## Editor Bios

**Francesco Musco** (1973), architect and urban planner, PhD in analysis and governance of sustainable development (Ca' Foscari, Venice), and Associate Professor in urban and regional planning at the Department of Design and Planning in Complex Environment, University Iuav of Venice.

Currently teaches "environmental planning," "territorial design for climate change," and "urban design."

Dean of the EU Erasmus Mundus Master Course on MSP Maritime Spatial Planning (2013–2019) and Director of the Master Program in Planning and

Policies for the City, Environment, and Landscape active at the University Iuav of Venice.

Supporter of a multidisciplinary approach to city and spatial planning, during the last years he finalized his research activity to the relationship between planning and sustainability, with particular attention to the implementation of bottom-up public policies to define sustainable development in local contexts.

He is responsible for international agreements of scientific collaboration with Drexel University (Philadelphia, USA), John Hopkins University (Baltimore, USA), University of Reading (UK), University of Seville (ES), and Future University (Khartum, Sudan).

Scientific coordinator of several projects granted by competitive bids, among these:

- City Action Plans in Climate Adaptation: A Global Comparative Analysis (2011) (Santander Foundation, Madrid)
- Natreg – Developing with Nature (EU South East Europe Program) (2009–2011)
- UHI development and application of mitigation and adaptation strategies and measures for counteracting the global urban heat island phenomenon in climate change scenario (EU Central Europe Program) (2011–2014)

- Sustainable development and new asset of territorial planning in the wetland area Humedal de Mantequilla (2012–2015) (UNDP and Veneto Region)
- ADRIPLAN (Adriatic Ionian Maritime Spatial Planning, 2013–2015) (EU DG Mare)
- URBAN_Wins (2016–2019) H2020 Waste 2015 6-b (EU Horizon 2020)

Department of Design and Planning in Complex Environments
IUAV University of Venice, Venice, Venezia, Italy
e-mail: francesco.musco@iuav.it

## Contributors

**Dominik Aleš** Faculty of Mathematics and Physics, Department of Meteorology and Environment Protection, Charles University Prague (CUNI), Prague, Czech Republic

**Brigitte Allex** Department of Landscape, Spatial and Infrastructure Sciences, Institute of Landscape Planning, University of Natural Resources and Life Sciences, Vienna, Austria

**Györgyi Baranka** Hungarian Meteorological Service, Budapest, Hungary

**Michal Belda** Faculty of Mathematics and Physics, Department of Meteorology and Environment Protection, Charles University Prague (CUNI), Prague (CUNI), Czech Republic

**Anna Błażejczyk** Bioklimatologia. Laboratory of Bioclimatology and Environmental Ergonomics, Warszawa, Poland

**Krzysztof Błażejczyk** Geoecology and Climatology Department, Institute of Geography and Spatial Organization, Polish Academy of Sciences, Warszawa, Poland

**Giovanni Bonafè** ARPA Emilia-Romagna, Bologna, Italy

**Filippo Bonazzi** Territorial Planning and Private Building office, Modena, Italy

**Lucio Botarelli** ARPA Emilia-Romagna, Bologna, Italy

**L. Bozó** Hungarian Meteorological Service, Budapest, Hungary

**Christiane Brandenburg** Department of Landscape, Spatial and Infrastructure Sciences, Institute of Landscape Planning, University of Natural Resources and Life Sciences, Vienna, Austria

**Filippo Busato** Department of Management and Engineering (DTG), University of Padova, Vicenza (VI), Italy

**Marcello Capucci** Urban Planning Department, Modena, Italy

**Svetlana Čermelj** Department of Environmental Protection, City of Ljubljana, Ljubljana, Slovenia

**Rok Ciglič** Anton Melik Geographical Institute, Research Centre of the Slovenian Academy of Sciences and Arts, Ljubljana, Slovenia

Geographical Institute ZRC SAZU, Ljubljana, Slovenia

**Claudia Dall'Olio** Territorial Planning and Mountain Development Service, Emilia Romagna Region, Italy

**Doris Damyanovic** Department of Landscape, Spatial and Infrastructure Sciences, Institute of Landscape Planning, University of Natural Resources and Life Sciences, Vienna, Austria

**Wojciech Dudek** Nofer Institute of Occupational Medicine, Łódź, Poland

**Stefan Emeis** Institute of Meteorology and Climate Research (IMK-IFU) of the Karlsruhe, Institute of Technology (KIT), Karlsruhe, Germany

**Joachim Fallmann** UK Met Office, Exeter

**Davide Fava** Democenter-Sipe Foundation and Emilia Romagna Region, Modena, Italy

**Davide Ferro** Department of Design and Planning in Complex Environments, IUAV University of Venice, Venice, Italy

**Federica Fiumi** Territorial Planning and Mountain Development Service, Emilia Romagna Region, Italy

**Jan Flegl** Prague Institute of Planning and Development, Prague, Czech Republic

**Giuliana Fornaciari** Department of Design and Planning in Complex Environments, IUAV University of Venice, Venice, Italy

**Laura Fregolent** Department of Design and Planning in Complex Environments, IUAV University of Venice, Venice, Italy

**Vladimír Fuka** Faculty of Mathematics and Physics, Department of Atmospheric Physics, Charles University Prague (CUNI), Prague, Czech Republic

**Birgit Gantner** Department of Landscape, Spatial and Infrastructure Sciences, Institute of Landscape Planning, University of Natural Resources and Life Sciences, Vienna, Austria

**Lodovico Gherardi** Territorial Cooperation Contact Point, Emilia Romagna Region, Italy

**Graziella Guaragno** Territorial Planning and Mountain Development Service, Emilia Romagna Region, Italy

**Tomáš Halenka** Faculty of Mathematics and Physics, Department of Atmospheric Physics, Charles University Prague (CUNI), Prague, Czech Republic

**Peter Huszár** Faculty of Mathematics and Physics, Department of Atmospheric Physics, Charles University Prague (CUNI), Prague, Czech Republic

**Radek Jareš** Prague Institute of Planning and Development, Prague, Czech Republic

**Rainer Kapp** Section of Urban Climatology, Office for Environmental Protection Municipality, Stuttgart, Germany

**Mária Kazmuková** Prague Institute of Planning and Development, Prague, Czech Republic

**Christine Ketterer** Albert-Ludwigs-University of Freiburg, Freiburg

iMA Richter & Roeckle, Freiburg, Germany

**Kristina Kiesel** Department of Building Physics and Building Ecology, Vienna University of Technology, Vienna, Austria

**O. Kocsis** Hungarian Urban Knowledge Centre Non-Profit Ltd, Budapest, Hungary

**Žiga Kokalj** Institute of Anthropological and Spatial Studies ZRC SAZU, Ljubljana, Slovenia

**Blaž Komac** Anton Melik Geographical Institute, Research Centre of the Slovenian Academy of Sciences and Arts, Ljubljana, Slovenia

Geographical Institute ZRC SAZU, Ljubljana, Slovenia

**Mária Kovács** Department of Climatology and Landscape Ecology, University of Szeged, Szeged, Hungary

**Beata Kręcisz** Nofer Institute of Occupational Medicine, Łódź, Poland

**Ilona Krüzselyi** Hungarian Meteorological Service, Budapest, Hungary

**Magdalena Kuchcik** Geoecology and Climatology Department, Institute of Geography and Spatial Organization, Polish Academy of Sciences, Warszawa, Poland

**Paolo Lauriola** ARPA Emilia-Romagna, Bologna, Italy

**Renato Lazzarin** Department of Management and Engineering (DTG), University of Padova, Vicenza (VI), Italy

**Alenka Loose** Energy Manager of the City of Ljubljana, Ljubljana, Slovenia

**Filippo Magni** Department of Design and Planning in Complex Environments, IUAV University of Venice, Venice, Italy

**Ardeshir Mahdavi** Department of Building Physics and Building Ecology, Vienna University of Technology, Vienna, Austria

**Denis Maragno** Department of Design and Planning in Complex Environments, IUAV University of Venice, Venice, Italy

**Stefano Marchesi** ARPA Emilia-Romagna, Bologna, Italy

**Letizia Martinelli** Chair of Environmental Meteorology, Albert-Ludwigs-University, Freiburg, Germany

Department of Planning, Design and Technology of Architecture, Sapienza University, Rome, Italy

**Davide Martinucci** Department of Design and Planning in Complex Environments, IUAV University of Venice, Venice, Italy

**Andreas Matzarakis** Albert-Ludwigs-University of Freiburg, Freiburg, Germany

Research Center Human Biometeorology, German Meteorological Service, Freiburg, Germany

**Paweł Milewski** Geoecology and Climatology Department, Institute of Geography and Spatial Organization, Polish Academy of Sciences, Warszawa, Poland

**Ulrich Morawetz** Department of Economics and Social Sciences, Institute for Sustainable Economic Development, University of Natural Resources and Life Sciences, Vienna, Austria

**Francesco Musco** Department of Design and Planning in Complex Environments, IUAV University of Venice, Venice, Italy

**Marco Noro** Department of Management and Engineering (DTG), University of Padova, Vicenza (VI), Italy

**R. Ongjerth** Hungarian Urban Knowledge Centre Non-Profit Ltd, Budapest, Hungary

**Krištof Oštir** Institute of Anthropological and Spatial Studies ZRC SAZU, Ljubljana, Slovenia

**Cezary Pałczyński** Nofer Institute of Occupational Medicine, Łódź, Poland

**Miha Pavšek** Anton Melik Geographical Institute ZRC SAZU, Ljubljana, Slovenia

**Jürgen Preiss** Vienna Environmental Protection Department, Municipal Department 22, Unit of Spatial Development, Vienna, Austria

**Florian Reinwald** Department of Landscape, Spatial and Infrastructure Sciences, Institute of Landscape Planning, University of Natural Resources and Life Sciences, Vienna, Austria

**Jaroslav Ressler** Institute of Computer Science, The Czech Academy of Sciences, Prague, Czech Republic

**Ulrich Reuter** Section of Urban Climatology, Office for Environmental Protection Municipality, Stuttgart, Germany

**Rayk Rinke**  Section of Urban Climatology, Office for Environmental Protection Municipality, Stuttgart, Germany

**Catia Rizzo**  Urban Design and Transformation office, Modena, Italy

**Michele Zanelli**  Urban Quality and Residencial Policy Department, Emilia Romagna Region, Italy

**Stefano Zauli Sajani**  ARPA Emilia-Romagna, Bologna, Italy

**Petr Skalák**  Department of Climatology, Czech Hydrometeorological Institute, Prague, Czech Republic

**F. Szkordikisz**  Hungarian Urban Knowledge Centre Non-Profit Ltd, Budapest, Hungary

**Jakub Szmyd**  Geoecology and Climatology Department, Institute of Geography and Spatial Organization, Polish Academy of Sciences, Warszawa, Poland

**Rodica Tomozeiu**  ARPA Emilia-Romagna, Bologna, Italy

**Maja Topole**  Anton Melik geographical Institute ZRC SAZU, Ljubljana, Slovenia

**Milena Vuckovic**  Department of Building Physics and Building Ecology, Vienna University of Technology, Vienna, Austria

**Sven Wagner**  Institute of Meteorology and Climate Research (IMK-IFU) of the Karlsruhe, Institute of Technology (KIT), Karlsruhe, Germany

**Pavel Zahradníček**  Department of Climatology, Czech Hydrometeorological Institute, Prague, Czech Republic

**Michal Žák**  Department of Climatology, Czech Hydrometeorological Institute, Prague, Czech Republic

Faculty of Mathematics and Physics, Department of Atmospheric Physics, Charles University Prague (CUNI), Prague, Czech Republic

**Ondřej Zemánek**  Prague Institute of Planning and Development, Prague, Czech Republic

**Gabriella Zsebeházi**  Hungarian Meteorological Service, Budapest, Hungary

# List of Figures

# List of Tables

# Planning and Climate Change: Concepts, Approaches, Design

Francesco Musco, Federica Appiotti, Irene Bianchi,
Michele Dalla Fontana, Elena Gissi, Giulia Lucertini,
Filippo Magni, and Denis Maragno

**Abstract** Reflect on the present, on the dynamics and the conditions that built it, and look forward at the same time, in search of a prospect to improve the future. Since Howard (1850–1928) and Geddes (1854–1932), this has been the dominant logic supporting the work of all those (architects, urban planners, planners, landscape architects, etc.) who grappled with city and territorial management and planning. However, from the 1970s, territorial planning has been confronted with new concepts – such as sustainable development, environmental sustainability and social equity – and more recently, new challenges – such as the ones linked to climate change, which led to the need to redefine territorial planning in disciplinary and operational terms. For some years now, the planner's new role is under discussion, especially in relation to the challenges posed by climate change. Sustainability, mitigation, adaptation, renewable energy, low-carbon transition, ecosystem approach and post-disaster planning are just some of the new keywords surrounding the discussion on territorial management and planning. This chapter aims to present rationally, what it means to re-organize and re-think the city, in a long-term perspective. It wants to show how it is possible, and above all is a duty to integrate the new concepts mentioned above in urban planning, to deal with the effects of climate change. The Urban Heat Islands contrast enters fully into the feasible experimentation with appropriate innovations in territorial planning. The paper draws attention to the Italian situation, in the light of the European reference framework.

**Keywords** Climate change • Adaptation • Policy • Urban heat islands • Urban planning & design

F. Musco • F. Appiotti • I. Bianchi • M.D. Fontana • E. Gissi • G. Lucertini
F. Magni • D. Maragno
Department of Design and Planning in Complex Environments,
IUAV University of Venice, S.Croce, 1957, 30135 Venice, Italy

# Introduction

Climate change has undoubtedly emerged as a crucial issue since the beginning of the twenty-first century. According to IPCC predictions, the phenomena associated with climate variability will intensify in the coming decades (2007), and climate-related extreme events will constitute an increasing risk on a social and ecological level (2012). Over the past 20 years, the need to address the dynamics of climate change on an urban scale has been recognized at the institutional, academic and operational levels.

In this context, the challenges posed by the changing climate scenario require a redefinition of the urban and territorial planner's role, as well as revising the planner's skills and planning tools. In fact traditionally, planning has been based on the assumption that human activities are planned and implemented in an "unchanging" context, characterized by stable regional and environmental conditions. The compressed environmental dynamics set in motion by environmental change and – more generally – the social, economic and environmental impacts related to climatic phenomena that occur in urban settings, even under emergency conditions, require the adoption of a new perspective and new tools, able to increase the adaptive capacity of cities compared to changes to the city, which are partially generated by the cities themselves.

The relationship between climate change and cities is rather complex and some of the challenges that planners will have to face, especially in terms of mitigation and adaptation, can be identified with the effects of climate change. The contrast with the Urban Heat Islands (UHI) is one of the most obvious, intensified by global warming, which in the coming years will also have to be addressed structurally by urban and territorial planning.

## Climate Change and the City: A Complex Relationship

### From Sustainability to Climate Change: Towards a New Approach

With respect to when cities and territories were built, conditions are changing radically. Urban planning, as a discipline, was developed in the late nineteenth and early twentieth century, mainly as a response to the crisis of the times, related to hygienic needs, clean water, decent housing, open spaces, efficient transport systems and social welfare. During the twentieth century however, urban planning expanded to meet the emerging challenges of environmental protection, sustainable urban development and international cooperation (Wheeler 2010).

To speak today of sustainability, in planning or in relation to territorial dynamics, is not easy and requires attention. A first element to consider is the lack of consensus that exists with respect to the concept of "sustainable city". A second consideration is related to the perception of sustainability, often viewed as an "abstract" goal, whose implementation is beset with difficulties. In addition, the perception of the city's decline has encouraged the integration of urban planning, economy and ecology moving more and more towards an understanding of social, political and environmental sustainability disciplines (Musco 2008).

If with the signing of the New Aalborg Charter in 2004, European local governments made specific commitments ranging from urban planning to new ways of life, from the economy to urban upgrading, it is with the Leipzig Charter on Sustainable Cities and the EU's Territorial Agenda (2007) that the strategies and principles for sustainable urban development policies in Europe were defined.

Nowadays, cities are facing a new crisis, which therefore requires a new perception of all the principles related to sustainability. Climate change goes beyond any previous human challenge, as it requires an integrated and dynamic approach.

Currently, the international scientific community recognizes climate change as a major challenge for the development and sustainability of the twenty-first century (UNDP 2005, 2010; OECD 2009; World Bank 2012; UN-Habitat 2011a, b), for the revitalization of urban areas, and it recognizes two main aspects: (i) the difficulty of reaching a shared consensus for the reduction of greenhouse gas emissions (GHG) in international negotiations and (ii) the growing international consensus on the urgent need to build strategies to adapt to climate change on a national, regional and local level (Musco and Magni 2014).

For this reason, during the last decade, urban areas have become central to the international debate on climate issues. The new geography of contemporary urbanization in fact identifies urban areas as a key element in the processes of globalization and transition to new land occupation models worldwide (Seto et al. 2010). Therefore, today as in the past, if the task of planning is to reduce the risks and negative externalities and help provide answers to the concerns and aspirations that people express with respect to their living environment, it is necessary to step back and critically reflect on the concepts that underlie the planning and reformulate them in the light of new urban scenarios.

## Mitigation and Adaptation in the European Agenda

The debate on climate change, supported by empirical evidence brought by the Stern Review (Carraro 2009), followed by regular reports from the IPCC (2007, 2013), the EU report on temperature increases and the EEA's (2012) report on "Urban adaptation to climate change" in Europe, has become increasingly important within the urban issues. Climate protection can be generally defined as a set of indirect policies for adaptation and mitigation aimed at reducing the impact of climate change on natural and anthropized systems to the reduction of environmental externalities that may favour the climate changes in the medium and long term (Musco 2009). This combined approach of policies to mitigate and adapt acquires a strategic value, since it allows different management levels, multiple policy areas and a number of actors to be held together, both in terms of top-down and bottom-up.

State of the art "climate protection planning" in Europe is far from consistent. Each country is characterized by a national indication (national mitigation and/or adaptation plans and strategies), and the presence of local initiatives in terms of climate plans and local authority tools or networks. The latter's status varies widely from case to case and only a few local authorities have introduced adaptation, mitigation and energy efficiency strategies in the existing territorial planning systems.

Although a growing part of the scientific community (Betsill and Bulkeley 2006; Biesbroek et al. 2009; Musco 2010), together with international institutions' research and policies (IPCC; EEA; EU White Paper, EC), recognizes the role that territorial planning can play in addressing both the causes and consequences of climate change, the explicit translation of CC-problems into territorial policy measures and actual management is far from being reached.

In 2006 the publication of the Green Paper on Energy, "An European Strategy for Sustainable, Competitive and Secure Energy" raised the issue of energy efficiency and exploitation of renewable energy sources. This tool was followed in 2007 by the proposal of an action plan for energy efficiency (2007–2012) and a SET Plan (Strategic Energy Technology Plan). With the so-called Climate and Energy package, the EU has finally set a solid and binding goal for the member countries: 20 % reduction in their greenhouse gas emissions (measured in $CO_2$ equivalent) by 2020 compared to 1990 levels, reduction in energy consumption by 20 % compared to a "business as usual" scenario and production of energy from renewable sources accounting for 20 % of final energy consumption. 2020 is not however a suitable timeframe for the resolution of problems related to the impacts of climate change. For this reason, the European Commission has already begun to explore the different scenarios ahead for post-2020. With the communication of 8th March 2011 ("A Roadmap for moving to a competitive low carbon economy in 2050"), the Commission states that this transition goes through stages involving a reduction of greenhouse gas emissions by 25 % by 2020, 40 % by 2030, 60 % by 2040 and 80 % by 2050 compared with 1990, thus surpassing the target set by the same package.

Although the implementation of policies and action plans is highly dependent on the national context and the various modes of urban governance, there are an increasing amount of experiences, programmes and projects that connect directly the local level, for the European Community, to the creation of new networks (Covenant of Mayors, GRaBS) or are based on already existing relations (Agenda 21, ICLEI, C40).

On this basis, local, regional and sometimes national authorities have begun to define, in many cases on an experimental basis, a series of plans aimed at protecting the areas from the effects of climate change.

# Towards Urban Adaptation

Adapting to CC can be considered a "new" theme on the planning stage. The need to address the CC from a point of view of adaptation and not just of mitigation represents a substantial leap in scale, from a global logic for mitigation, to an urban and strongly localized one for adaptation. Adaptation is an urban and local issue, since it is very specifically the cities and the people that must find their "way" to adapt to the effects of CC that impact them and there are no appropriate policies and adaptive measures that are suitable to be applied anytime and in all contexts. Adaptation is a complex mechanism that is based primarily on the geomorphologic specificities of the place and the local community that lives in it with its customs and traditions, but the economy, infrastructure and flows that characterize it must necessarily also be taken into account. Adaptation is therefore primarily a spatial, territorial concept, which cannot forcefully enter as a new standard in the elaboration of the theories and tools of the plan and the project of urban and territorial planning.

The need to face CC at an urban scale can be attributed to diverse considerations, that should be addressed in an integrative way. First, as partially highlighted in the previous section, CC became an issue in urban agendas in response to the necessity to face urban vulnerability, defined as "the degree to which people, places, institutions and sectors are susceptible to, and unable to cope with, climate change impacts and hazards" (UN-HABITAT 2014). The higher vulnerability of urban contexts can be attributed to a series of factors, such as "their heavy reliance on interconnected networked infrastructure, high population density, large numbers of poor and elderly people and major concentration of material and cultural asserts" (Carter et al. 2015: 4, see also EEA 2010). With this respect, a further consideration concerns climate change and risk perception: also due to the factors mentioned above, the impacts of climate change are mainly experienced at a urban and local scale. Secondly, the emergence of urban CC issues is related to the need to limit the urban drivers that cause pollution. Currently, cities are the main producers of greenhouse gases, and this incidence will steadily increase with the growing urbanization trend (UN 2008).

At the conceptual level, adaptation would adopt an integrated theoretical framework capable of integrating Climate Change Adaptation (CCA) and Disaster Risk Reduction (DRR) also considering their relevance for urban planning.

## A New Role for Planning

The marginalization of territorial and urban planning in recent years has become an objective and consolidated fact. The reasons for this have been identified in the inability to understand how the city and the territory in general were changing (yesterday and today), in the progressive loss of a complex design idea in which space and society, physical and socio-economical dimensions, general concepts and specific action plans, interactions between scales and times interact constantly

(Gasparrini 2015; La Cecla 2015; Benevolo 2012). Considering the above, planning can and must (re-) play an important role by sharing the challenges established by CC, by ecological issues, the geo-strategic and environmental re-appropriation of our territories and our cities. The spread of environmental issues and CC can reshape planning discipline by focusing on water, soil, energy, waste, accessibility/mobility, but also on concepts such as blue and green infrastructures, recovery and regeneration of marginal areas (*vague terrains*), the densely populated and widespread city. In addition, the issues of recovery and regeneration through environmental and ecological networks are closely linked to security (ANCE/CRESME 2012), which opens a new and important line of research and design on "post-disaster planning". The many risks, as well as their dynamic and cumulative interaction, require planning strategies guided by adaptive logic in order to rethink the space we live in structurally and not limit ourselves to making buildings "safe".

What territorial and urban planning must do is be more attentive to the physical and social realities of the places, going further than just looking at the individual events and embracing the extreme complexity of each territory and city. Planning must be more attentive to the spatial project to recognize the peculiarities and opportunities and to ensure not only quality urban landscapes, but also externalities and interdependencies that only efficient and safe cities and territories can provide (Gasparrini 2015). The great environmental and spatial challenges posed by CC require visions and relations on a super-local and a place-specific scale at the same time: a continuous multi-scale attitude that links resilience and recovery tactics and strategies. It seems obvious that all these issues require a rethinking of the shape and use of the territory and the city through the integrated enhancement of environmental components, to counter the effects of the CC and at the same time to rethink the contemporary city by looking for a sustainable balance.

## *New Concepts*

Adaptation to climate change, broadly defined by the IPCC in 2007, and subsequently analysed in its various meanings in a lot of literature can be divided into different types: (i) anticipatory, (ii) autonomous and (iii) planned. These three different aspects of the concept and adaptation strategies support a number of new slogans and tools that fill the discussion on territorial management and planning. If in recent decades, the concept of "sustainability" has become a key element of territorial urban development, and "adaptation" aims at laying the foundations for durability through specific strategies, measures and actions. Given the difficulty in predicting the change of climatic parameters on different scales and different natural and anthropic components, adaptation strategies must be regulated by seeking not just to ensure the system's functionality but also to take advantage of opportunities that may arise from the change. For this reason, in recent years, headway is being made in the idea of using an "ecosystem approach" (Grumbine 1994; Christensen et al. 1996 Millennium Ecosystem Assessment, 2005) to mitigate and adapt to climate change and its effects (Doswald and Osti 2011; Naumann et al. 2011).

The ecosystem approach concept is a way of thinking and acting in a science-based, ecological way, integrating the biological, social and economic conditions to achieve a socially and scientifically acceptable balance between the priorities of nature conservation, the use of resources and the division of benefits (sustainability). This approach attempts to remove the barriers between human economy, social aspirations and the natural environment, placing humans within the ecosystem models and aspiring to maintain the ecosystem's natural structures and functions, taking into consideration the emerging properties from the interaction of these systems. Given the holistic view, which sees man as an integral part of the natural system, and the aspiration to integrate policies and measures that affect the system, the use of this approach is proving to be a promising strategy to increase resilience of the cities and territories in response to growing pressures. In this perspective, the use of renewable energy sources and low-carbon transition does not just take on a role in mitigative strategies for reducing $CO_2$ emissions, but become key tools in adaptation strategies that follow an ecosystem approach. The measures and actions that are being taken at a territorial level from an adaptive viewpoint following an ecosystem approach are manifold. Examples of these are the planning and use of blue and green infrastructures, river corridors, overflow basins for storage of rain and river water, containment tanks for the management of river floods, becoming more frequent due to the change in extreme rainfall patterns, living roofs and reconstruction of ecological corridors.

In this perspective, a reflection should be done about the importance of defining adaptation plans totally integrated with mitigation strategies, as well as about the urgent need to provide cities with management and planning strategies to be adopted after extreme climate events (such as draughts, floods and urban heat waves).

In fact if mitigations reduce the causes affecting climate, adaptation plans are aimed at reducing the future vulnerability on cities and built environment, thus at anticipating the adverse effects of climate change and at reducing potential damages deriving from it. At the same, if potential impacts are not more avoidable, *post-disaster planning* and management seek to define long-term recovery strategies, and ultimately to transform cities in more sustainable and resilient places, also through the direct involvement of local communities.

Re-shaping cities in ways that enable to enhance their adaptive capacity does not mean to bring them back to the way in which they were before the change and/or the disaster, nor to modify their deepest nature and raison d'être. Redefining urban patterns in this contexts means to take the opportunities that are hidden behind the change and use them to rethink a more secure, sustainable and resilient future. For the development and implementation of adaptation and mitigation including post-disaster recovery strategies, cities must be considered in their complexity, and all their dimensions (spatial, geographical, environmental, social, economic and cultural) must be addressed.

The adoption of measures and adaptation actions should not, however, be a short- and medium-term response to the negative effects of climate change, but become part of a routine planning that recognizes in dynamic, changeable and resilient nature, a model to follow.

Controlling the effects of Urban Heat Islands is fully embedded in a new resilient planning aimed at reducing the impact of temperature change.

## Conclusions: Building Urban Adaptation – The Main Role of Planners

The dynamics of climate change require a thorough review, not only of the approaches but also, at the same time, of the Territorial Governance tools. Operating within a Climate Proof scenario, territorial planning will have to be able to identify territorial vulnerabilities and implement effective measures designed around the territorial characteristics of the vulnerable area. The local effectiveness of the adaptation action identified is not just attributable to its design but also to the forms under which it is implemented.

Planning on all scales has so far only partially considered regulation of the relationship between climate, urban vulnerability and territorial planning, leaving room for activities and/or projects of a voluntary nature. The growing attention to these processes, however, has not yet led to suitable policy responses. It is more than ever evident that "climate protection" presents rather disjointed situations with cases in which adaptation plans and strategies have been introduced, and, on the other hand, realities where the risks and impacts are still undervalued despite the relevance of the phenomena in progress. In most Italian urban contexts, the impact of a changing climate is still just relegated to the civil protection. The main reasons can be traced to a shared lack of public awareness on climate variability and its territorial impact, to a slow response to extreme weather conditions due to lack of preparation and resources and a lack of public policies and regulations relating to urban and environmental planning designed to manage climate change.

It seems evident that adaptation, although by its nature being developed locally, needs to be supported by processes to integrate the different project and planning scales closely related to mitigation policies and efficiency of the urban scale.

## References

ANCE/CRESME. (2012, October). Lo stato del territorio Italiano 2012. Insediamento e rischio sismico e idrogeologico. Primo rapporto ANCE/CRESME.

Bart, I. (2011). Municipal emission trading: Reducing transport emission through cap and trade. *Climate Policy, 11*, 813–828.

Benevolo, L. (2012). Il tracollo dell'urbanistica italiana, Edizioni Laterza.

Betsill, M., & Bulkeley, H. (2006). Cities and multilevel governance of global climate change, In *Global governance: A review of multilateralism and international organizations* (Vol. 12, N° 2). Boulder: Lynne Rienner Publishers.

Biesbroek, G. R., Swart, R. J., & Van der Knaap, W. (2009). The mitigation-adaptation dichotomy and the role of spatial planning. *Habitat International, 33,* 230–237.

Carraro, C. (2009). La Stern Review: Tra Scienza e politica dei cambiamenti climatici. In N. Stern (Ed.), *Clima è vera emergenza.* Milano: Francesco Brioschi editore.

Carter, J. G., Cavan, G., Connelly, A., Guy, S., Handley, J., & Kazmierczak, A. (2015). Climate change and the city: Building capacity for urban adaptation, *Progress in Planning, 95,* 1–66.

Christensen, N. L., Bartuska, A., Brown, J. H., Carpenter, S., D'Antonio, C., Francis, R., Franklin, J. F., MacMahon, J. A., Noss, R. F., Parsons, D. J., Peterson, C. H., Turner, M. G., & Moodmansee, R. G. (1996). The report of the Ecological Society of America Committee on the scientific basis for ecosystem management. *Ecological Applications, 6,* 665–691.

Doswald, N. & Osti, M. (2011). *Ecosystem-based adaptation and mitigation: Good practice examples and lessons learnt in Europe.* BfN Skripten. Available online at: https://www.bfn.de/fileadmin/MDB/documents/service/Skript_306.pdf

EEA. (2012). *Urban adaptation to climate change in Europe: Challenges and opportunities for cities together with supportive national and European policies.* Copenhagen: European Environment Agency.

Gasparrini, C. (2015). *In the city on the cities. Sulla citta nelle città.* Trento: LIStLab.

Grumbine, R. E. (1994). What is ecosystem management? *Conservation Biology, 8*(1), 27–38.

IPCC. (2007). *Climate change 2007: Mitigation on climate change. Contribution of working group III to the fourth assessment report of the intergovernmental panel on climate change.* Cambridge University Press

IPCC. (2007). Climate change 2007. AR4 synthesis report: Contribution of Working groups I, II and III of the IPCC. Cambridge: Cambridge University Press.

IPCC. (2007). Fourth assessment report: Climate change. Geneva.

IPCC. (2007). *Summary for policymakers di climate change 2007: Impact, adaptation and vulnerability.* Contribution of working group II to the fourth assessment report of the intergovernmental panel on climate change. Cambridge University Press.

IPCC. (2012). Managing the risk of extreme events and disasters to advance climate change adaptation. A special report of working groups I and II of the IPCC. Cambridge: Cambridge University Press.

Krause, R. M. (2012). An assessment of the impact that participation in local climate networks has on cities' implementation of climate, energy and transportation policies. *Review of Policy Research, 29,* 585–603.

La Cecla, F. (2015). *Contro l'urbanistica.* Einaudi.

Millennium Ecosystem Assessment. (2005). *Ecosystems and human well-being: Current state and trends.* Washington, DC: Island Press.

Musco, F. (2008). Cambiamenti Climatici, Politiche di Adattamento e Mitigazione: una Prospettiva Urbana. *ASUR,* N. 93.

Musco, F. (2010). Policy design for sustainable integrated planning: From local agenda 21 to climate protection. In Van Staden, & F. Musco (Eds.), *Local goverments & climate change.* New York: Springer.

Musco, F., & Magni, F. (2014). Mitigazione ed adattamento: le sfide poste alla pianificazione del territorio. In F. Musco, & L. Fregolent (a cura di), *Pianificazione urbanistica e clima urbano. Manuale per la riduzione dei fenomeni di isola di calore urbano,* Il Poligrafo, Padova (forecoming)

Naumann, S., Anzaldua, G., Berry, P., Burch, S., McKenna, D., Frelih-Larsen, A., Holger, G., & Sanders, M. (2011), *Assessment of the potential of ecosystem-based approaches to climate change adaptation and mitigation in Europe. Final report to the European Commission,* DG Environment, Contract no. 070307/2010/580412/SER/B2, Ecologic institute and Environmental Change Institute, Oxford University Centre for the Environment

Seto, K., Sanchez Rodriguez, R., & Fragkias, M. (2010). The new geography of contemporary urbanization and the environment, In *Annual review of environment and resources* (Vol. 35). Palo Alto, California.

UN Habitat. (2011a). *Planning for climate change. A strategic values based approach for urban planners*. Nairobi.

UN Habitat. (2011b). *Global report on human settlements 2011: Cities and climate change*. Nairobi.

UN-Habitat. (2014). Planning for climate change: A strategic, values-based approach for urban Planners. UN-Habitat: Cities and Climate Change Series.

UNDP. (2005). *Adaptation policy framework for climate change: Developing strategies, policies and measures*. New York: United Nations.

UNDP. (2010). *Designing climate change adaptation initiatives. A UNDP toolkit for practitioners*. New York: United Nations.

Wheeler, S. (2010). A new conception of planning in the era of climate change. *Berkeley Planning Journal, 23*(1), 19–16

World Bank. (2012). *Building urban resilience: Principles, tools and practice*, Washington, DC, (2011).

Zahran, S., Grover, H., Brody, S. D., & Vedlitz, A. (2008). Risk, stress, and capacity: Explaining metropolitan commitment to climate protection. *Urban Affairs Review, 43*, 447–474.

# Introduction

**Paolo Lauriola**

**Abstract** Urban heat island (UHI) is micro-climatic phenomenon which occurs within urban areas and consists of generally warmer temperature than rural sur-roundings. The current development of cities together with FORESEEN urban DEVELOPMENT makes this phenomenon of fundamental importance also for stakeholders and urban planners.

The attention to these items is proved by the funding of the project "Development and application of mitigation and adaptation strategies and measures for counteracting the global Urban Heat Islands phenomenon".

The main objective of this project is to establish a trans-national attention for the prevention, adaptation and mitigation of the natural and anthropogenic risk arising from the urban heat island phenomenon. The partnership is basically twofold, with technical institutions as well as local stakeholders, both coming from each of the regions involved in the project.

The direct participation to the project of local stakeholders guarantees the possibility of an effective impact of UHI project objectives into planning strategies, with a specific emphasis on the human bio-meteorological factors that are relevant for the urban planning process.

**Keywords** Micro-climate • Trans-national    policies • Counteracting    UHI • European Territorial Cooperation

P. Lauriola
Regional Agency for Environmental Protection (ARPA),
Emilia Romagna Region, Via Begarelli 13, 41100, Modena, Italy
e-mail: plauriola@arpa.emr.it

# Introduction

The term "heat island" describes a micro-climatic phenomenon that occurs in urban environments. It consists of a relevant increase of air temperature within urban areas which are thus generally warmer than the surrounding rural neighbourhoods. Usually, the temperature difference is more relevant during the night than during the day and it is most apparent when winds are weak.

At seasonal level, urban heat island phenomenon occurs in winter as well as in summer, when it is more severe for the population living in the urban environment. The UHI threat to human health within cities have to be carefully tackled: in fact, the high summer temperatures heavily affect the quality of life in the cities producing a lot of negative impacts that may be summarized as a relevant deterioration of human health with bioclimatic discomfort, as well as an increase in energy consumption (for example, because of the need of air conditioning) which in turn determines higher emission of air pollutants and greenhouse gases. These aspects make urban agglomerations increasingly vulnerable to climate change. Although UHI phenomenon is not a direct consequence of climate change, it is expected to exacerbate due to the predicted overall warming in the framework of climate change scenario for the second half of the century, when mean temperature are generally likely to increase.

UHI phenomenon has been shown to be directly linked with the size of urban areas and with population living inside. In this respect, it is likely to become more severe in the forthcomings years due to the constant growing of the number of people living in urban areas. In fact, global population is increasingly concentrating in cities: since 2007 more than a half of the human beings is living inside urban areas. In Europe, about 75 % of people is currently living in urban areas, with a projected increase up to 80 % in 2020. As for the whole globe, in 2050 the share of the urban population will reach almost 70 % of the total, implying that about 6.3 billion people will live in urban areas.

There are a number of reasons why UHI phenomenon affects urban areas. The main cause is related to the physical characteristics of materials composing urban surfaces absorbing rather than reflecting solar radiation (concrete and asphalt among the others) so that the surplus heating of the surfaces determines the emission of a large amount of long-wave radiation, especially during the night. In addition, urban areas generally contains small portions of natural surfaces (vegetation in parks, gardens, etc.) which could contribute to maintain a stable energy balance. A further factor related to the increase of temperature is the waste heat generated by energy consumption (heating and cooling plants, industrial activities, transports, etc.).

Our cities and urban areas in general are facing many challenges – economic, social, health and environmental. However, the proximity of people, business and services associated with cities also creates opportunities to improve resource efficiency. Indeed, well-designed and well-managed urban settings offer great opportunities for sustainable living; partnership and coordination from the local to the European level can support their improvement. Climate change has the potential to influence all the components in the urban environment and to raise new and complex challenges for the quality of life, health and human biodiversity inside urban areas. Poor urban design can worsen the impacts of climate change.

The attention of the European Union to the problems related to UHI is proved by the funding of the transnational cooperation project within the Central Europe Programme "Development and application of mitigation and adaptation strategies and measures for counteracting the global Urban Heat Islands phenomenon" (3CE292P3). This project is coordinated by the Regional Agency for Environmental Protection in Emilia-Romagna, Italy, and involves 17 partners within Central Europe area. The partnership is basically twofold, since it is characterized by the presence of technical institutions, as well as local stakeholders, both coming from each of the regions involved. Partners are listed in the following:

- Regional Agency for Environmental Protection (Arpa), Emilia-Romagna, Italy
- Emilia-Romagna Region, General Directorate Territorial and Negotiate Planning, Agreements, Italy
- Veneto Region, Spatial Planning, Italy
- CORILA, Italy
- Karlsruhe Institute of Technology (KIT), Germany
- Municipality of Stuttgart, Germany
- Meteorological Institute, University of Freiburg, Germany
- Institute of Geography and Spatial Organization, Polish Academy of Science, Poland
- NOFER Institute of Occupational Health, Poland
- Department of Building Physics and Building Ecology, Vienna University of Technology, Austria
- Environmental Protection Department, Municipal Department 22 (MA 22), Vienna, Austria
- Hungarian Meteorological Service, Hungary
- Faculty of Mathematics and Physics, Charles University Prague, Czech Republic
- City Development Authority of Prague, Czech Republic
- Czech Hydrometeorological Institute, Czech Republic
- Scientific Research Center of the Slovenian Academy of Sciences and Arts, Slovenia
- Department of Environmental Protection, Municipality of Ljubljana, Slovenia

The main objective of UHI project is to establish a trans-national attention for the prevention, adaptation and mitigation of the natural and anthropogenic risk arising from the urban heat island phenomenon. UHI project is organized through six Work Packages, namely:

1. Project management and coordination
2. Communication, knowledge management and dissemination
3. Framework analysis
4. Transnational network and UHI assessment's tools
5. Mitigation and adaptation strategies

6. Pilot and capitalization actions for limiting UHI's effect

The first step consists in gaining a more detailed insight of the phenomenon and of potentially correlated risks, starting from a deep analysis of the current knowledge both from a scientific and a legislative point of view. In other words, to collect scientific planning and legislative experiences throughout Central Europe in order to influence some suitable and sustainable actions in urban land use. This state-of-the-art analysis is carried out with the traditional micro-meteorological techniques and is specifically designed to develop mitigation and adaptation as well as management strategies. UHI intensity has been measured within each of the main urban areas within Central Europe comparing meteorological data obtained in a monitoring station located inside cities with data obtained in a station located in the rural surroundings using a common methodology. In addition, a web atlas has been implemented, after the development of a GIS database where project partners uploaded meteorological and air-quality data, as well as maps of satellite images, soil use and DEM referring to the different areas involved in UHI-project.

Communication and dissemination within UHI project represent a very relevant aspect of the developed activities, with a sharing of the competence and experiences not only in the framework of the partnership itself but also with all the other European Institutions and stakeholders in general. Moreover, also people living in urban areas should be addressed by the communication and dissemination activities through the largest possible number of mass-media. Among the other dissemination activities, local events are worth of some words, since they are designed to facilitate the interaction between partners that are developing pilot activities and local stakeholders that can contribute to them. The necessary interaction in a trans-national network will then ensure a widespread audience for the dissemination and exploitation of methods and findings from the project activities. A number of local events have been organized by UHI project partners with a certain success in terms of participation and debate among the various subjects involved, not only related to the development of a trans-national debate on the UHI theme but also to the interactions with climate change.

Another important aspect is related to the elaboration of scenarios for UHI phenomenon taking into account its relationship with climate change. In this respect, the implementation of appropriate mitigation and adaptation strategies as well as the integration of these strategies inside urban planning tools is of outstanding importance. Mitigation strategies are related to the adoption of urban and land planning models that are able to counteract the development of UHI within urban environments, while adaptation strategies aim at reducing the impact of those phenomenon related to UHI, such as summer bioclimatic discomfort.

Pilot actions constitute the final phase of UHI project: they started as feasibility studies in order to evaluate how city space can be developed taking into account the mitigation and adaptation strategies developed in the course of UHI project. The development of pilot actions in the framework of mitigation and adaptation strategies represents another very important aspects of UHI project. The most relevant metropolitan areas in Central Europe are the pilot areas where pilot actions are developed (Budapest, Ljubljana, Modena, Padua, Prague, Stuttgart, Warsaw and Wien). In fact, UHI project aims at adopting urban and land planning models in order to prevent urban heat island effect and to reduce its impact. Pilot areas have been identified by project partners inside each of the afore-mentioned metropolitan areas in Central Europe.

Trans-national Focus Groups (TFGs) are one of UHI project's operative tools to ensure maximum synergy between various partners and local stakeholders involved in the pilot actions. The rationale of TFGs is to encourage the exchange of ideas and best practices in a trans-national and multi-disciplinary context. Experts taking part to TFGs are non necessary limited to the partnership of UHI project, rather it has been considered an enlargement of the expertise during the course of the UHI project.

Trans-national focus groups (TFGs) are trans-national thematic meetings where experts with a sound experience in UHI related topics, such as meteorological, climatic and biometeorological aspects, architectural techniques and urban planning debate on the issues related to UHI phenomenon. TFGs took place in conjunction with the formal project meetings established during the course of UHI project.

TFGs are conceived to manage the knowledge shared among partners and stakeholders, the debate developed in the TFG framework faced general and scientific issues that could support the activities carried out within the whole UHI project.

The items chosen as drivers of the debate are the following:

- Urban planning: urban sustainability, regeneration and sprawl limitation policies
- Environmentally driven consent: policy and communication (pro-active strategies aiming at an environmentally significant UHI accounting behaviour – attitude and context – addressing to citizens, planners, policy makers, researchers, etc.);
- Urban health: bioclimatic discomfort, human health
- Urban meteorology: micro- and macro-scale analysis of the phenomenon

TFGs can thus be considered as parallel insight that contribute to define the technical scenarios linked with UHI phenomenon. The methodological approach is based on the definition of general issue (for example, urban planning) developed in different sub-topics acting as a pathway of the debate (for example, regeneration and sprawl limitation policies).

The discussion within TFGs dealt with generally developed problems and implemented scientific activities carried out in parallel in the context of the whole project: working groups may in fact be considered as a cross-cutting approach that helps to define technical scenarios associated with the phenomenon of urban heat islands.

A general objective of UHI project aimed at drawing a trans-national attention, as well as policies and practical actions, in order to prevent risks deriving from UHI, both from natural and anthropogenic origin. In particular, UHI project includes the review of a wide range of possible mitigation and adaptation actions for lowering the negative UHI effects within cities. Most of the actions that are commonly employed can be divided into three main types of intervention, namely buildings, pavements and vegetation.

As for buildings, the mitigation of their effects on urban heat islands are primarily related to the changes of material properties, as well as on the geometry of the urban settings created by buildings themselves (street sections and urban canyons). The first of the aforementioned strategies mainly deals with the thermal performance of buildings, while the other is mainly related with the way in which air currents can remove excess heat from areas between buildings (streets, passageways, etc.). Pavements in turn play a very important role in the formation of the UHI phenomenon, since conventional paving materials (concrete and asphalt among the others) tend to absorb large amounts of solar radiation during daytime and to release it to the cooler surrounding air at night. Another important property of paving materials is their limited permeability to water, which prevents water absorption in the ground, thus reducing the potential evaporation of the ground surface which may contribute to the reduction of air temperatures. Last but not least, trees and vegetation in general reduces ambient air temperature mainly by evapo-transpiration and shading and is therefore expected to help in the mitigation process of UHI intensity. The common practices within this scope are the planting of trees and vegetation in an existing urban fabric (city streets and car-parkings) or the creation and preservation of wider green areas (parks, groves, lawns, etc.) within the urban fabric.

UHI is obviously a common problem for Europe on a continental scale. Pilot actions brought together the most relevant metropolitan areas in Central Europe for a kind of shared study of the urban heat island phenomenon and for a joint experimentation of countermeasures.

UHI project structure deals also with the comparison of the impact of potential mitigation measures of the urban heat island through the use of different modelling tools in order to give a quantitative evaluation of the reduction of UHI intensity implementing the mitigation strategies. These estimate may also be an innovative strategy in order to support local stakeholders thus contributing to bridge the gap between two traditionally unrelated disciplines such as meteo-climatology and urban planning. The development of traditional meteorological models and the analysis of relative outputs can in turn contribute to the definition of specific strategies to guide the choice of urban development and renewal.

The core of UHI project can be stated as an effort to create a positive relationship between knowledge and actions. Policy-making is a very complicated process partly due to the wide range of topics and uncertainty in the scientific results. Research in UHI project should aim at gaining a greater understanding of the complexities of meteorological issues with respect to mesoscale interactions, primarily at the urban boundary layer. One of the goal of this attention is certainly paid to put in place energy-efficiency and energy-saving approaches in urban and territorial planning.

In addition, systematic and interdisciplinary applied research can help policy makers to gather intelligence and to monitor and evaluate the efficacy of their approaches. On the other hand, policy makers in urbanized and urbanizing regions can create opportunities to reduce the coupled impacts associated with rapid urban-ization and changing urban climates as exemplified by the UHI effect.

That is policy makers at all levels should be able to craft policies, incentives and regulations matching economic, social and environmental imperatives. All these issues are the directions towards which UHI project moved. Consistently, the part-nership is a balanced (and quite strong) mix of policy makers on one side (namely "institutional partners"), and environmental monitoring agencies and university on the other side (namely "scientific partners"). All of them act within the orbit of the most relevant metropolitan areas in Central Europe.

# Part I
# The Urban Heat Island: Evidence, Measures and Tools

# Chapter 1
# Forecasting Models for Urban Warming in Climate Change

**Joachim Fallmann, Stefan Emeis, Sven Wagner, Christine Ketterer, Andreas Matzarakis, Ilona Krüzselyi, Gabriella Zsebeházi, Mária Kovács, Tomáš Halenka, Peter Huszár, Michal Belda, Rodica Tomozeiu, and Lucio Botarelli**

**Abstract** Defining UHI phenomenon required and interdisciplinar approach using both simulation models and climate data elaborations at regional and metropolitan level. In particular the WP 3 of UHI project provided a detailed survey on the main studies and practices to counteract urban heat islands in different European areas;

J. Fallmann (✉)
UK Met Office, Exeter
e-mail: Joachim.Fallmann@metoffice.gov.uk

S. Emeis
Head of Research Group "Regional Coupling of Ecosystem-Atmosphere", Karlsruhe Institute of Technology (KIT), Institute of Meteorology and Climate Research, Atmospheric Environmental Research (IMK-IFU), Garmisch-Partenkirchen, Germany

S. Wagner
Institute of Meteorology and Climate Research (IMK-IFU) of the Karlsruhe Institute of Technology (KIT), Karlsruhe, Germany

C. Ketterer (✉)
Albert-Ludwigs-University of Freiburg, Werthmannstr. 10, D-79085, Freiburg, Germany

iMA Richter & Roeckle, Eisenbahnstrasse 43, 79098, Freiburg, Germany
e-mail: Christine.Ketterer@meteo.uni-freiburg.de

A. Matzarakis
Albert-Ludwigs-University of Freiburg, Werthmannstr. 10, D-79085, Freiburg, Germany

Research Center Human Biometeorology, German Meteorological Service, Stefan-Meier-Str. 10, D-70104, Freiburg, Germany

I. Krüzselyi (✉) • G. Zsebeházi
Hungarian Meteorological Service, Budapest, Hungary
e-mail: kruzselyi.i@met.hu

M. Kovács
Department of Climatology and Landscape Ecology, University of Szeged, Szeged, Hungary

© The Author(s) 2016
F. Musco (ed.), *Counteracting Urban Heat Island Effects in a Global Climate Change Scenario*, DOI 10.1007/978-3-319-10425-6_1

discussed climate models at regional level; simulated the evaluation of urban warming in the different cities involved in the project, providing locally proper measuring and analysis in connection with the specific urban forms.

**Keywords** Forecasting UHI • Climate models • Regional scale • Climate scenario

## 1.1 General Introduction

The working package WP 3 collects technical and scientific definitions and state of the art about the urban heat island phenomenon and further presents strategies to simulate future scenarios by using modelling systems. The knowledge review is a core output of action WP 3.1, giving a complete overview over the problem. It discusses methods to mitigate and adapt to the intensification of the UHI in Central Europe (CE) and beyond and further provides background information for local authorities related to urban planning, building and land use regulations in compliances with EU rules. The review is developed with the contribution of all PPs, scientific and institutional, in order to have main examples of excellences, best practices, innovative regulations and intervention put in act to face the UHI phenomenon.

Working Package WP 3.2 discusses regional climate model simulations and tries to give an estimation of future climate conditions (temperature, humidity, precipitation, wind speed, cloud cover, etc.) which may serve as outer conditions for the assessment of the UHI phenomenon in the cities of CE. The suitable simulations can be made e.g. with WRF or RegCM, especially when urban land use parameterizations involved, and statistical output on means and standard deviations of the meteorological variables can be supplied. The regional climate model uses available boundary conditions provided by existing global climate models. Statistical downscaling techniques can be used as well. Within the city structures, other microscale models are necessary to provide estimates of the local conditions, e.g. in street canyons.

The project partners being involved in this action try to show possibilities to simulate the effect of the Urban Heat Island and analyse its characteristic features for their city of interest. They try to set up models with different backgrounds also to account for simulation of mitigation scenarios counteracting this urban climate

T. Halenka (✉) • P. Huszár
Faculty of Mathematics and Physics, Department of Atmospheric Physics,
Charles University Prague (CUNI), Prague, Czech Republic
e-mail: tomas.halenka@mff.cuni.cz

M. Belda
Faculty of Mathematics and Physics, Department of Meteorology and Environment
Protection, Charles University Prague (CUNI), Prague, Czech Republic

R. Tomozeiu (✉) • L. Botarelli
ARPA Emilia-Romagna, Bologna, Italy
e-mail: rtomozeiu@arpa.emr.it

phenomenon, to work out plans about sustainable strategies for future urban planning together with the local stakeholders. The operators of these models contributing to this report are manifold. On the one hand side there are meteorological services like the HMS (Hungarian Meteorological Service) and research institutes like the KIT (Karlsruhe Institute of Technology. On the other side there are Universities like Prague and Freiburg or territorial alliances like ARPA Emilia Romagna. In the following, the models used by the project partners trying to forecast the UHI are listed.

In the following, a broad range of different tools and studies are presented which have been carried out by the project partners in the course of the activities in working package WP3. The studies range from climate change projections for central European cities with the regional climate model (RCM) WRF, to regional climate modelling with RegCM and statistical downscaling approaches. Further, the microclimatic model RayManis used to assess climate change on street scale and another study investigates urban effects by coupling a town energy model (TEB) to the surface modelling platform SURFEX.

All the information is collected by the working package leader PP5 – Karlsruhe Institute of Technology and presented according to the requirements introduced in the WP3 methodological document.

## 1.2 Overview of Models and Tools

Collection of models for investigating the extend of Urban Heat Islands and the impact of climate change for Central European urban regions in the course of WP 3 – activities
EnviMet (http://www.envi-met.com/)

- Commonly agreed to serve as primary model for simulating urban climatology and mitigation scenarios
- three-dimensional microclimate model designed to simulate the surface-plant-air interactions in urban environment
- typical resolution of 0.5–10 m in space and 10 s in time.
- ENVI-met is a **Freeware program** based on different scientific research projects
- ENVI-met is a prognostic model based on the fundamental laws of fluid dynamics and thermo- dynamics. The model includes the simulation of:

  - Flow around and between buildings
  - Exchange processes of heat and vapour at the ground surface and at walls
  - Turbulence
  - Exchange at vegetation and vegetation parameters
  - Bioclimatology
  - Particle dispersion and pollutant chemistry

- Applied by TU Vienna, University of Friburg and others

  WRF (Weather Research and Forecasting Model)

- Developed by the National Center of Atmospheric Research (NCAR)
- Mesoscale, numerical weather prediction model, which also can be used for climate modeling
- Nested to global circulation model ECHAM5/MPI-OM http://www.mpimet.mpg.de/en/science/models/echam.html
- Open source, code downloadable from the web
- http://www.mmm.ucar.edu/wrf/users/
- http://www.wrf-model.org/index.php
- Applied by KIT, Germany

  Statistical downscaling approach

- Using STREAM 1 simulations from ESEMBLES-Project (http://www.ensembles-eu.org/)
- Methodology and forcing that were defined by CMIP3 simulations contributing to IPCC AR4; CMIP3 (Coupled model Intercomparison Project)
- http://www-pcmdi.llnl.gov/ipcc/about_ipcc.php
- Applied by ARPA-Emilia Romagna, Italy

  Micro-Climatic Model RayMan to assess climate change on city scale

- Boundary conditions from ENSEMBLE model RT2B (http://ensembles-eu.metoffice.com) and REMO regional climate model (http://www.remo-rcm.de)
- Calculation of the Physiological Equivalent Temperature (PET)
- RayMan: calculation of short- and long-wave radiation fluxes affecting the human body and takes complex urban structures into account
- calculated mean radiant temperature, required for the human energy balance
- meteorological and thermo-physiological data as input
- open source: http://www.mif.uni-freiburg.de/rayman/intro.htm
- Applied by University of Freiburg

  SURFEX combined with TEB (Town Energy Model)

- **SURFEX** (Surface Externalisée) is the surface modelling platform developed by Meteo-France
- computes averaged fluxes for momentum, sensible and latent heat for each surface grid box → boundary condition for meteorological model
- input land cover information from ECOCLIMAP database
- TEB: computes energy balance considering canyon concept
- ALADIN-Climate RCM as atmospheric forcing
- http://www.cnrm-game.fr/spip.php?article145&lang=en
- Applied by Hungarian Meteorological Service

  Regional Climate Model RegCM (http://users.ictp.it/RegCNET/regcm.pdf)

- Boundary conditions from GCM CNRM-CM5 (http://www.enes.org/models/earthsystem-models/cnrm-cerfacs/cnrm-cm5)

- Community Land Surface Model v3.5 (CLM3.5) as an optional land surface parameterization
- Urban surface treated by coupling with Single Layer Urban Canopy Model linked to SUBBATS surface scheme
- Applied by Charles University, Prague

  CLMM (Charles University Large Eddy Microscale Model)

- LES tool for simulation of the flow in microscales with complex terrain or structures solving CFD problems
- In addition to flow equations it includes transport equation for scalars like temperature, moisture and passive pollutants
- Applied by Charles University, Prague

## 1.3 Case Studies

### 1.3.1 Projections of Climate Trends for Urban Areas in Central Europe Using WRF

Joachim Fallmann
UK Met Office, Exeter
Joachim.Fallmann@metoffice.gov.uk

Stefan Emeis
Head of Research Group "Regional Coupling of Ecosystem-Atmosphere",
Karlsruhe Institute of Technology (KIT), Institute of Meteorology and Climate
Research, Atmospheric Environmental Research (IMK-IFU), Garmisch-
Partenkirchen, Germany

Sven Wagner
Institute of Meteorology and Climate Research (IMK-IFU) of the Karlsruhe
Institute of Technology (KIT), Karlsruhe, Germany

#### 1.3.1.1 Introduction

In 2050 the global fraction of urban population will rise to a level up to almost 70 %, which means that around 6.3 billion people are expected to live in urban areas. Next to that development a rise of the global temperature of about 0.2 K per decade for the twenty-first century is projected within the range of the SRES scenarios of the Intergovernmental Panel on Climate Change (Intergovernmental Panel on Climate Change – IPCC) for Europe (Wagner et al. 2013).

It's predicted, that extreme events are to increase in the future, which means more and heavier storms, precipitation events and thus increased danger of flooding, occurrence of heat waves or days with high air pollution, especially dangerous in combination with high temperature periods (Beniston et al. 2007).

This note describes how scenario simulations for Central Europe have been performed with a regional climate model based on global climate model scenario simulations.

### 1.3.1.2 Data and Methods

Results from high resolution, multi-ensemble regional climate models are an essential input for many climate impact studies. In the course of the CEDIM (Center for Disaster Management and Risk reduction Technology) project 'Flood Hazards in a Changing Climate' (Wagner 2013) a multi model ensemble of high resolution 7 km regional climate simulations for a present (1971–2000) and a near future (2021–2050) time period were conducted. To assess the climate change on regional scales, regional climate models (RCMs) were nested into coarser global circulation models (GCMs). For the bulk of the simulations the ECHAM5/MPI-OM Model in T63 resolution (horizontal grid spacing of approximately $140 \times 210$ km at mid-latitudes) served as GCM. ECHAM5 is the fifth-generation atmospheric general circulation model developed at the Max Planck Institute for Meteorology, in that case it was coupled to the Max Planck Institute ocean model (MPI-OM). IPCC SRES (Special Report on Emissions Scenario) A1B forcing scenario served as boundary condition.

The spatial resolution of RCM simulations has steadily increased over the last decades. In the past, several larger ensembles were carried out to assess climate change, like for example in PRUDENCE (Christensen and Christensen 2007) with a resolution of 50 km or ENSEMBLES (Hewitt CD 2005) with a spatial resolution of 25 km. To get more information on this, please refer to the respective literature.

Using WRF (Weather Research and Forecasting Model) as regional climate model was one part of the contribution of the Institute of Meteorology and Climate Research (IMK-IFU) to CEDIM. To set up WRF, different steps had to be conducted to make reliable forecasts. Thus three different runs had to be carried out: one past climate run, one validating reanalysis run and the final future climate scenario run. Each of these runs had a calculation time of approximately 3 month (Wagner 2013). Covering Germany and the near surroundings it was possible to extract modeling results for urban areas ($7 \times 7$ km grid cells) contributing to the CENTRAL Europe Project. Thus, WRF is used in the following to illustrate the effect of climate change on urban regions within the area of central Europe

The regional climate model WRF followed a double nesting procedure, where the coarse nest covered an area of entire Europe with a resolution of about 50 km, whereas the fine nest consisted of Germany and the near surroundings (Fig. 1.1).

The fine model domain of 174 by 174 grid cells covers an area between 1.5 to 17.5° E and 44.5 to 54.5° N. The model resolution of 7 km implies that every urban area of interest is covered by at least one grid cell 40 vertical levels where used for both nests. For further specifications on model physics and modeling proceedings refer to Berg et al. (2013) and Wagner et al. (2013).

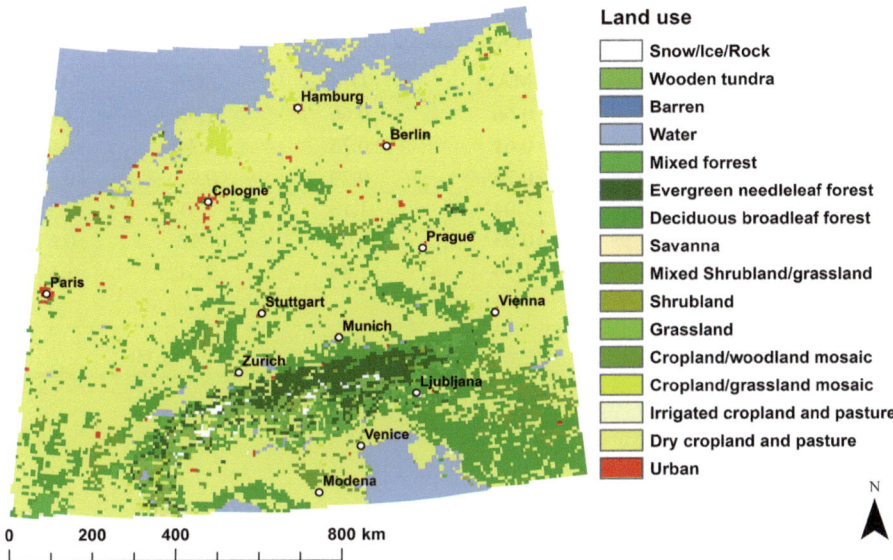

**Fig. 1.1** WRF nested domain with 7 km horizontal resolution, showing USGS 2006 classified land use, projected on a LAT/LON grid with the coordinate system WGS 84 Zone 32 N. The urban areas of interest are marked

### 1.3.1.3   Results

All simulation results project a significant warming throughout the model domain between 0.8 and 1.1 K. All ECHAM5 driven regional climate models predict an increase of annual precipitation in the range of 2 to 9 % (average of 3 % for Germany), with higher values in winter and autumn months. Using WRF as RCM reveals changes of temperature and precipitation (Wagner 2013). Differences in 2 m temperature between future and past regional climate model run, extracted for certain urban areas located in the model domain and in the project region Central Europe reveal the following trends (Table 1.1).

Monthly mean temperatures are extracted for one grid cell in the centre of a selected urban region to create probability density functions in order to statistically compare the modelling results. By calculating the values to fall below the 5 % confidence interval, the tendency towards extreme values is to be analysed on the basis of the comparison of the future (2021–2050) with the reference period (1971–2000). Probability density functions (PDF) for 4 selected cities are presented in Fig. 1.2. Following the expectations given by the IPCC AR4, temperatures will develop in direction to the extremes (Beniston et al. 2007). The PDFs indicate a compression and widening of the future curve compared to that one of the past. The shift in the 95th percentile reflects the climate change signal.

**Table 1.1** Projected fine nest seasonal and annual temperature changes [°C] between 1971 and 2000 and 2021–2050 for the WRF simulation averaged for urban area

|            | DJF  | MAM  | JJA  | SON  | Annual |
|------------|------|------|------|------|--------|
| Ljublijana | 1.47 | 0.66 | 0.66 | 1.35 | 1.03   |
| Modena     | 1.11 | 0.61 | 0.75 | 1.24 | 0.93   |
| Padua      | 0.86 | 0.26 | 0.29 | 0.9  | 0.58   |
| Vienna     | 1.92 | 1.04 | 1.13 | 1.91 | 1.5    |
| Prague     | 1.43 | 0.05 | 0.07 | 1.13 | 0.67   |
| Stuttgart  | 2.05 | 1.36 | 1.86 | 2.31 | 1.89   |

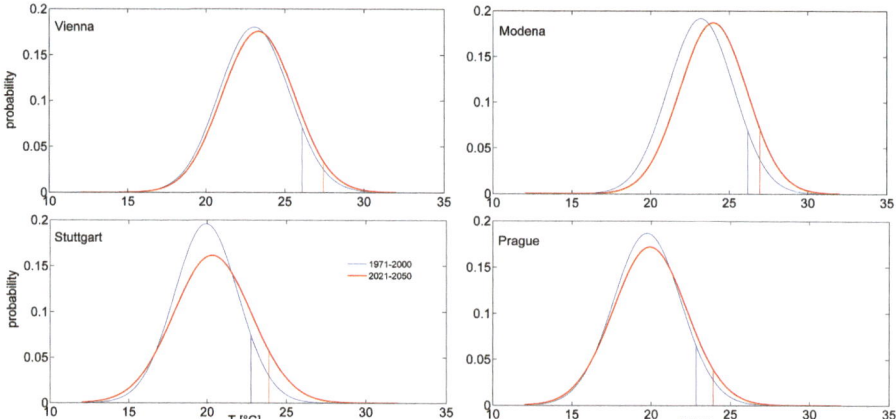

**Fig. 1.2** Probability density functions (PDFs) extracted for the central $7 \times 7$ km pixel of a selected urban area. The *blue line* indicates the probability density curve for extracted monthly mean temperatures in the past (1971–2000), the *red line* shows the same for the future (2021–2050) period. The *vertical lines* illustrate the 95th percentile for each plot and time frame

### 1.3.1.4   Conclusion

Delivering climate runs with higher resolution, which may fit better to the urban scale will be very cost intensive. Nevertheless, results from projects like the above-mentioned one can be used as boundary conditions for high resolution city-scale models to conduct scenario runs (e.g. different urban planning strategies) for future climate conditions and region of interest. For further studies on impact of climate change on urban settlements please refer also to the project Ensembles-Based Predictions of Climate Changes and Their Impacts – ENSEMBLES (Hewitt CD 2005). This study can be used for other working packages dealing with mitigation and adaptation strategies, with the background that climate change will amplify Urban Heat Islands and future problems for urban inhabitants coming along with that phenomenon. Specific measures like urban greening, changing radiative properties of building materials or restructuring of city quarters, are not discussed in this report, rather should the results serve as basis for referring the problem of UHI's to a more to raise public awareness on a different level.

# References

Beniston, M., Stephenson, D., Christensen, O., Ferro, C., Frei, C., Goyette, S. P., Halsnaes, K., Holt, T., Jylhae, K., Koffi, B., Palutikof, J., Schoell, R., Semmler, T., & Woth, K. (2007). Future extreme events in European climate: an exploration of regional climate model projections. *Climatic Change, 81*(1), 71–95. Available from: http://dx.doi.org/10.1007/s10584-006-9226-z

Berg, P., Wagner, S., Kunstmann, H., & Scheadler, G. (2013). High resolution regional climate model simulations for Germany: Part I validation. *Climate Dynamics, 40*(1–2), 401–414. Available from: http://dx.doi.org/10.1007/s00382-012-1508-8

Christensen, J., & Christensen, O. S. (2007). A summary of the PRUDENCE model projections of changes in European climate by the end of this century. *Climatic Change, 81*(1), 7–30. Available from: http://dx.doi.org/10.1007/s10584-006-9210-7

Hewitt, C. D. (2005). The ENSEMBLES project: Providing ensemble-based predictions of climate changes and their impacts. *EGGS Newsletter, 13*, 22–25.

Jacobson, M. Z. & Ten Hoeve, J. E. (2011). Effects of urban surfaces and white roofs on global and regional climate. *Journal of Climate, 25*(3), 1028–1044. Available from: http://dx.doi.org/10.1175/JCLI-D-11-00032.1. Accessed 17 Jan 2013.

Wagner, S., Berg, P., Schaedler, G., & Kunstmann, H. (2013). High resolution regional climate model simulations for Germany: Part II projected climate changes. *Climate Dynamics, 40*(1–2), 415–427. Available from: http://dx.doi.org/10.1007/s00382-012-1510-1

## 1.3.2 Human-Biometeorological Assessment of Changing Conditions in the Region of Stuttgart in the Twenty-First Century

Christine Ketterer
Albert-Ludwigs-University of Freiburg, Werthmannstr. 10, D-79085, Freiburg, Germany

iMA Richter & Roeckle, Eisenbahnstrasse 43, 79098, Freiburg, Germany
Christine.Ketterer@meteo.uni-freiburg.de

Andreas Matzarakis
Albert-Ludwigs-University of Freiburg, Werthmannstr. 10, D-79085, Freiburg, Germany

Research Center Human Biometeorology, German Meteorological Service, Stefan-Meier-Str. 10, D-70104, Freiburg, Germany

### 1.3.2.1   Data and Methods

Regional climate simulations of the ENSEMBLE RT2B model (for more information see http://ensembles-eu.metoffice.com/) with daily resolution are used as the
data basis for the analysis of climate change in the greater region of Stuttgart for
1960–2100. The RT2B model focuses on the SRES A1B scenario. The regional
climate model REMO (for more information see http://www.remo-rcm.de/) in
hourly resolution and 10 km spatial resolution focusing on SRES A1B and B1 scenario is used to analyze the human thermal comfort conditions.

   Therefore, the thermal index Physiologically Equivalent Temperature (PET;
Höppe 1993; Mayer and Höppe 1987; Matzarakis et al. 1999) is calculated with the
help of the micro-climate model RayMan (Matzarakis et al. 2007, 2010). PET is
used to quantify especially the frequency and intensity of heat stress. Thereby,
PET between 18 °C and 23 °C was assessed to be comfortable, while PET above
35 °C (Nastos and Matzarakis 2012) stands for strong heat stress (Matzarakis and
Mayer 1996).

   The dataset is used to calculate the number of climatologically event days and
their change until the end of the twenty-first century.

### 1.3.2.2   Results

Table 1.3 shows the frequency of climatological event days. The frequency of hot
days (Tamax $\geq$ 30 °C) and summer days (Tamax $\geq$ 25 °C) will increase to 174 % and
140 % in the period 2021–2050 and 280 % and 157 % until end of the twenty-first
century. On the other hand the number of frost (Tamax $\leq$ 0 °C) and ice days
(Tamax $\leq$ 0 °C) per year will decrease to 33 % in the period 2071–2100.

   The average annual air temperature might rise by 1.5 °C from 1961–1990 to
2021–2050 and 3.5 °C until the end of the twenty-first century (Table 1.3). Thereby,
the increase in air temperature is strongest during summer and winter and weakest
in spring.

   The number of days per year with heat stress (PET > 35 °C) at 14:00 MEZ will
increase by 6 days from 1961 to 1990 to 2021–2050 and by 28 days until the end of
the twenty-first century according to the REMO data A1B scenario. The B1 data
shows no increase until the mid of the twenty-first century, but an increase of 4 %
(16 days) until 2071–2100 (Table 1.2).

   In the early morning (6:00 MEZ), the number of days with (extreme) cold stress
will significantly decrease by 15 (10) days according to the A1B (B1) scenario until
2021–2050 and by 48 (29) days until 2071–2100 (Fig. 1.3). In contrast, the number
of days with PET > 29 °C will rise by 5 (15) days until the mid (end) of the twenty-
first century according to the A1B scenario.

**Table 1.2** Analysis of the number of climatological event days in the greater area of Stuttgart using the ENSEMBLE model RT2B for three different time periods 1961–1990, 2021–2050 and 2071–2100

| Event | | 1961–1990 | 2021–2050 | 2071–2100 |
|---|---|---|---|---|
| Extreme hot days | $Ta_{max} \geq 39\ °C$ | $0 \pm 1$ | $1 \pm 2$ | $3 \pm 5$ |
| Hot days | $Ta_{max} \geq 30\ °C$ | $7 \pm 9$ | $12 \pm 14$ | $20 \pm 24$ |
| Summer days | $Ta_{max} \geq 25\ °C$ | $27 \pm 19$ | $37 \pm 22$ | $42 \pm 38$ |
| Frost days | $Ta_{min} \leq 0\ °C$ | $104 \pm 29$ | $79 \pm 27$ | $35 \pm 30$ |
| Ice days | $Ta_{mix} \leq °C$ | $30 \pm 13$ | $18 \pm 10$ | $7 \pm 8$ |
| Extreme cold days | $Ta_{max} \leq -10\ °C$ | $1 \pm 1$ | $\pm 0$ | $\pm 0$ |

**Table 1.3** The average annual air temperature simulated by the ENSEMBLE models RT2B and the standard deviation in the greater region of Stuttgart from 1961–1990 to 2021–2050 and 2071–2100

| Air temperature | 1961–1990 | 2021–2050 | 2071–2100 |
|---|---|---|---|
| Yearly | $7.9 \pm 1.7$ | $9.4 \pm 1.8$ | $11.4 \pm 2.1$ |
| Spring | $6.7 \pm 2.0$ | $7.8 \pm 2.3$ | $9.6 \pm 2.0$ |
| Summer | $16.5 \pm 2.5$ | $18.0 \pm 2.7$ | $20.5 \pm 3.4$ |
| Autumn | $8.3 \pm 1.8$ | $9.7 \pm 1.8$ | $11.8 \pm 2.4$ |
| Winter | $0.1 \pm 2.1$ | $1.8 \pm 2.1$ | $3.8 \pm 2.1$ |

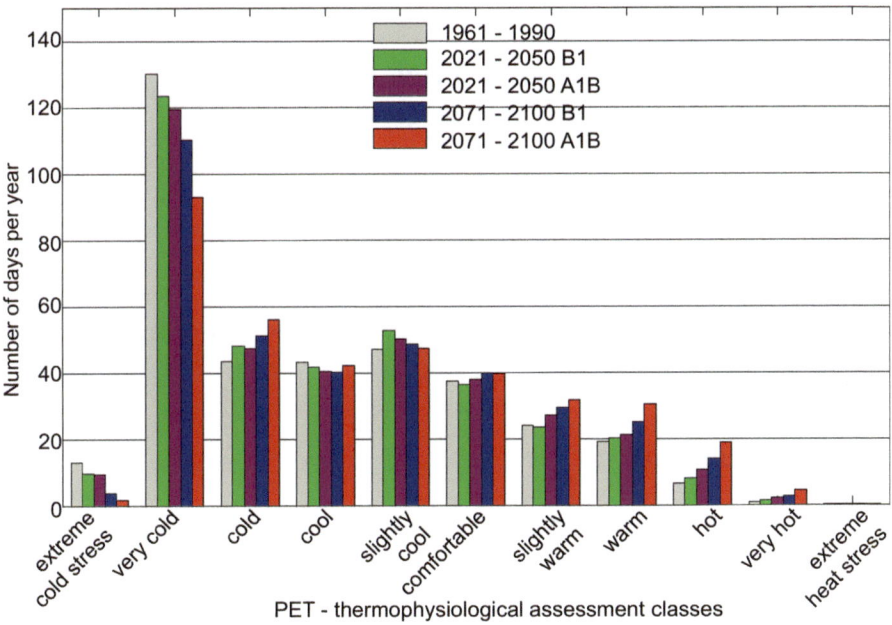

**Fig. 1.3** Number of days with PET assessment classes in the climate normal period 1961–1990 compared to the frequency in 2021–2050 and 2071–2100 at 6:00 a.m. Data basis: REMO A1B and B1 scenario

# References

Höppe, P. R. (1993). Heat balance modelling. *Cellular and molecular life sciences*, 49(9), 741–746.

Matzarakis, A., & Mayer, H. (1996). Another kind of environmental stress: Thermal stress. In *WHO collaborating centre for air quality management and air pollution control* (Vol. 18, pp. 7–10).

Matzarakis, A., Mayer, H., & Iziomon, M. G. (1999). Applications of a universal thermal index: Physiological equivalent temperature. *International Journal of Biometeorology, 43*, 76–84.

Matzarakis, A., Rutz, F., & Mayer, H. (2007). Modelling radiation fluxes in simple and complex environments – Application of the RayMan model. *International Journal of Biometeorology, 51*(4), 323–334.

Matzarakis, A., Rutz, F., & Mayer, H. (2010). Modelling radiation fluxes in simple and complex environments: Basics of the RayMan model. *International Journal of Biometeorology, 54*(2), 131–139.

Mayer, H., & Höppe, P. (1987). Thermal comfort of man in different urban environments. *Theoretical and Applied Climatology, 38*, 43–49.

Nastos, P. T., & Matzarakis, A. (2012). The effect of air temperature and human thermal indices on mortality in Athens, Greece. *Theoretical and Applied Climatology, 108*(3–4), 591–599.

## 1.3.3 *Urban Climate Modelling with SURFEX/TEB at the Hungarian Meteorological Service*

Ilona Krüzselyi and Gabriella Zsebeházi
Hungarian Meteorological Service, Budapest, Hungary
kruzselyi.i@met.hu

Mária Kovács
Department of Climatology and Landscape Ecology, University of Szeged, Szeged, Hungary

### 1.3.3.1  Introduction

Half of the world's population lives in cities nowadays, which are continuously growing and have significant effects on local climate. Moreover, consequences of climate change in cities might be enhanced by the impact of urban surfaces. Thus, to make adaptation strategies to climate change, investigating these impacts is especially important. Therefore urban climate modelling activity started besides regional climate modelling at the Hungarian Meteorological Service (HMS) in 2010. To portray the interactions between the atmosphere and the urban areas, SURFEX surface model is applied. Main objective of using SURFEX is downscaling the regional projections for the future over Hungarian cities, and this paper is focusing on the first step of this, i.e. the validation of the surface model.

### 1.3.3.2  Methodology

The SURFEX (SURFace EXternalisée; Le Moigne 2009) surface model consists of four schemes for urban surface, sea, inland water and nature. Amongst these schemes the Town Energy Balance (TEB) model (Masson 2000) describes interactions between urban surface and atmosphere by simulating turbulent fluxes. It follows local canyon approach, where canyon represents the road with buildings on the sides. TEB considers three surfaces (roof, wall, road) with different energy budgets. It takes several processes into account which are important in urbanized areas, e.g., it treats water and snow interception by roofs and roads, fog, runoff, radiative trapping, momentum and heat fluxes. The anthropogenic heat and moisture fluxes derived from traffic, industry and domestic heating are also considered.

As input, SURFEX needs information about the atmospheric conditions, i.e. the atmospheric forcing, which can be supplied either by measurements or an atmospheric model. The atmospheric model may be coupled with SURFEX and thus it can get feedback from the surface scheme, but SURFEX running in offline mode (i.e., without feedback) is feasible as well. It is noted that advection is not taken into account in SURFEX, thus there is no interaction between grid points in offline mode, which is only possible through the atmospheric model.

At HMS, the SURFEX studies started in 2010 (Vértesi 2011) for modelling urban heat island (UHI) effect in Budapest. Some 10-year long experiments were achieved over Budapest and Szeged. The atmospheric forcing was obtained from ERA-40 re-analysis (Uppala et al. 2005) produced by ECMWF (European Centre for Medium-range Weather Forecast). Re-analyses are three-dimensional climate databases, which are created with data assimilation technique using as many observations as possible plus short-range weather forecasts. ERA-40 is a global dataset at ca. 125-km horizontal resolution, which was downscaled by ALADIN-Climate regional climate model (Csima and Horányi 2008) to a 10-km resolution domain covering the Carpathian Basin for 1961–2000. These results were interpolated by a special configuration of the model to two smaller areas around Budapest and Szeged at 1 km resolution (Fig. 1.4) for the investigated periods. These served as inputs for SURFEX, which was run in offline mode at also 1 km resolution. The information for the fine surface coverage and physiography was derived from the ECOCLIMAP database (Masson et al. 2003).

The first experiment was conducted over Budapest for 1961–1970. ECOCLIMAP was created in 2006, thus it might not describe the surface characteristics of the given period realistically, as several houses have been built since the 60s, especially in the outskirt. Therefore, the experiment was repeated for 1991–2000 to see

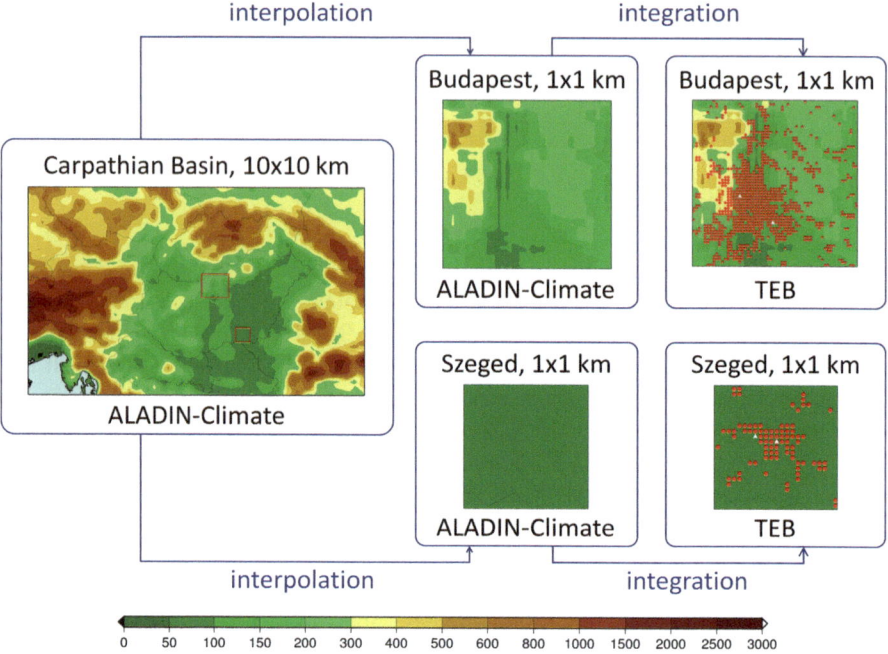

**Fig. 1.4** Flow chart about the use of SURFEX with topography [m] and the gridpoints that include urban surface (*right panels; red*); *white dots* indicate the observational stations in the *right panels*

whether these changes modify the results. For this latter period a simulation was achieved over Szeged as well to investigate the effect of urbanization in another location.

The results were validated against observations of HMS stations. In both cities two stations were selected: one in the centre and one in the outer part of the town. In case of Budapest the inner station is situated in Kitaibel Street, close to the Buda Hills, the other one operates in Pestszentlőrinc, in the outskirts (Fig. 1.4). However, ECOCLIMAP considers both points as temperate suburban (it is composed of 60 % town and 40 % nature). Szeged is located in the Southern Great Plain region, where higher elevated orographic objects (hills or mountains) cannot be found in the vicinity of the town. The surroundings of its inner observational site are also categorized as temperate suburban in ECOCLIMAP, while the outer point is actually a rural point without buildings. This paper summarizes the performance of the SURFEX model for these three experiments. Climate change assessment is not the aim of this study, since 10-year periods are insufficient for such investigations.

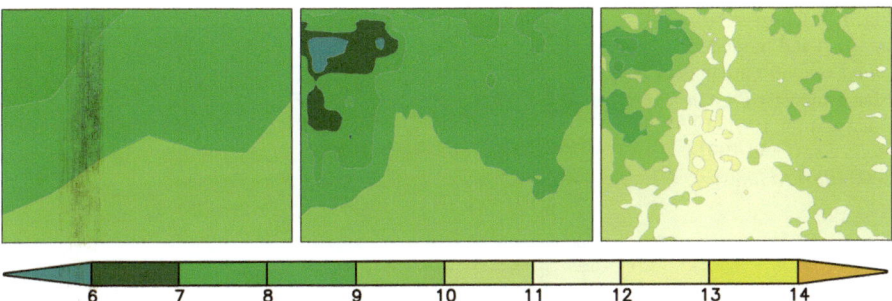

**Fig. 1.5** Spring 2-m mean temperature (°C) of ALADIN-Climate (at 10 and 1 km resolution – *left* and *middle panel*) and SURFEX (at 1 km resolution – *right panel*) over Budapest for 1961–1970 (Vértesi 2011)

**Fig. 1.6** Difference of monthly mean temperature (°C) between SURFEX and observation in Kitaibel street (*filled squares*) and Pestszentlőrinc (*open squares*) for 1961–1970 (*solid lines*) and 1991–2000 (*dashed lines*)

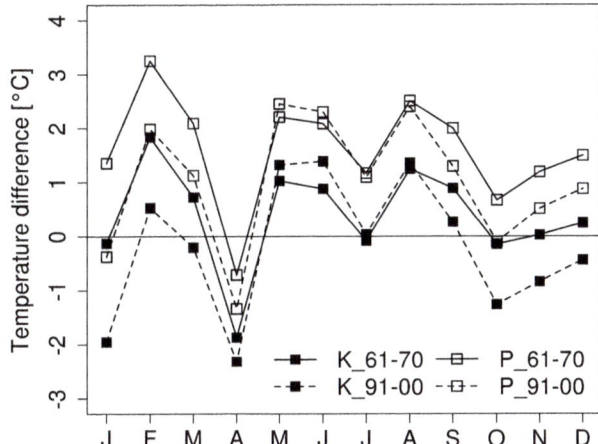

### 1.3.3.3  Results

#### 1.3.3.3.1  Budapest

Figure 1.5 shows how the interpolation and SURFEX integration modify the original temperature field of ALADIN-Climate. The applied interpolation takes into account the 1-km resolution topography, therefore, some new orographic features appear in the middle panel, like the cooler Buda hills. As a result of the sophisticated surface schemes in SURFEX, its temperature field (right panel) shows much more detailed information, Danube becomes slightly visible and temperature excess appears over the heart of the city.

In both reference points and periods mean intra-annual temperature differences between the SURFEX results and observations are very similar (Fig. 1.6), and mainly reflect the behaviour of the bias of ALADIN-Climate (not shown). From May to September the model exaggerates the temperature, and in April very strong

**Fig. 1.7** Observed (*open squares*) and modelled (*filled squares*) urban heat island intensity (°C) in Budapest for 1961–1970 (*solid lines*) and 1991–2000 (*dashed lines*)

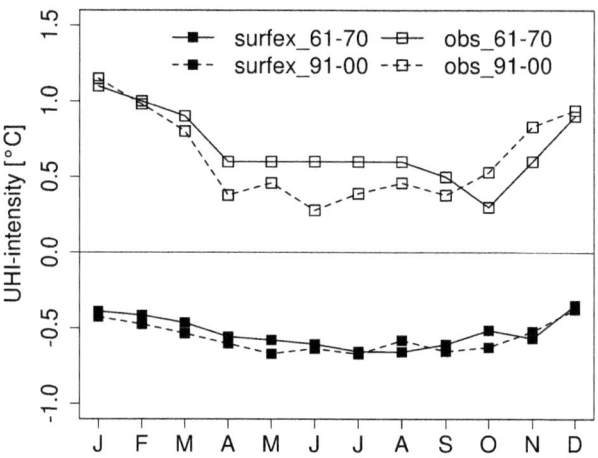

underestimation can be seen. In the concordantly overheated months, the results of Kitaibel Street are better, while in the rest of the year (except for April) this statement is valid only for the earlier period. Since ECOCLIMAP database provides information about recent surface characteristics, the reduced overestimation in Pestszentlőrinc in 1991–2000 compared to the results of 1961–1970 might be caused partly by the more realistic coverage description. (However, the fact that the improvements cannot be detected in every months indicates the key role of different atmospheric forcings in the two periods, especially in summer.)

In contrast with the temperature measurements, urban heat island cannot be noticed in any periods of the year (Fig. 1.7), which means that the inner point is colder than the outer one in the model. This already appears in the ALADIN-Climate results, and SURFEX cannot improve this, especially because the two points are characterised with the same cover type in the ECOCLIMAP. Moreover, Kitaibel Street locates on higher elevation than Pestszentlőrinc, and the neighbouring of the Buda Hills to the inner site might cause too strong cooling in ALADIN-Climate compared to the observations (recall that SURFEX does not simulate interactions between the neighbouring grid cells).

In the two reference points the results does not indicate good performance, but if a larger area is taken, SURFEX captures the daily cycle of UHI (Fig. 1.8). In daytime the air temperature of the city centre does not differ from the reference point in the suburban area; however after dusk (in winter already at 18:00, in summer at 21:00 UTC) UHI appears and its maximum intensity can be seen 5–6 h after sunset. The physical reason is that energy supply by solar radiation ends after sunset and upward longwave radiation is much more effective over natural surfaces than in the densely built-in urban area due to the smaller heat capacity of soil and the trapping of radiation in urban canyons (Basically the same conclusions were drawn for 1991–2000).

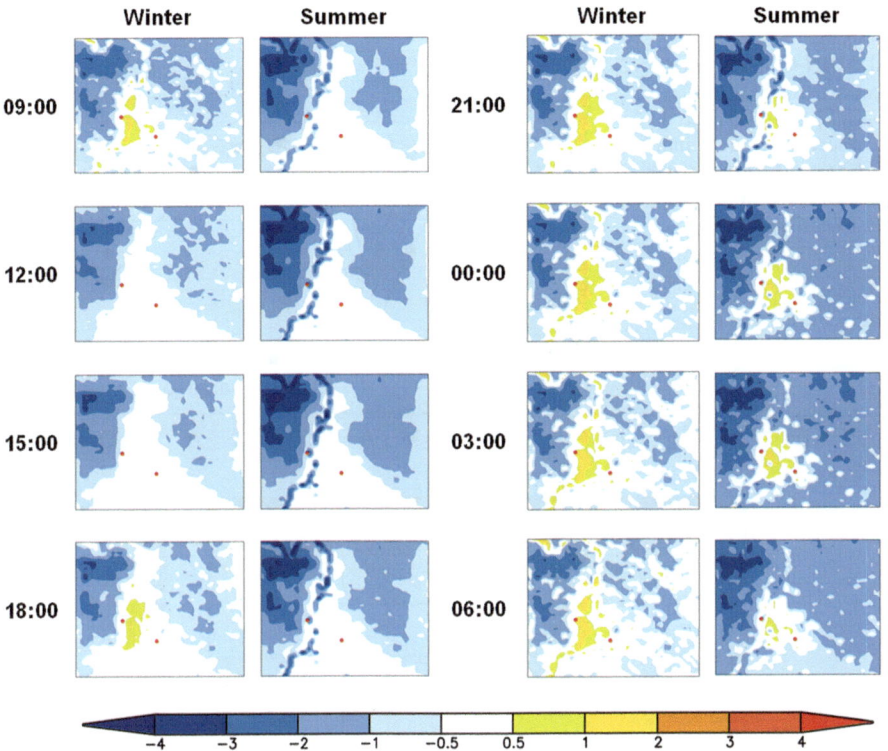

**Fig. 1.8** Difference of simulated mean temperature (°C) from the value at Pestszentlőrinc (the outskirt point) in winter and summer for 1961–1970 (the two *red points* represent the selected stations in Kitaibel Street and Pestszentlőrinc)

### 1.3.3.3.2  Szeged

The meteorological measurements in the inner site of Szeged have started in May 1998, thus the validation was limited for 1999–2000. Figure 1.9 presents the monthly average biases of ALADIN-Climate and SURFEX in the two reference points in this period. The annual cycles of the biases are similar to the result for Budapest. In general, SURFEX is giving more heat to the temperature fields of ALADIN-Climate, and due to the representative locations of the reference points (being in a flat area, and the outer point is situated in natural environment), the inner site gains larger warming to the extent that the difference between the bias of the two points almost diminishes. This implies that the monthly average UHI intensity is positive in all months (Fig. 1.10) and the magnitudes are represented adequately, as well.

The average annual and daily cycle of UHI simulated by SURFEX in 1991–2000 (Fig. 1.10) follows the theoretical pattern, namely the largest intensity occurs in the nocturnal hours from May to September with a peak of 1.8–2.2 °C. In contrast, in the late mornings of summer and autumn the outer point can be warmer than the inner one, since they warm slowly due to the larger heat capacity of urban surfaces

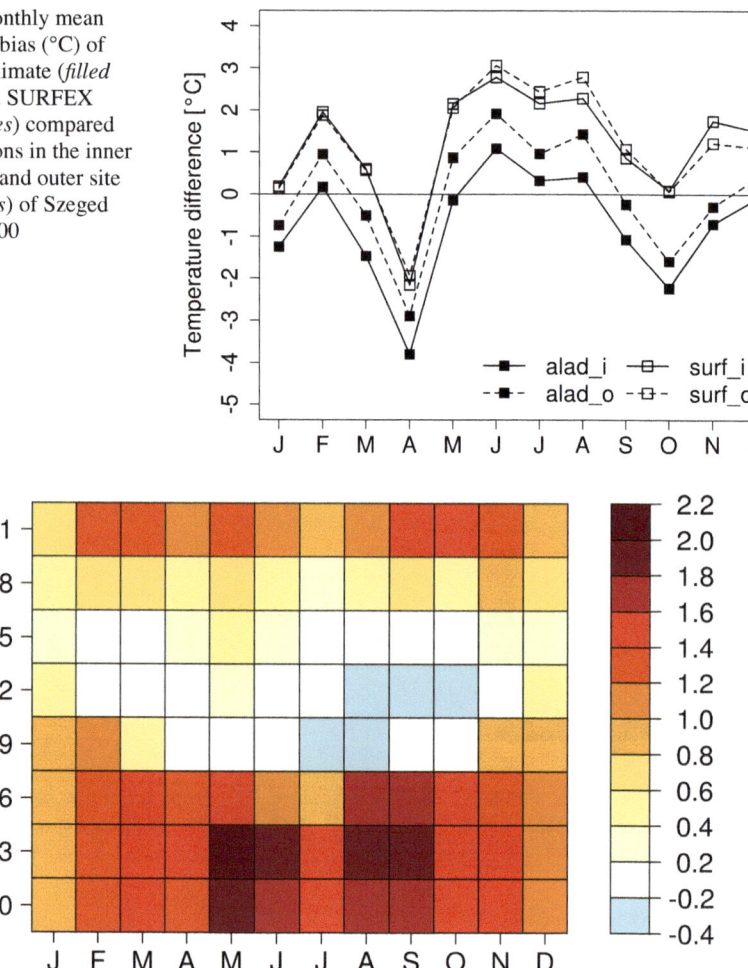

**Fig. 1.9** Monthly mean temperature bias (°C) of ALADIN-Climate (*filled squares*) and SURFEX (*open squares*) compared to observations in the inner (*solid lines*) and outer site (*dashed lines*) of Szeged for 1999–2000

**Fig. 1.10** Annual and daily cycle of UHI intensity (°C) in 1991–2000 between the two selected points in Szeged simulated by SURFEX

and the urban canyons form obstacles against low sun-rays. In December and January UHI feature can be seen all day, however the maximum intensities are lower than in summer.

### 1.3.3.4 Summary

At HMS SURFEX dynamical scheme is applied to describe interactions between the atmosphere and the urban surfaces. Some 10-year long test simulations were accomplished with the model over Budapest and Szeged. In this paper preliminary results of these validation experiments are performed. Based on the results, SURFEX captures the main characteristics of the urban climatology: temperature excess in the heart of the city and daily cycle of the urban heat island. However, the evaluation indicated also some discrepancies: in the reference points SURFEX overestimates temperature in most months and in Budapest the central point is colder than the outer one, contrary to the observations. These errors derived partly from the atmospheric forcings and the inaccurate coverage description of ECOCLIMAP. Some additional investigations and test experiments are planned in order to examine the behaviour of the model in detail and develop its performance. Further goal with SURFEX is the dynamical downscaling of climate projections over towns, furthermore, supplying a reliable basis for impact studies in the cities for the future.

## References

Csima, G. & Horányi, A. (2008). Validation of the ALADIN-Climate regional climate model at the Hungarian Meteorological Service. *Időjárás, 112*, 155–177.

Le Moigne, P. (2009). SURFEX scientific documentation. Note de centre (CNRM/GMME), Météo-France, Toulouse, France.

Masson, V. (2000). A physically-based scheme for the urban energy budget in atmospheric models. *Boundary-Layer Meteorology, 94*, 357–397.

Masson V., Champeaux, J.-L., Chauvin, F., Meriguet, C. & Lacaze, R. (2003). A global database of land surface parameters at 1 km resolution in meteorological and climate models. *Journal of Climate, 16*, 1261–1282.

Uppala, S. M., Kallberg, P. W., Simmons, A. J., Andrae, U., da Costa Bechtold, V., Fiorino, M., Gibson, J. K., Haseler, J., Hernandez, A., Kelly, G. A., Li, X., Onogi, K., Saarinen, S., Sokka, N., Allan, R. P., Andersson, E., Arpe, K., Balmaseda, M. A., Beljaars, A. C. M., van de Berg, L., Bidlot, J., Bormann, N., Caires, S., Chevallier, F., Dethof, A., Dragosavac, M., Fisher, M., Fuentes, M., Hagemann, S., Hólm, E., Hoskins, B. J., Isaksen, L., Janssen, P. A. E. M., Jenne, R., McNally, A. P., Mahfouf, J.-F., Morcrette, J.-J., Rayner, N. A., Saunders, R. W., Simon, P., Sterl, A., Trenberth, K. E., Untch, A., Vasiljevic, D., Viterbo, P., & Woollen, J. (2005). The ERA-40 re-analysis. *Quarterly Journal of Research Meteorology Society, 131*, 2961–3012.

Vértesi, Á. É. (2011). Modelling possibilities of the urban heat island effect in Budapest (in Hungarian). Master Thesis, ELTE, Budapest, Hungary.

### *1.3.4  Regional Climate Modelling Considering the Effect of Urbanization on Climate Change in Central Europe*

Tomáš Halenka and Peter Huszar
Faculty of Mathematics and Physics, Department
of Atmospheric Physics, Charles University Prague (CUNI),
Prague, Czech Republic
tomas.halenka@mff.cuni.cz

Michal Belda
Department of Meteorology and Environment Protection,
Faculty of Mathematics and Physics, Charles University, Prague (CUNI),
Prague, Czech Republic

## 1.3.4.1  Introduction

Big cities or urban aglomerations can significantly impact both climate and environment. The emissions of large amount of gaseous species and aerosols, which affect the composition and chemistry of the atmosphere (Timothy et al. 2009), can have adverse effect on the environment in the cities and their vicinity. Moreover, this can negatively impact the population (Gurjar et al. 2010). In addition, this pathway can result in indirect impact on the meteorology and climate as well, due to radiation impact of the atmospheric composition on the thermal balance and thus affect the temperature as well. Especially within the canopy layer in the cities, the changes can be quite significant.

However, the primary reason for temperature increase within the cities or urban aglomerations with respect to the rural vicinity, is the effect of so called urban heat island (UHI, Oke 1973), which is mainly due to construction elements within the urban environment. This is extensively covered by artificial objects, buildings, using by large stone, bricks or concrete, and by quite large spaces often paved. This kind of surface clearly differs from natural surfaces (e. g. grassland, forest) by mechanical, radiative, thermal, and hydraulic properties, therefore, these surfaces represent additional sinks and sources of momentum and heat, affecting the mechanical, thermodynamical, and hydrological properties of the atmosphere (Lee et al. 2010). Nevertheless, the changes of meteorological conditions within the urban areas due to UHI can further affect the air-quality. This has been studied recently by e.g. Ryu et al. (2013), they found significant impact on the ozone day and night-time levels especially due to circulation pattern changes for the Seoul metropolitan area.

For WP3 we have focused on the aspects of climate conditions changes in urban environment, yet especially on those with strong potential to impact the air-quality, based on the experiment setup described below. For the region of Central Europe, we investigate the impact of the urban environment by means of its introducing into the regional climate model. As the spatial scale of the meteorological influence due to the cities is much smaller than the scale resolved by the mesoscale model, inclusion of urban land-surface requires additional parameterizations. The most common parameterizations considering the urban effects are the slab models (bulk parameterization), where the urban surface constants (e.g., surface albedo, roughness length, and moisture availability) can vary to better describe those of the urban surfaces. This treatment however ignores the three-dimensional character of the urban meteorological phenomena, moreover, in feasible resolutions the urban environment cannot be well resolved. Therefore, a more accurate approach is provided using urban canopy models (single layered – SLUCM, or multi-layered MLUCM) coupled to the driving mesoscale model (Chen et al. 2011). Our study describes in more details the implementation of such a SLUCM into our regional climate chemistry modelling system.

### 1.3.4.2   Background of Modelling for Europe

For the purpose of the climate simulations for UHI we follow two lines. First one aims to prepare up-to-date background information on the changing conditions in Europe using Euro-CORDEX rules. Euro-CORDEX is European part of the CORDEX initiative under WCRP which intend to provide downscaled information on climate change for individual continents around the world. It supposes to perform the coordinated simulations (i.e. for coordinated domains and periods), which results in creating large ensemble of model simulations, both for validation based on ERA-Interim reanalysis and/or historical runs with a GCM from CMIP5 driving an RCM. Especially for the historical runs, as well as the subsequent transient future runs, the matrix of GCMs and RCMs is expected to be quite large. While the resolution 0.44° (50 km) has been selected as standard for the CORDEX, for Europe the resolution of 0.11° (12 km) is emphasised as well. We participate in this activity with the model RegCM, both in evaluation experiment driven by ERA-Interim for 1989–2008 and transient historical and future run covering the period 1960–2100, at 50 km resolution, using scenario RCP4.5 for the future changes. As driving GCM, CNRM-CM5 is used. Figures 1.11 and 1.12 show the validation of temperature for the simulations driven by ERA-Interim and CNRM model, respectively. Further, Figs. 1.13 and 1.14 present climate change temperature signal for near future (2021–2050) and far future (2071–2100), respectively.

   The regional climate model we use is the model RegCM version 4.2 (hereafter referred to as RegCM4.2) from The International Centre for Theoretical Physics (ICTP), which is a three-dimensional mesoscale model. In terms of physical parameterizations it is based on RegCM3 (Pal et al. 2007) with many additional options. Major changes in the model from version 3 to version 4.2 include the following: the inclusion of the Community Land Surface Model v3.5 (CLM3.5) as an optional land surface parametrization, a new optional parametrization for diurnal SST variations, and a major restructuring (modularization) of the code base. RegCM4.2 and its evolution from RegCM3 is fully described by Giorgi et al. (2012).

   RegCM4.2 includes a two land-surface models: BATS and the CLM model. Both land-surface models can work in mosaic-type mode where the model grid is divided into sub-grid boxes for which the calculation of fluxes is carried out separately and the fluxes are then aggregated back to the large scale model gridbox (for BATS scheme refered as SUBBATS, see Pal et al. 2007). While in Europe scale we used standard BATS scheme, for the second line aiming to get high resolution downscaling with effect of urban parameterization included we have selected the SUBBATS on 1 km grid, which enable to identify clearly urban and suburban types of the land-use.

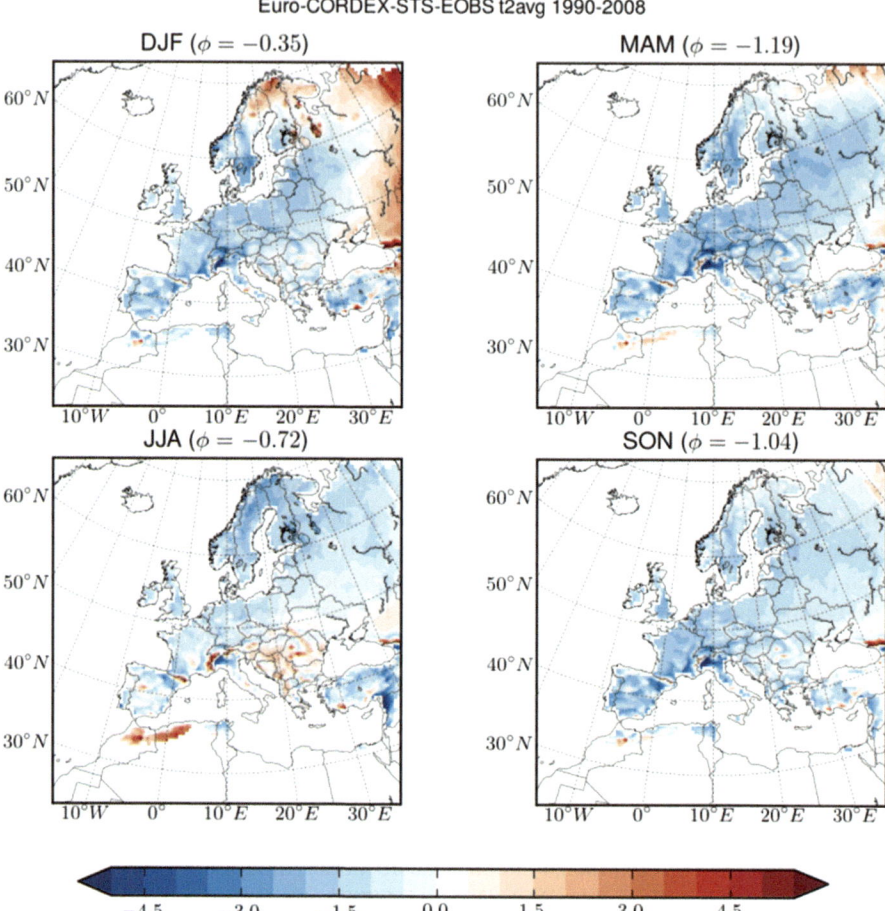

**Fig. 1.11** The validation of model mean temperature in terms of the difference of the ERA-Interim driven simulation against E-OBS data, for temperature and individual seasons

### 1.3.4.3 Urban Parameterization and Experimental Setup

Cities affect the boundary layer properties thus having direct influence on the meteorological conditions and therefore on the climate. The urban surface is covered by large number of artificial object with complex 3 dimensional structure and considerable vertical size. Specific characteristics in urban morphology can be involved in complicated physical processes such as increased momentum drag, radiation trapping between buildings (effect of vertical surfaces), and heat conduction by the artificial surfaces. There had been many field measurements in cities that found characteristic features of mean flow, turbulence and thermal structures in the urban boundary layer (e.g. Allwine et al. 2002; Rotach et al. 2005).

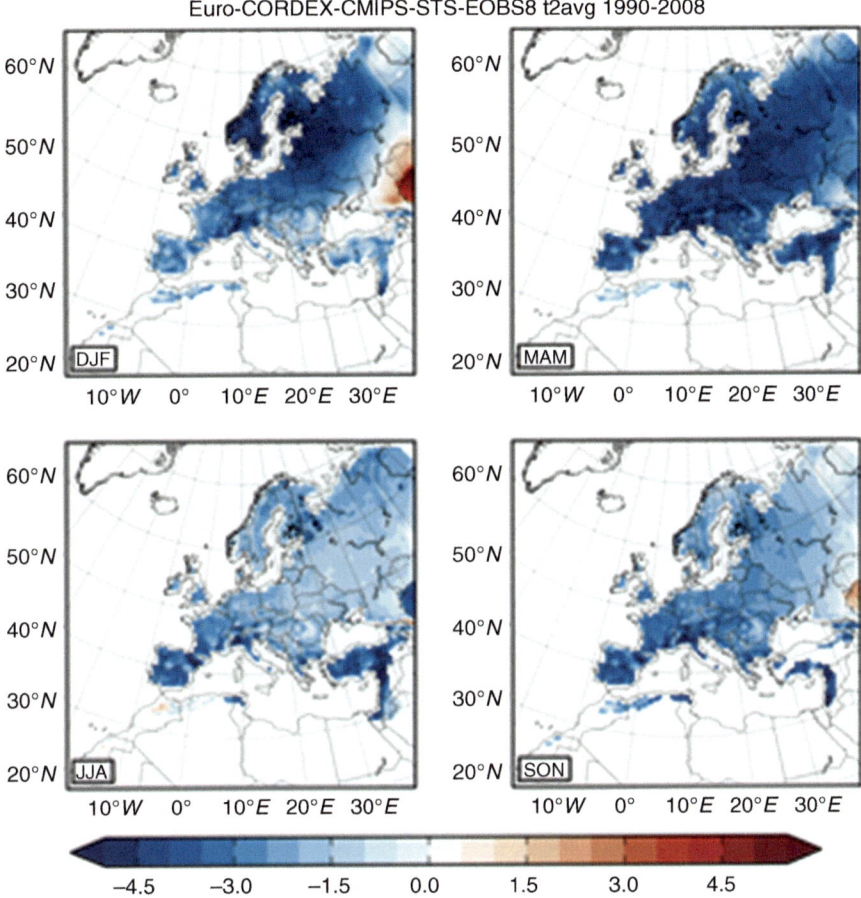

**Fig. 1.12** The validation of model mean temperature in terms of the difference of the CNRM-CM5 driven simulation against E-OBS data, for temperature and individual seasons

Although there is a trend (enabled by the faster computational resources) to increase the spatial resolution of the mesoscale models, regional weather prediction and climate models still fail to capture appropriately the impact of local urban features on the mesoscale meteorology and climate without special sub-grid scale treatment. This accelerated the implementation and application of urban canopy sub-models (Chen et al. 2010 or Lee et al. 2010). For the regional climate model RegCM4 we have chosen the Single Layer Urban Canopy Model (SLUCM) developed by Kusaka et al. (2001) and Kusaka and Kimura (2004); this scheme is proven to perform well in simulating the urban environment and it is less demanding in computational resources unlike its multi-layer counterparts (Lee et al. 2010).

SLUCM considers the urban surface in a realistic way: it assumes street canyons with a certain width; in the street canyon, shadowing, reflection and trapping of

Euro-CORDEX-CMIP5-STS-2021-2050 t2avg 1961-1990

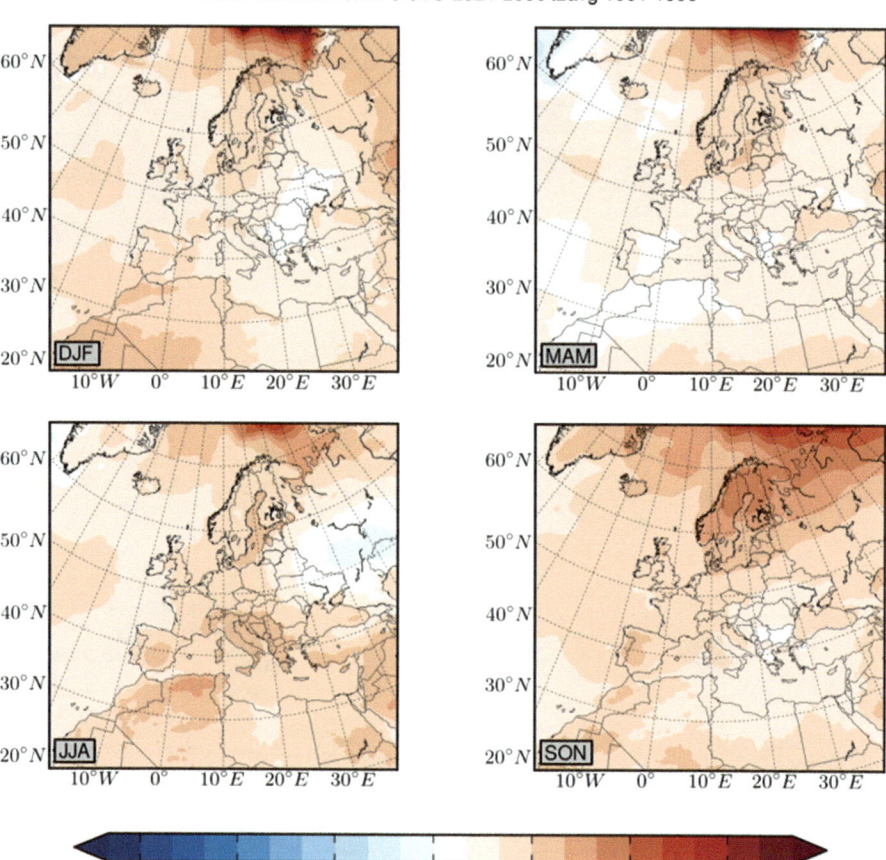

**Fig. 1.13** The climate change signal for near future in terms of the difference of the CNRM-CM5 driven simulation 2021–2050 against 1961–1990, for temperature and individual seasons under RCP4.5

radiation are considered. An exponential wind profile is prescribed. SLUCM treats surface skin temperatures at the roof, wall, and road and temperature profiles within roof, wall and road layers as prognostic variables. The heat fluxes from each surfaces are calculated using the Monin-Obuchov similarity theory and finally the canyon drag coefficient and friction velocity is computed using a similarity stability function for momentum. Figure 1.15 presents the conceptual design of SLUCM with the fluxes between street canyon air and the surrounding surfaces (road and walls) and the fluxes from/to the building roofs.

For high resolution downscaling with dynamic resolution of 10 km for the Central Europe we use SUBBATS scheme at 1 km resolution as already mentioned above. However, as the parameterization used till now in RegCM4 did not recognised the urban effects, an improvement can be achieved by implementing more

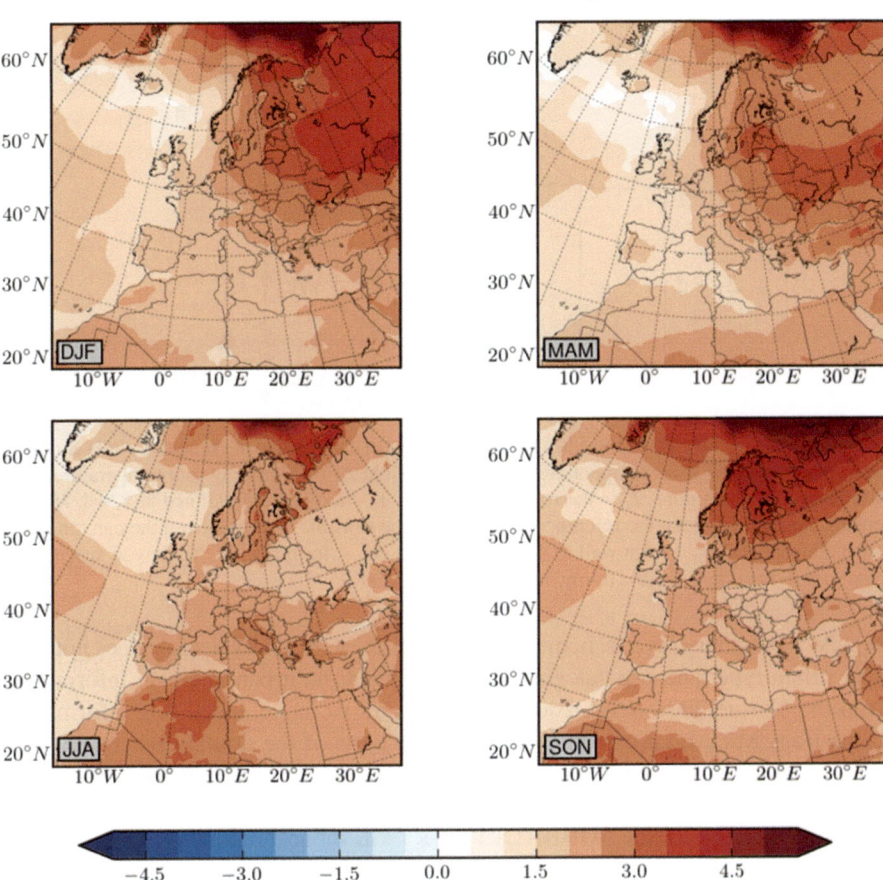

**Fig. 1.14** The climate change signal for near future in terms of the difference of the CNRM-CM5 driven simulation 2071–2100 against 1961–1990, for temperature and individual seasons under RCP4.5

sophisticated urban parameterizations lying under these land-surface models that better represent for the urban land-use type most urban features like building morphology, street geometry, variability of the properties of artificial surfaces, as well as the description of radiation trapping in the street canyon. For this purpose, Chen et al. (2010) provide a Single Layer Urban Canopy Model (SLUCM), originally developed by Kusaka et al. (2001) and applied in Kusaka and Kimura (2004).

This SLUCM model has been implemented into RegCM4.2 by linking it to the BATS surface scheme, applying SUBBATS with 1 km × 1 km sub-grid resolution. SLUCM is called within SUBBATS wherever urban land-use categories are recognized in the land-use data supplied. The scheme returns the total sensible heat flux from the roof/wall/road to BATS, as well as the total momentum flux. The total friction velocity is aggregated from urban and non-urban surfaces and passed to

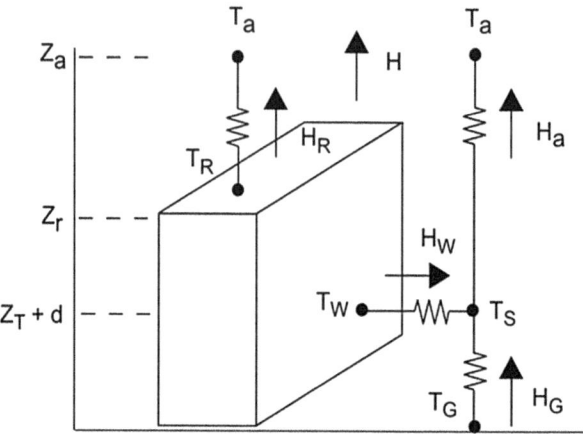

**Fig. 1.15** Energy fluxes in the SLUCM between the street canyon and the road and walls and from the buildings roof ($T_a$ – air temperature at reference height $z_a$, $T_R$ – building roof temperature, $T_W$ – building wall temperature, $T_G$ – the road temperature, $T_S$ – temperature defined at $z_T$+ d, $H$ – the sensible heat exchange at the reference height, $H_a$ is the sensible heat flux from the canyon space to the atmosphere, $H_W$ – from wall to the canyon space, $H_G$ – from road to the canyon space, $H_R$ – from roof to the atmosphere) (following Kusaka et al. 2001)

RegCM's boundary layer scheme. However, as RegCM4.2 by default does not consider urban type land-use categories, we extracted the urban land-use information from the Corine 2006 (EEA 2006) database and we have added this information to the RegCM4.2 land-use database. In those parts of the domain where this was not available in Corine data, the GLC2000 (GLC 2000) database was used. We considered two categories, urban and suburban. See Fig. 1.16 for the urban land-use coverage for the SUBBATS 1 km × 1 km subgrid module.

The domain for the present study has been selected to cover most of Central Europe with a spatial resolution of 10 km × 10 km. It is divided into 23 vertical levels reaching up to 5 hPa. For convection, we have used the Grell scheme (Grell 1993). RegCM4.2 is initialized and driven by the ERA-Interim reanalysis (Simmons et al. 2007). The time step for the integration is 30 s.

#### 1.3.4.4 Results

Figure 1.17 presents the change of selected meteorological parameters between experiments SLUCM (the urban canopy model turned on) and NOURBAN (urban canopy not considered) averaged over years 2005–2009. Shaded areas represent significant changes on the 95 % confidence level. We show only winter (left panels) and summer (right panels) seasons, actually, the effect is well expressed in spring and autumn as well, but summer signal is stronger.

For temperature, there is an evident increase with urban canopy introduced in summer, for winter only slight signal can be seen for big cities like Berlin and

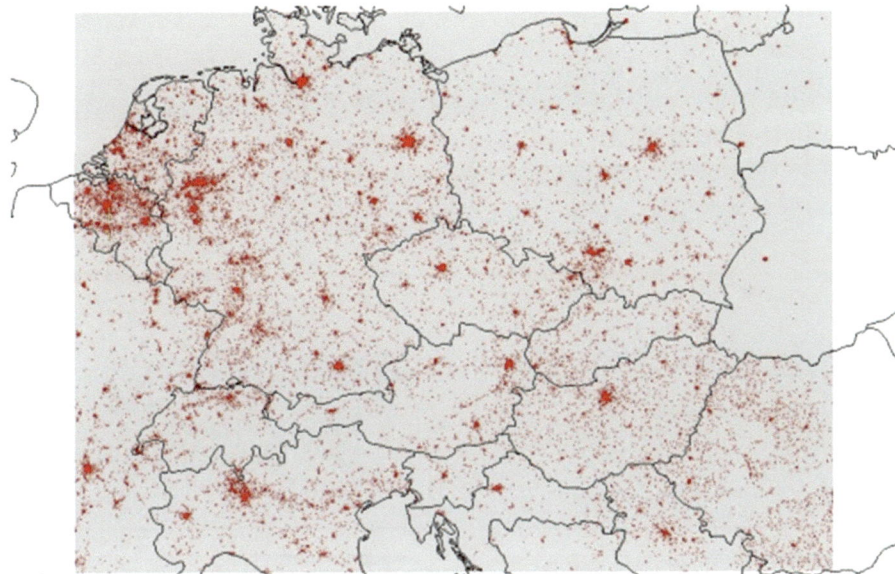

**Fig. 1.16** Urban and suburban land-surface categories at 2 km × 2 km resolution

Vienna, similarly for urban and industrial areas like Rhine-Ruhr region and Po-valley. In summer, this temperature increase can be of 1 K over urbanized areas (effect of cities like Budapest, Vienna, Prague, Berlin are well seen), but it is statistically significant elsewhere with up to 0.4 K increase even over non-urban areas. Opposite effect can be seen for specific humidity. Urban surfaces can absorb less water vapor than other surfaces and they represent a sink for the precipitated water as well. Therefore the evaporation from the urban surfaces is reduced as well which leads to the lower humidity over urban areas as seen in Figure 1.17. Again, this decrease is highest above cities (up to -0.8 g/kg), but significant decrease is simulated over non-urbanized areas as well, up to -0.3 – -0.4 g/kg. Signal is quite strong in summer, but similar patterns, although much slighter, can be seen in winter. For wind speed, introducing the urban canopy parameterization leads to stronger wind over the surface (Fig. 1.13). This increase is limited mainly over urban areas where it can reach 0.4–0.6 m.s$^{-1}$ in summer, much less it is expressed in winter, when for Po-valley there is even decrease. However, the signal is rather small in winter and not so much significant in all the domain. The increase above the cities in summer has to be further studied, one possible reason might be support of convection above the city with stronger winds in the bottom. Finally, we assess the effect of urban canopy parameterization on the height of planetary boundary layer from the model, which leads to statistically significant increase in summer above most of the domain, with quite strong signal above the cities and industrial regions (Fig. 1.17) of about 100–150 m, mostly negligible and not significant in winter.

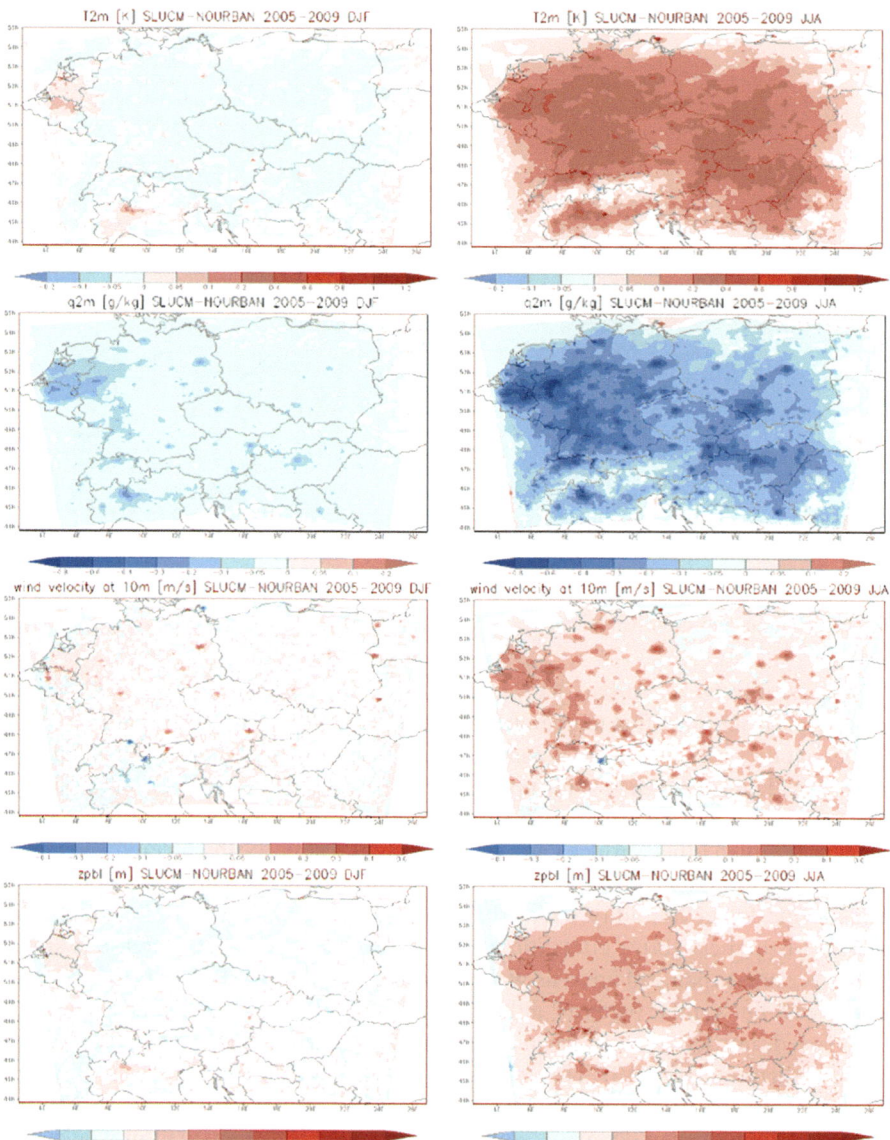

**Fig. 1.17** The mean differences of meteorological parameters between experiments with SLUCM against NOURBAN averaged over 2005–2009 for winter (*left panels*) and summer (*right panels*): from the top – temperature at 2 m (K), specific humidity at 2 m (g.kg$^{-1}$), wind speed at 10 m (m.s$^{-1}$), and planetary boundary height (m). *Shaded areas* represent significant changes on the 95 % level of confidence

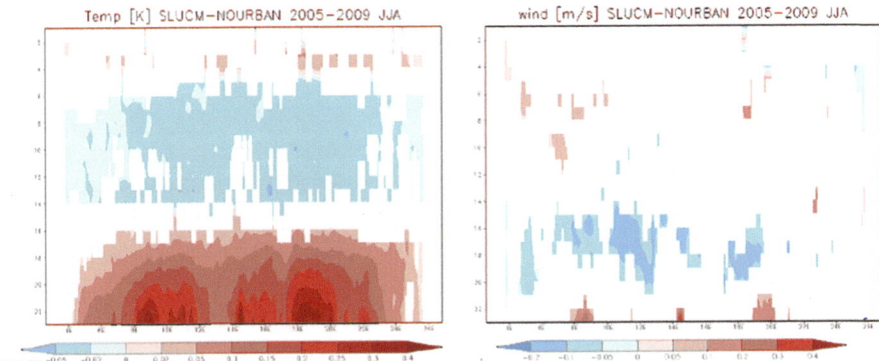

**Fig. 1.18** The mean differences in summer between experiments with SLUCM against NOURBAN averaged over 2005–2009 for vertical cross-section on 50 N of temperature (K, *left panel*) and wind speed (m.s⁻¹, *right panel*). *Shaded areas* represent significant changes on the 95 % level of confidence

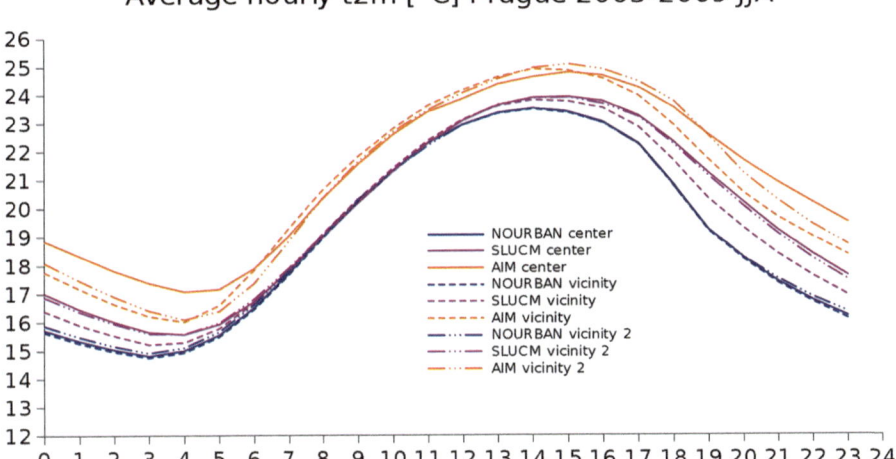

**Fig. 1.19** Daily course of 2 m temperature for Prague city center and two points in vicinity, where number 2 is rather suburban. The simulation without urban effect included is shown in *blue*, with the effect included in *violet* and observations are in *orange*. Summer season is shown

Figure 1.18 presents the more detailed analysis of significant patterns of vertical structure of the urban parameterization effects in summer. The increase of temperature in the boundary layer is accompanied with temperature decrease above, concerning the humidity there is no effect in the free atmosphere (not shown). Stronger wind can be seen only at surface level, the effect throughout the boundary layer is rather negative.

In Fig. 1.19, the daily courses of surface temperature (2 m) are shown for Prague, with more detailed analysis of the simulations with respect to the observation data.

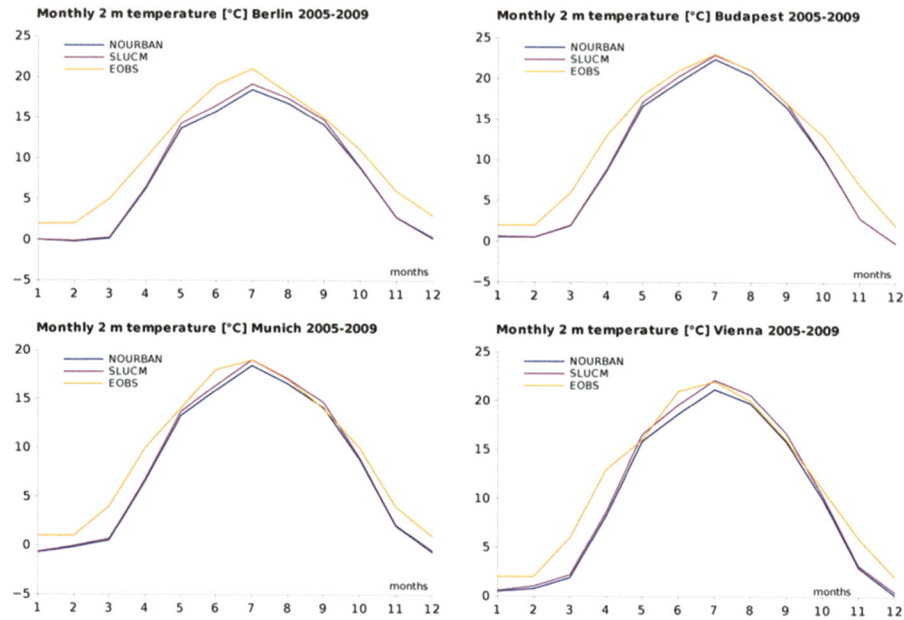

**Fig. 1.20** Annual course of 2 m temperature for selected cities based on the simulation without urban effect included (*blue curve*) and with the effect included (*violet*), compared to observations (*orange*)

In summer the strongest effect is shown, with clear extension of high temperature in evening hours both in city center data and in the simulation with urban effect included, to some extent as well in case of the second vicinity point, which has rather suburban character and it seems to be interpreted by the model with similar patterns as urban. Underestimation of the temperature in both simulations is evident, but especially during afternoon and night hours the urban effect contribute toward the bias reduction. Similar effect is clearly seen in comparison of simulations for other selected cities of the Central Europe in Fig. 1.20 in annual course, with E-OBS data shown as well. Again, introduction of the urban effect results in bias reduction in warm part of a year.

### 1.3.4.5   Conclusions

We have prepared the modelling stream for the assessment of the urban environment patterns and their changes due to both climate and other land-use or city structure changes. We have completed the decade 2021–2030 high resolution downscaling with respect to the pilot action in Prague, where city authorities are interested rather in closer future. Having both only BATS and SUBBATS results we can assess the effect of the urban environment. The analysis of other Euro-CORDEX results can provide the information on the uncertainity or the spread of the local changes. Finally, we are running the CTM CAMx coupled with the RegCM4 to assess the impact of the changes on air-quality. The changes of emissions will be taken into account as well. Moreover, CLMM model is used to get effects in street canyons.

We successfully implemented a single layer urban canopy parametrization into the regional climate model RegCM4.2. Preliminary assessment is based on present day conditions simulation for 5 year long period with and without urban canopy parameterization included. Our simulations have shown that the impact on meteorological parameters is significant not only over urbanized areas but also over rural ones far from cities. The most important impact is the increase of surface temperature (up to 1 K), decrease of humidity, increase of surface wind speed, decrease of precipitation (not shown here) and increase of boundary layer height.

# References

Giorgi, F., Coppola, E., Solmon, F., Mariotti, L., Sylla, M., Bi, X., Elguindi, N., Diro, G. T., Nair, V., Giuliani, G., Cozzini, S., Guettler, I., O'Brien, T. A., Tawfik, A., Shalaby, A., Zakey, A., Steiner, A., Stordal, F., Sloan, L., & Brankovic, C. (2012). RegCM4: Model description and preliminary tests over multiple CORDEX domains. *Climate Research, 52*, 7–29.

GLC. (2000). Global Land Cover 2000 database. European Commission, Joint Research Centre, 2003. http://bioval.jrc.ec.europa.eu/products/glc2000/glc2000.php

Grell, G. (1993). Prognostic evaluation of assumptions used by cumulus parameterizations. *Monthly Weather Review, 121*, 764–787.

Gurjar, B. R., Jain, A., Sharma, A., Agarwal, A., Gupta, P., Nagpure, A. S., & Lelieveld, J. (2010). Human health risks in megacities due to air pollution. *Atmospheric Environment, 44*, 4606–4613.

Kusaka, H., Kondo, H., Kikegawa, Y., & Kimura, F. (2001). A simple single-layer urban canopy model for atmospheric models: Comparison with multi-layer and slab models. *Boundary-Layer Meteorology, 101*, 329–358.

Kusaka, H., & Kimura, F. (2004). Coupling a single-layer urban canopy model with a simple atmospheric model: Impact on urban heat island simulation for an idealized case. *Journal of the Meteorological Society of Japan,, 82*, 67–80.

Lee, S.-H., Kim, S.-W., Angevine, W. M., Bianco, L., McKeen, S. A., Senff, C. J., Trainer, M.S., Tucker, C., & Zamora, R. J. (2010). Evaluation of urban surface parameterizations in the WRF model using measurements during the Texas Air Quality Study 2006 field campaign. *Atmospheric Chemistry and Physics Discussions, 10*, 25033–25080.

Oke, T. R. (1973). City size and the urban heat island. *Atmospheric Environment* (1967), 7(8),769–779.

Pal, J. S., Giorgi, F., Bi, X., Elguindi, N., Solomon, F., Gao, X., Francisco, R., Zakey, A., Winter, J., Ashfaq, M., Syed, F., Bell, J. L., Diffenbaugh, N. S.,

Karmacharya, J., Konare, A., Martinez, D., da Rocha, R. P., Sloan, L. C., & Steiner, A. (2007). The ICTP RegCM3 and RegCNET: Regional climate modeling for the developing world. *Bulletin of the American Meteorological Society, 88*, 1395–1409.

Ryu, Y.-H., Baik, J.-J., Kwak, K.-H., Kim, S., & Moon, N. (2013). Impacts of urban land-surface forcing on ozone air quality in the Seoul metropolitan area. *Atmospheric Chemistry and Physics, 13*, 2177–2194.

Simmons, A., Uppala, S., Dee, D., & Kobayashi, S. (2007). ERAinterim: New ECMWF reanalysis products from 1989 onwards. *Newsletter, 110*(Winter 2006/07), ECMWF, Reading.

Timothy, M. & Lawrence, M. G. (2007). The influence of megacities on global atmospheric chemistry: A modeling study. *Environment and Chemistry, 6*, 219–225.

## 1.3.5  Statistical Downscaling Techniques Applied to ENSEMBLES GCMs: Bologna-Modena Case Study

Rodica Tomozeiu and Lucio Botarelli
ARPA Emilia-Romagna, Bologna, Italy
rtomozeiu@arpa.emr.it

### 1.3.5.1  Introduction

Another tool used by the scientific community in order to construct future climate projections is the statistical downscaling techniques (SDs). One of the main advantages of this technique is that it produces information at local scale, station or grid points and it is not expensive in terms of computational time. One major problem for all tools that produce climate change scenario is to quantify and reduce the uncertainties that appear in modelling processes. Particular attention has been paid on this problem and many projects have been focused on this issue. One of this is Ensembles project (http://www.ensembles-eu.org/), where it was recommended use of a range of models over the same area and construction of an ensemble mean (EM). This technique has been applied in the present work, in order to produce climate change scenario over Bologna-Modena case study selected in the project.

### 1.3.5.2  Data and Methods

The SDs model developed by ARPA-SIMC, is a multivariate regression based on Perfect–Prog approach, built using observed local fields at station level, and large scale fields derived from re-analysis data set. A set of 75 stations distributed over N-Italy, including Bologna station, that measure minimum and maximum temperature and the large scale fields from ERA-40 reanalysis, over the period 1960–2002 has been used in order to set-up SDs model. Once the most skilful model is selected for each season and predictand, this is then applied to the predictors simulated by AOGCMs experiments in the framework of A1B emission scenario, such as to evaluate the local future scenarios of seasonal temperature. The SDs scheme here proposed use as predictors a selection of fields between mean sea level pressure (MSLP), 500 hPa geopotential height (Z500), and temperature at 850 hPa (T850), already tested over Emilia-Romagna and N-Italy region (Tomozeiu et al. 2013). These fields (predictors) derived from ERA40 re-analysis (http://www.ecmwf.int/products/) have a spatial resolution of $1.125° \times 1.125°$, cover the window 90°W–90°E and 0°–90°N and are referred to the period mid-1957 (September) to mid-2002 (August). As regards the AOGCMs predictors from the ENSEMBLES –STRAEM1 (Van der Linden and Mitchell 2009) runs had been used, over the period 1961–1990 (control-run) and 2021–2050 (A1B scenario). These fields are archived in the Climate and Environmental Retrieval and Archive (CERA data base) of the World Data Center System for Climate (WDC) and the access at the data is given by http://ensembles.wdc-climate.de. The STREAM1 simulations (http://www.ensembles-eu.org/), used in the present work have been performed with the methodology and the forcing that were defined for the CMIP3 simulations contributing to the IPCC AR4 assessment. Thus, the experiments were done using a common set of agreed forcing for historical simulations over the period 1860–2000, and for the three IPCC scenario A1B, A2, B2, over the twenty-first century. The scenarios were started from an initial condition obtained for year 2000 in the historical simulation. Several runs, produced by the following modelling groups have been take into account in the present work: INGV, NERSC, FUB, IPSL, METOHC (2 runs), MPIMET+DMI. The statistical downscaling scheme (CCAReg scheme) was applied to each seasons and each predictands (seasonal minimum and maximum temperature). The presence of different AOGCMs gives the opportunity to construct an Ensemble Mean (EM) of climate projections. The results obtained by applying the outputs of the AOGCMs to the CCAReg scheme at Bologna station are presented bellow.

### 1.3.5.3   Results

The future changes are presented in terms of probability density functions (PDFs) of seasonal minimum and maximum temperature, which provide a good estimation of changes not only in the mean but also in the extreme values. As could be noted from Fig. 1.21, that presents the PDFs of changes in winter minimum temperature as projected by the CCAReg scheme applied to each AOGCM, all the outputs emphasizes an increase in the winter Tmin between 0.7 (BCCR and ECHAM5 models) up to 1.8 °C (IPSL and EGMAM run2), over the period 2021–2050 with respect to 1961–1990.

As concerns the other seasons, the Ensemble Mean of changes in minimum temperature computed taking into account all runs, reveals an increase of temperature in all seasons (see Fig. 1.22), around 1.5 °C during spring and autumn and 2.5 °C during summer.

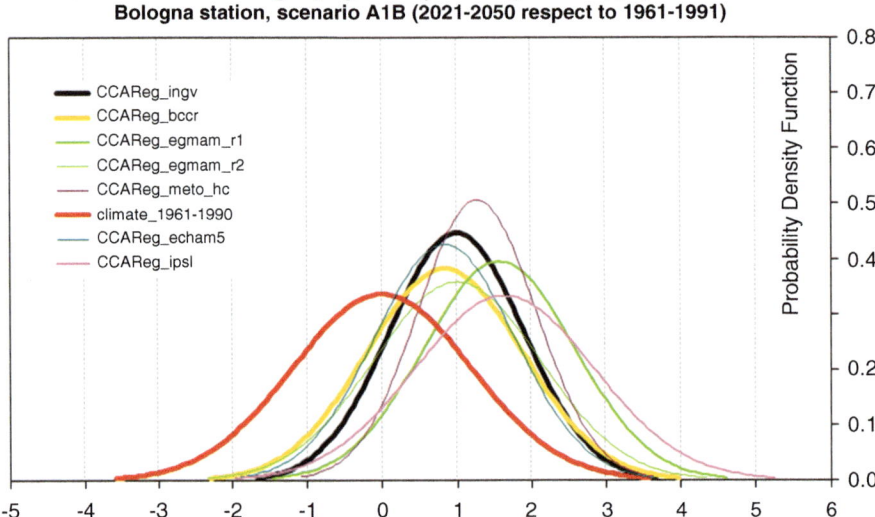

**Fig. 1.21** Climate change projections of winter minimum temperature-Bologna station, scenario A1B, 2021–2050

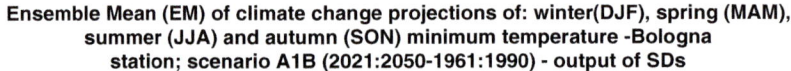

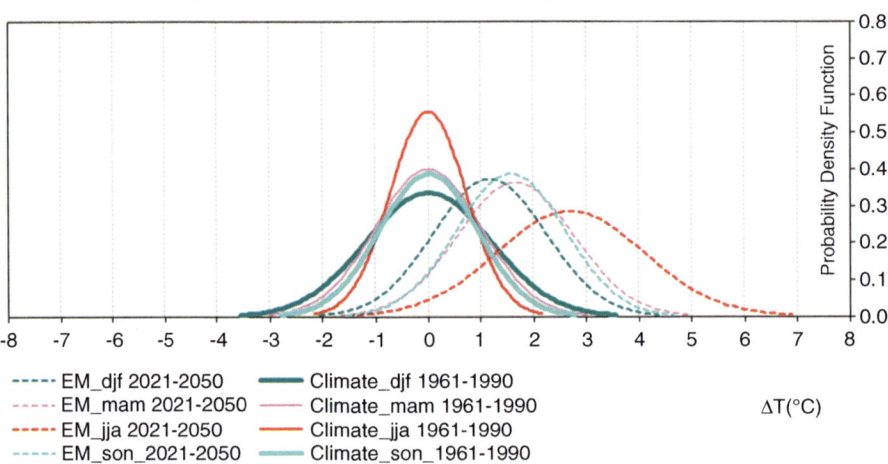

**Fig. 1.22** Ensemble Mean (EM) of seasonal changes of minimum temperature projected at Bologna station (CCAReg model), scenario A1B, 2021–2050 with respect to 1961–1990

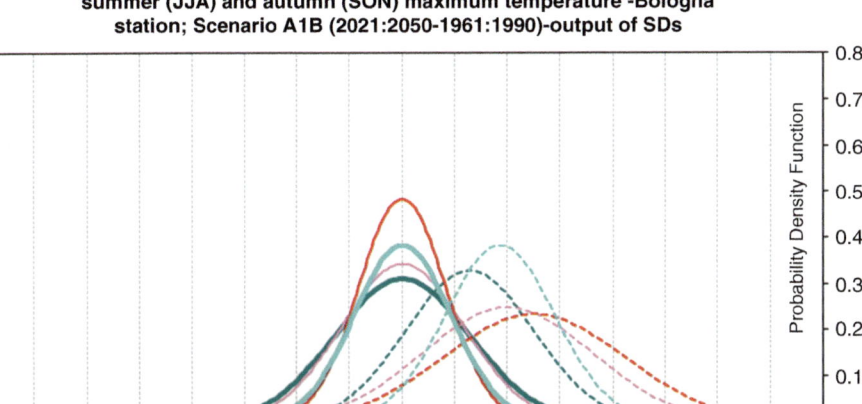

**Ensemble Mean (EM) of climate change projections of: winter(DJF), spring (MAM), summer (JJA) and autumn (SON) maximum temperature -Bologna station; Scenario A1B (2021:2050-1961:1990)-output of SDs**

**Fig. 1.23** Ensemble Mean (EM) of seasonal changes of maximum temperature projected at Bologna station (CCAReg model), scenario A1B, 2021–2050 with respect to 1961–1990

A similar signal of warming has been projected in seasonal maximum temperature. As it could be noted from Fig. 1.23, that presents the Ensemble Mean of changes in seasonal maximum temperature, the projected warming is around 1 °C during winter, similar with those projected in minimum temperature. During spring and autumn the maximum temperature is projected to increase with 2 °C (central moment of the distribution) with respect to 1961–1990. The peak of warming is projected to appear during summer season when the changes will be around 2.5 °C (central moment of the distribution).

Analyzing the PDFs of changes in seasonal minimum and maximum temperature (Figs. 1.22 and 1.23) it could be noted that all projected distributions (dashed curves) tend to shift to warm values with respect to present distributions (continuous curves). In addition significant changes could be noted not only in the central moment of the distribution but also in the tails, more significant in the upper tails of summer minimum and maximum temperatures when the 90th percentile could reach changes of 5 °C. A signal of increasing could be noted also in 10th percentile of minimum temperature, especially during summer. This could connect to an increase in the heat waves (days with Tmax > 90th percentile of Tmax) and number of tropical nights (Tmin > 20 °C) over Bologna during the period 2021–2050 with respect to 1961–1990. In fact, the downscaling o the heat wave index for each season emphasis a possible increase in the future 2021–2050 with respect to 1961–1990 Fig. 1.24 presents the climate scenarios of seasonal heat wave at Bologna, present

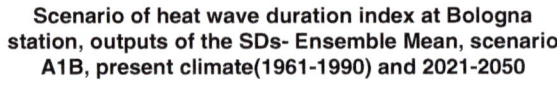

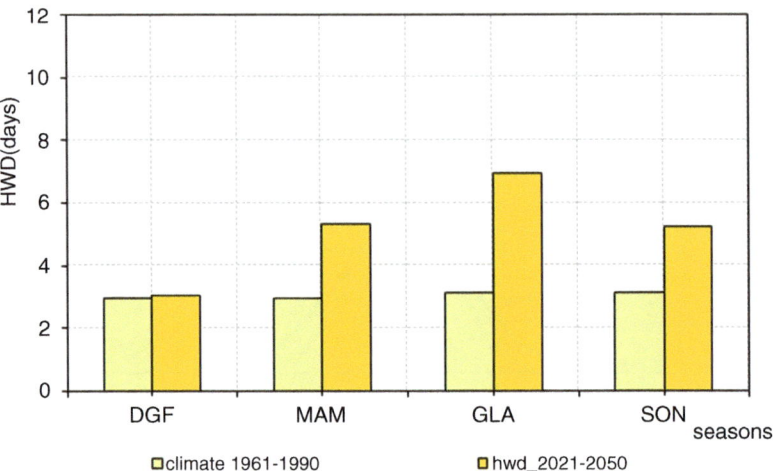

**Fig. 1.24** Ensemble Mean (EM) of seasonal heat waves projected at Bologna station (CCAReg model), scenario A1B, 2021–2050 and present climate (1961–1990)

and future climate. As could be noted significant increase is projected during summer season, followed by spring and autumn.

### 1.3.5.4 Conclusion

The future scenarios constructed through the statistical downscaling scheme applied to several GCMs show a possible increase in the seasonal minimum and maximum temperature over Bologna, around 2 °C over the period 2021–2050 with respect to 1961–1990. The signal is more intense during summer (around 2.5 °C the central moment of the distribution) when an increase of the frequency of heat waves is projected. The results are in agreement with those obtained by the regional climate models (Van der Linden and Mitchell 2009).

**Acknowledgement** The ENSEMBLES data used in this work was funded by the EU FP6 Integrated Project ENSEMBLES (Contract number 505539) whose support is gratefully acknowledged.

# References

Tomozeiu, R., Agrillo, G., Cacciamani, C., & Pavan, V. (2013). Statistically downscaled climate change projections of surface temperature over Northern Italy for the periods 2021–2050 and 2070–2099. Natural Hazards. doi:10.1007/s11069-013-0552-y

Van der Linden, P., & Mitchell, J. F. B. (2009). ENSEMBLES: Climate change and its impacts: Summary of research and results from the ENSEMBLES project, Met Office Hadley Centre, UK, 160 pp.

# Chapter 2
# Urban Heat Island Gold Standard and Urban Heat Island Atlas

## Gold Standard for UHI Measurements and Introduction of The Central-European Urban Heat Island Atlas

**Györgyi Baranka, L. Bozó, Rok Ciglič, and Blaž Komac**

**Abstract** Choosing appropriate measurement sites for further investigation is one of the crucial points in determining the UHI intensity. To obtain comparable values among different cities we should use measurements obtained in similar circumstances. In the context of the deployment of an urban climate network, this guideline contains recommendations for meteorological measurement and data processing for data users who are not professionals in the field of climate measurement in urban environments. This paper presents a classification of urban measuring sites, lists steps for choosing representative stations, and provides standardized methods developed for urban climate stations and network design. Finally, these standardized approaches and guidelines are useful for making recommendations for the future deployment of urban climate networks. The final parts include the new communication and data transmission techniques for urban observation networks.

**Keywords** Urban climate observation • Local Climate Zone • Urban site location • Exposure of instruments • Metadata • Data management

---

The original version of this chapter was revised. An erratum to this chapter can be found at http://dx.doi.org/10.1007/978-3-319-10425-6_15

G. Baranka (✉) • L. Bozó
Hungarian Meteorological Service, Kitaibel u. 1, 1024 Budapest, Hungary
e-mail: baranka.gy@met.hu

R. Ciglič • B. Komac
Anton Melik Geographical Institute, ZRC SAZU, Novi trg 2, 1000, Ljubljana, Slovenia

© The Author(s) 2016
F. Musco (ed.), *Counteracting Urban Heat Island Effects in a Global Climate Change Scenario*, DOI 10.1007/978-3-319-10425-6_2

## 2.1   Introduction

Before any type of standard recommended by standardization committees is
employed in practice, the following points should be emphasized:

- The application of standards is not compulsory.
- Obeying standards is usually voluntary.
- These statements are supplied as recommendations and suggestions during the
  investigation and studying of urban climate.

The original aims in elaborating these guidelines were:

- to review the current status of urban observations and the establishment of urban
  meteorological networks;
- to identify sampling infrastructures to be developed in urban areas;
- to provide a reference system for cities which lack a monitoring network but
  intend to deploy one;
- to demonstrate a prototype for a new monitoring system that is already
  operational.

Any city operating measurement sites for UHI detection is strongly encouraged
to adopt the recommendations of the Gold Standard to its measurement and evaluation systems in order to obtain better coverage of UHI phenomena over the city.

Finally, these standardized approaches and guidelines are useful for making recommendations for the future deployment of urban climate networks. The final chapter includes the new communication and data transmission techniques for urban
observation networks.

In this era, environmental observations are not the sole privilege of hydrological
and meteorological institutions. There are several examples of different networks
working together, with the output of each supplementing the others. It remains true
that reliable data are needed; networks and measuring systems need to meet various
requirements (for example ISO 2009:2008). Complying with these regulations and
laws can make the operation of such measuring systems more expensive.

The description presented below shows that any kind of observation requires a
significant financial effort, technical background, and maintenance capacity for any
agency, municipality, or national institution undertaking this activity. New in-situ
observatories based on citizens' personal devices reduce investment in and running
costs of in-situ observations and monitoring applications. One method called
"Citizen Observatories" is based on devices such as smart phones, tablets, laptops,
and social media, and it can strengthen environmental monitoring capabilities.

## 2.2   Concepts

The heterogeneous natural and artificial surfaces of urban environments imply that
atmospheric observations require a dense measuring network to resolve the local
climate adequately. In this document, the current state of urban meteorological sites
and networks is shown. Afterward, this document presents suggestions for better

descriptions and representations of the surroundings of stations, and it also outlines better documentation of network characteristics, standardized approaches, and recommended guidelines for urban observations to follow.

Before the deployment of a meteorological station, logistics and plans are needed. The establishment a new urban climate network requires a huge financial commitment from the developer; underscoring the need for forward-looking planning. Prior to implementation, the investigator should make efforts to find suitable meteorological sites and instruments; determine their measurement programs; and describe the frequency of maintenance and calibration procedures required by quality assurance and quality control (QA/QC) systems for any kind of measurement.

The observation of urban climate can proceed through mobile measurements and stationary sites. Because of the relatively high cost and difficulty in siting equipment for fixed meteorological monitoring stations, their deployment and maintenance ultimately results in sparse data coverage for urban areas. In this document, recommendations are made for the operation of fixed meteorological stations and networks, which are the only way to gain sufficient, detailed information from urban regions. These observation methods can be complemented by remote sensing techniques used for interpolation, but even these do not allow for the appropriate spatial and temporal resolution or a sufficiently wide range of observed variables.

Traditionally, meteorological measurements have not been taken in urban areas but in open areas representing larger regions, as the WMO (2008) prescribed in siting synoptic measuring stations, and as it was also presented in Chapter 5 of the UHI Assessment Manual. However, on many occasions it is impossible or makes no sense to conform to these guidelines. This document recommends some principles that will help in such circumstances, even though it is not possible to anticipate all eventualities. The recommendations presented here remain in agreement with general objectives set out by the WMO (see the chapter "Exposure of instruments"). Many urban stations have been placed over short grass in open locations (parks, playing fields), and as a result they are actually monitoring an environment of the type modified rural. In many respects, the generally accepted standards for the exposure of meteorological instruments set out in WMO (2008) applies to urban sites. All details, including deviations from guidelines, should be logged thorough "metadata" (additional information about the whole network), which is essential in order to provide a data end-user with the information required to process and use the network's data adequately.

This paper aims to provide instruction on how to obtain integrated and harmonized measurement data in relation to UHI phenomenon, by standardizing and unifying urban microclimate data. Toward these ends, the following items will be discussed:

• Defining urban climate zones
• Site placement
• Network design
• Instrumentation
• Operational definition
• Reporting of data.

Complex, morphologically heterogeneous urban environments must be studied in fine detail, so it as necessary to obtain a better understanding of weather and climate interactions and impacts in these areas. Robust planning, design, field documentation, installation, management, quality assurance (QA), and maintenance are essential parts of any successful network of sensors.

Prior to the establishment of a new monitoring network in an urban environment, the following logistical steps should be undertaken:

1. Collect information and consult with experts about the climate observations and surveying or measuring activities in advance.
2. In the planning phase, write a detailed task list. Determine the aims and tasks of the measurement program.
3. List the measurement climate elements (see the UHI Assessment Manual, Chapter 6) and define the technical requirements of measurement as range, reported resolution, required measurement uncertainty, sensor time constant, output averaging time, and achievable measurement uncertainty (see the UHI Assessment Manual, Chapter 7, for a detailed description).
4. Process orders to obtain the most appropriate equipment.
5. Choose the sites and install the equipment.
6. Prepare sensors and data logging and metadata files (see the UHI Assessment Manual, Chapter 8, for a description of metadata).
7. Record the calibration and maintenance requirements of the climate network. These steps are summarized in Fig. 2.1. After their implementation, reliable climate data will be available for the further evaluation and statistical analysis of climate signals in a given urban area.

The original aims in elaborating these guidelines were

– to review the current status of urban observations and the establishment of urban meteorological networks;
– to identify sampling infrastructures to be developed in urban areas;

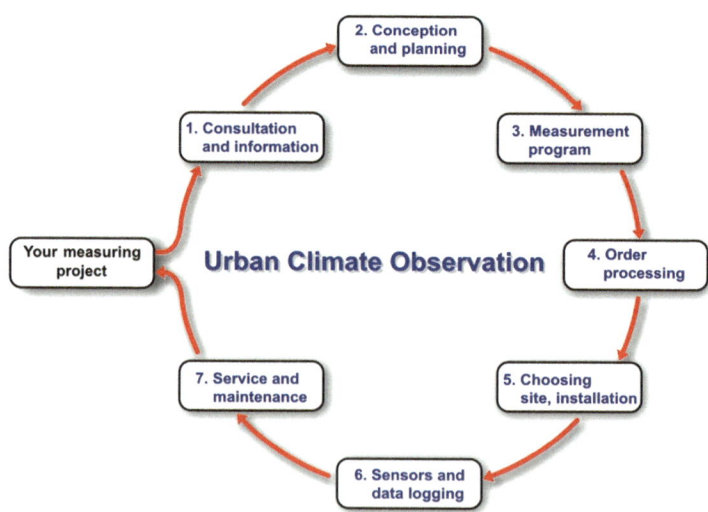

**Fig. 2.1** Process chain for measurement project planning, management and service

- to provide a reference system for cities which lack a monitoring network but intend to deploy one;
- to demonstrate a prototype for a new monitoring system that is already operational.

Any city operating measurement sites for UHI detection is strongly encouraged to adopt these recommendations to its measurement and evaluation systems in order to obtain better coverage of UHI phenomena over the city.

## 2.3   Planning a Representative Urban Climate Station Network

The simple classification of measuring sites contains categories for urban, downtown, suburban, and rural sites. For correct and precise heat island evaluation, far more detailed descriptions of urban areas are necessary, as Stewart and Oke's method (2012) suggests applying. Local Climate Zones were introduced for characterizing the building and vegetation structures of a compact urban region.

Stewart and Oke formally defined Local Climate Zones (LCZs) as regions of uniform surface cover, structure, material, and human activity that span hundreds of meters to several kilometers in the horizontal scale. The LCZ system comprises 17 zone types at the local scale (100–1000 m). Each type is unique in its combination of surface structure, cover, and human activity. Classification can be made on the basis of built-up area (10 classes) and types of land cover (7 classes). Classification of sites into appropriate LCZs requires basic metadata and surface characterization. The zone definitions provide a standard framework for reporting and comparing field sites and their climate observations.

LCZs are a widely used classification system (Unger et al. 2011). It provides useful information for

- defining UHI magnitude,
- establishing comparable values originating from different sites,
- modeling climate and analyzing temperature,
- detecting features that influence microclimates.
- Determination of LCZs means defining urban areas characterized by the same structure, a similar built-up ratio, and relatively similar building heights (Balázs et al. 2009). Bordering is essential for finding ideal and representative sites for urban climate observation(Unger et al. 2011). Typically, a "three-step process" is suggested by Stewart and Oke (2012) "to users when classifying field sites into LCZs" (p. 1889):

  "Step 1: Collect site metadata. Users must collect appropriate site metadata to quantify the surface properties of the source area (as defined in Step 2) for a temperature sensor. This is best done by a visit to the field sites in person to survey and assess the local horizon, building geometry, land cover, surface wetness, surface relief, traffic flow, and population density [...]. If a field visit is not possible, secondary sources of site metadata include aerial photographs, land cover/land use maps, satellite images (e.g., Google Earth©), and published tables of property values (e.g., Davenport terrain roughness lengths)" (Stewart and Oke, 2012, pp. 1889–1890).

"Step 2: Define the thermal source area. The thermal source area for a temperature measurement is the total surface area «seen» by the sensor [...]." (Stewart and Oke, 2012, p. 1890) "Sources will include upwind buildings, the walls and floor of an upwind street, and perhaps a branching network of more distant street canyons" (p. 1890).

"Quantifying the surface properties for field sites and source areas located on or near the border of two (or more) zones is problematic. If the location of the sensor can be moved, it should be placed where it samples from a single LCZ. [...]. If the location of the sensor cannot be moved, temperature data retrieved from that site should be stratified first according to wind direction, then to LCZ. [...] A site with a split classification is less ideal for heat island studies because changes in airflow and stability conditions interfere confuse the relation between surface form/cover and air temperature. It is recommended that transitional areas be avoided when siting meteorological instruments" (Stewart and Oke, 2012, p. 1891).

"Step 3: Select the local climate zone. Metadata collected in Step 1 should lead users to the best, not necessarily exact, match of their field sites with LCZ classes. Metadata are unlikely to match perfectly with the surface property values of one LCZ class. If the measured or estimated values align poorly with those in the LCZ datasheets, the process of selecting a best-fit class becomes one of interpolation rather than straight matching. Users should first look to the surface cover fractions of the site to guide this process. If a suitable match still cannot be found, users should acknowledge this fact and highlight the main difference(s) between their site and its nearest equivalent LCZ" (Stewart and Oke, 2012, p. 1891).

Stewart and Oke (2012) note that "updating LCZ designations is crucial for all sites, particularly those used in long-term temperature studies" (p. 1893). They add that "sites located on the edges of cities where urban growth and environmental change are rapid, or in the cores of cities where land redevelopment and large-scale greening projects are taking place, should be surveyed and classified annually. For sites used in mobile or short-term stationary surveys, the frequency of updates is dictated largely by day-to-day variations in weather and soil moisture [...]" (p. 1893).

## 2.4  Exposure of Instruments

There are numerous possibilities for setting up urban climate stations, such as screen level, on a roof, in a street canyon, which vary depending on the LCZs where they are situated.

A primary survey of study area can document the presence of obstacles close to the measurement site. The main discrepancies are caused by unnatural surfaces and shading:

(a) Obstacles around the screen influence the irradiative balance of the screen. A screen close to a vertical obstacle may be shaded from the solar radiation, "protected" against the night radiative cooling of the air by receiving the warmer infrared radiation from this obstacle, or otherwise influenced by reflected radiation.
(b) Neighboring artificial surfaces may heat the air. Reflective surfaces (e. g., buildings, concrete surfaces, car parks) and water sources (e. g., ponds, lakes, irrigated areas) should be avoided.

Each climate parameter being measured at a site carries its own considerations. The following requirements represent a good urban observation station, acceptable for air temperature and humidity measurements (Fig. 2.2):

(a) measurement point situated at least 10 m from artificial heat sources and reflective sources (buildings, concrete surfaces, car parks, etc.) or expanses of water (unless indicative of the region) occupying

  (i) less than 50 % of the surface within a circular area of 10 m around the screen,
  (ii) less than 30 % of the surface within a circular area of 3 m around the screen.

(b) station away from all projected shade when the sun is higher than 20°.
(c) station within ground covered with natural and low vegetation (<25 cm) representative of the region.

Within these guidelines, the recorded temperature and humidity data contain additional uncertainty of up to 2 °C compared to a station placed on flat, horizontal land surrounded by open space.

The choice of a site for representative precipitation measurement in an urban area should fulfill the following minimum requirements (Fig. 2.3):

(a) The land is surrounded by an urban area, on a slope of less than 30°.
(b) Possible obstacles must be situated at a distance greater than one half the height of the obstacle. An obstacle represents an object with an angular width of 10° or more.

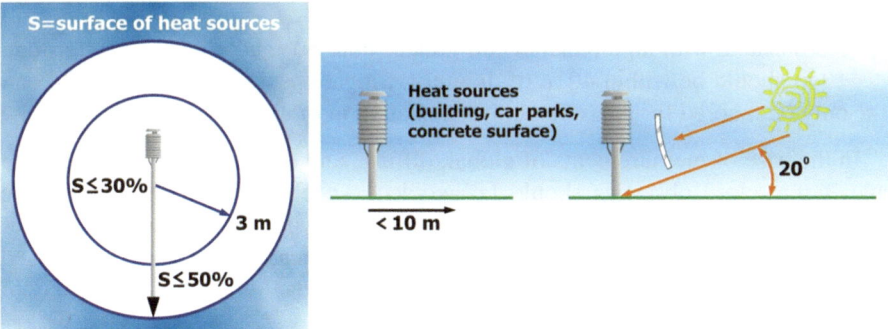

**Fig. 2.2** Suitable location of air temperature and humidity sensors in urban environments (Based on WMO Guide, 2008)

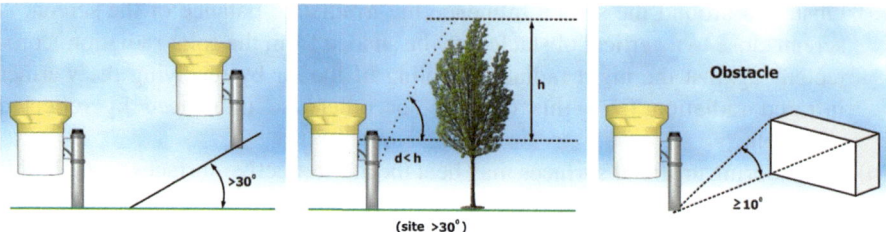

**Fig. 2.3** Representative precipitation measurement in an urban area (Based on WMO Guide, 2008)

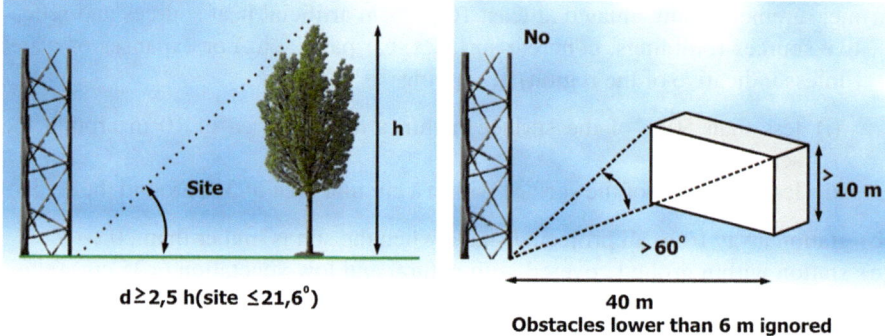

**Fig. 2.4** Reliable location for measuring mast (Based on WMO Guide, 2008)

By following these guidelines, uncertainty of precipitation measurements remains within a 25 % deviation from standard observations taken in a flat and open area.

(c) Where numerous obstacles (buildings, trees, etc.) are present, it is recommended that sensors be placed above the average height of the obstacles to minimize the influence of adjacent obstacles. A measuring mast can serve as a reliable location (Fig. 2.4).

(d) The mast should be placed away from obstacles, at a distance of at least 2.5 times the height of surrounding obstacles;

(e) No obstacle with an angular width greater than 60° and a height greater than 10 m should be within 40 m of the mast. Single obstacles lower than 6 m can only be ignored for measurements taken at 10 m or above.

In the case of measurements of global, diffuse, and direct radiation, the general rule is that close obstacles should be avoided. An obstacle is considered to be reflecting if its albedo is greater than 0.5. The reference position for elevation angles is the sensitive element of the instrument. It is recommended that shade be allowed to project into the sensor during no more than 30 % of daylight hours for any day of the year (Fig. 2.5).

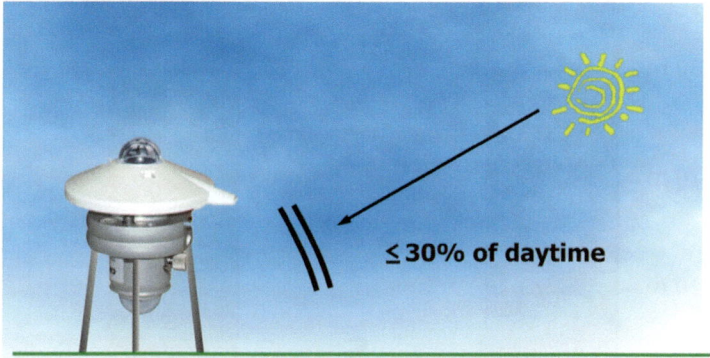

**No shade projected for more than 30% of daytime**

**Fig. 2.5** Recommendation for representative solar radiation observation (Based on WMO Guide, 2008)

If the site does not meet all the requirements above, the specific environment or application should be recorded in a metadata file. The general template of a station metadata file is presented in Sect. 2.6.

## 2.4.1   Temperature

In different LCZs, the following suggestions should be considered in finding suitable locations for air temperature measurement:

- The sensor should be relatively far from warm surfaces, walls, roads, or vehicles with hot engines, and it might receive reflected heat from glassed surfaces.
- In very densely built-up LCZs (such as LCZ 1, LCZ 2, LCZ 4, and LCZ 5) sensors should be set up 5–10 m from buildings of height 20–30 m, as shown in Fig. 2.6.
- Sometimes in slightly greater source areas, sensor damage can be prevented by placing the sensor away from the path of vehicles, thereby avoiding exhaust heat and dust contamination. Measurement heights of 3 m or 5 m are accepted as a standard.
- There is no simple, general scheme for extrapolating air temperature horizontally inside the Urban Canopy Layer (UCL).

The following recommendations are made for choosing the most reliable height for a thermometer:

- The recommended screen height is between 1.25 m and 2 m above ground level for urban sites (similar to non-urban stations), but it is better to allow a greater height in densely built-up areas (such as LCZ 1, LCZ 2, LCZ 4, and LCZ 5).

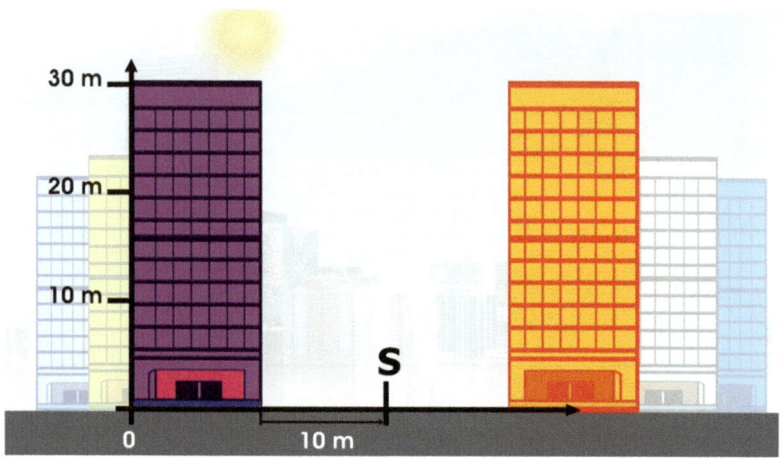

**Fig. 2.6** Thermal and humidity measures on the horizontal scale in a densely built-up area

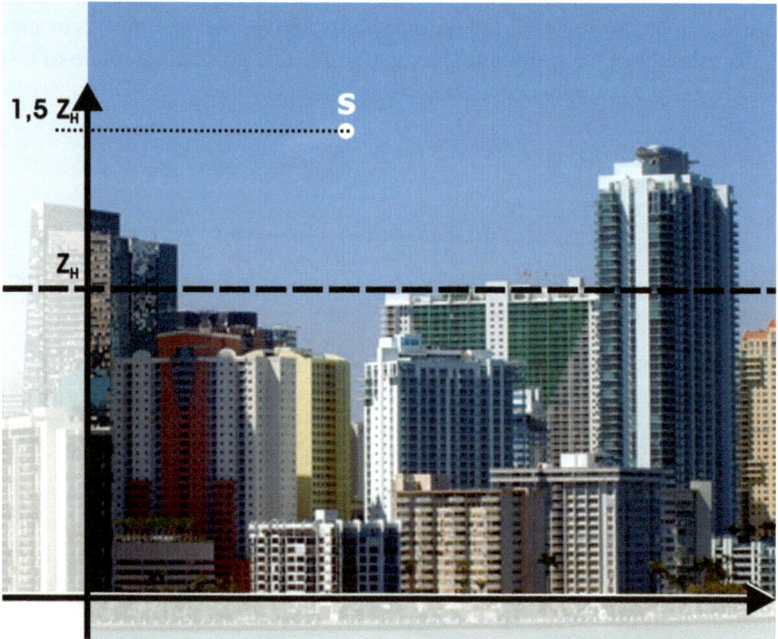

**Fig. 2.7** The height scale of thermal measurements in a densely built-up area

• In the Urban Canopy Layer there is little variation in temperature with height, while there is a discontinuity near roof level both horizontally and vertically. Consequently, if a meaningful spatial average is sought, sensors should be well above mean roof level. The recommended height is $>1.5z_H$ (where $z_H$ is mean height of the roughness elements), so that the mixing of roof and canyon air is accomplished (see Fig. 2.7).

- For an urban environment, there is currently no standard method for extrapolating air temperature data down towards screen level (2 m in height).

Other facts to consider during the placement of a thermometer in urban environments:

- Careful attention to radiation shielding and ventilation is highly recommended.
- Use of the same sensor assemblies (with/without shields and ventilation) inside a network is strongly advised in order to avoid inter-site differences.

Surface temperature is not commonly measured at urban stations; its measurement is only possible using infrared remote sensing or a downward-facing pyrgeometer, or by employing one or more radiation thermometers for which the combined field of view covers a representative sample of the urban district.

## 2.4.2   Humidity

The instruments normally used for measuring humidity can be used in urban areas. The siting and exposure of air temperature sensors presented in the previous chapter apply equally to humidity sensors in the Urban Canopy Layer and above the Roughness Sublayer (see Figs. 2.6 and 2.7). For humidity measurement, the same sensor is used as in temperature sampling. Because urban environments are far dirtier than rural sites (in terms of dust, oils, and pollutants), thermometers and hygrometers require increased maintenance and frequent service. Yearly changing and calibration are strongly advised. The provision of shielding from extraneous sources of solar and long-wave radiation is also recommended.

## 2.4.3   Wind Speed and Direction

Wind speed and direction are very sensitive to flow distortion by obstacles including

- effects of local relief due to hills, valleys, and cliffs,
- sharp changes in roughness or in the effective surface elevation ($z_d$: zero-plane displacement length),
- perturbation of flow around clumps of trees and buildings,
- obstacles in the form of individual trees and buildings (Fig. 2.8),
- disturbances induced by the physical bulk of the tower or mounting arm to which the instrument is attached.

In choosing the height for wind measurement, some basic principles should be considered.

- The standard height for rural wind observations is 10 m above ground, at a horizontal distance from obstructions of at least 10 times the height of the obstacle. Following this guideline is difficult in typical urban districts; where a patch of at least 100 m radius would be required around 10 m-high buildings and trees.
- In a densely built-up area where the effects of individual roughness elements persist, the top of roughness sublayer is about $1.5z_H$, meaning that the recommended anemometer height is at least 15 m if the closest buildings are 10 m high.

Problems of turbulence and vortex flow, resulting from inappropriately placing anemometers behind a building, can be avoided by mounting the sensors, as shown in Fig. 2.8. The recommended height for airflow detection in urban environments is summarized below, with distinctions among different LCZs:

- In urban districts with low element height and density (such as LCZ 6, LCZ 7, LCZ 8, and LCZ 9), it may be possible to use a site where the "open country" standard exposure guidelines can be met. To use the 10 m height, the closest obstacles should be at least 10 times their height from the anemometer and no more than about 6 m high on average.
- In more densely built-up districts (such as LCZ 1 and LCZ 2) with relatively uniform element height and density (buildings and trees), wind speed and direction measurements should be taken with the anemometer mounted on a mast of open construction at a minimum height of 1.5 times the mean height of the elements.

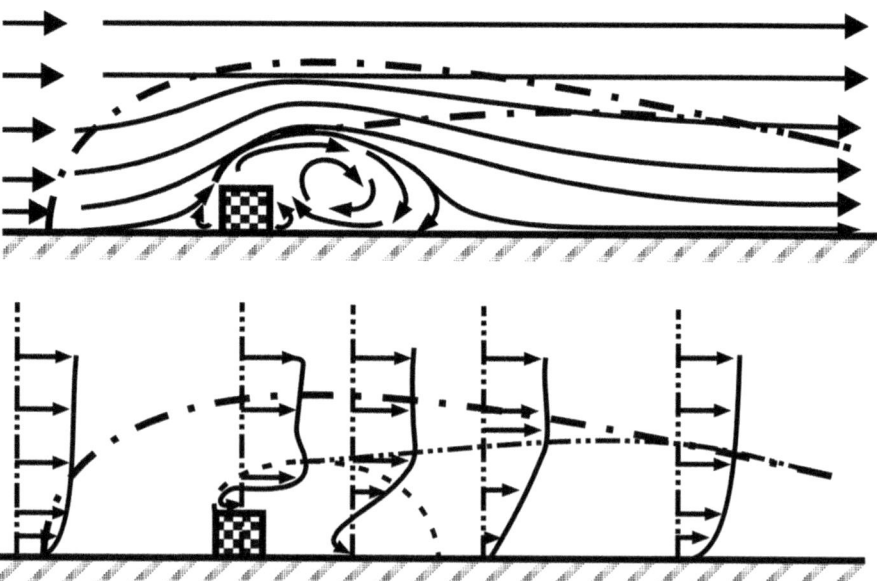

**Fig. 2.8** Typical two-dimensional airflow and a vertical wind structure around a building (Figures after Halitsky 1963)

- In urban districts (such as LCZ 4 and LCZ 5) with scattered tall buildings, the recommendations are as in previous point, but with special attention toward avoiding the wake zone of the tall structures.

According to Harman I.N., 2003, setting up an urban climate station in a street canyon requires consideration of different airflows, depending on canyon sizes. Horizontal air currents above the roof level create eddies, caused by flow blockage from buildings. If the street canyon is wide enough, the current system within the street canyon consists of two parts: one part that has closed circulation, and another part which is blown through. The ratio of these two, easily discernible parts depends on the buildings and the width of the street.

If the width of the street is at least three times the height of the buildings (Fig. 2.9a), then the eddying air currents have no effect whatsoever on the neighboring building, and the street canyon has free airflow.

If the width of the street is gradually decreased within the model, then the closed eddy will dominate the entire width of the street, resulting in the characteristic feature that the direction of the prevailing wind blowing at street level is the opposite of the direction of the currents at roof level (Fig. 2.9b).

If the width of the street is narrowed further, so that the ratio of the height of the buildings and the width of the street exceeds 2/3 (Fig. 2.9c), then the entire street canyon will be characterized by a closed eddy that prevents airing through, causing hot air to become trapped and allowing air pollutants produced by motor vehicles to accumulate.

### 2.4.4 Precipitation

In urban areas, instruments and methods for the measurement of precipitation are the same as an open site. The measurement of precipitation (such as rain or snow) is very susceptible to changes in airflow in the vicinity of the measurement.

In urban environments, measurement errors are associated with the following main causes:

- the interception of precipitation during its trajectory to the ground by nearby collecting surfaces, such as tress and buildings,
- hard surfaces near the gauge which may cause splash-in into the gauge, and overhanging objects which may drip precipitation into the gauge,
- the spatial complexity of the wind field around obstacles in the LCZ, causing significant localized concentration or absence of rain- or snow-bearing airflow,
- the gustiness of the wind in combination with the physical presence of the gauge itself, causing anomalous turbulence around it and leading to under- or over-catch.

In open country, standard exposure requires that obstacles should be no closer than two times their height. In some ways, this guideline is less restrictive than for

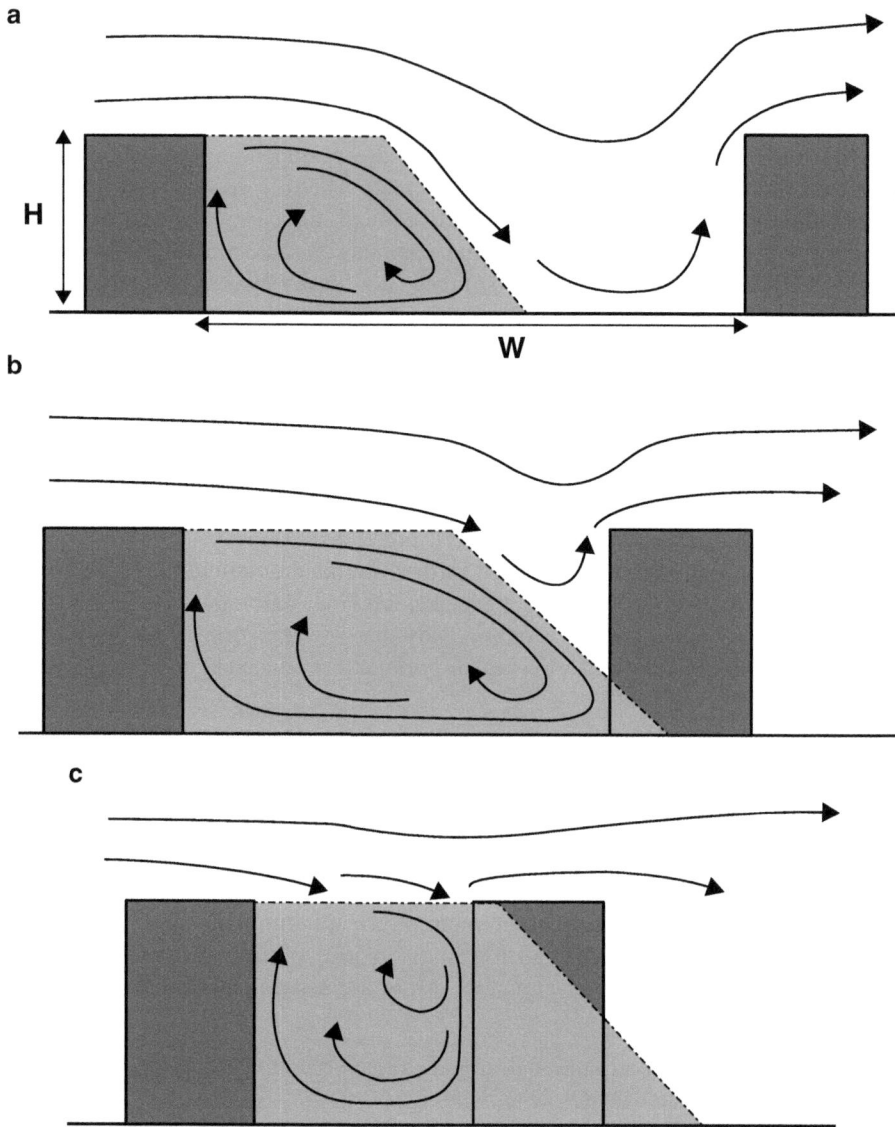

**Fig. 2.9** Typical air flow in a street canyon (**a**) H/W ratio is 1/3 (**b**) H/W ratio around 1/2 (**c**) a narrow street

temperature, humidity, or wind measurements. However, in the UCL the turbulent activity created by flow around sharp-edged buildings is more severe than that around natural obstacles, and it may persist for greater distances in their wake. Again, the highly variable wind speed and direction encountered on the roof of a building make rooftops sites to be avoided.

With regard to precipitation gauges in urban areas, the following guidelines are recommended:

- Gauges should be located in open sites within the city, where the standard exposure criteria can be met (for example, playing fields, open parkland with a low density of trees, urban airports).
- Gauges should be located in conjunction with wind instruments if a representative exposure for them is found.
- Gauges should not be located on the roofs of buildings unless they are exposed at a sufficient height to avoid the wind envelope of the building.
- The measurement of snowfall depth should be taken at an open site or, if made at developed sites, a large spatial sample should be obtained to account for the inevitable drifting around obstacles. Such sampling should include streets oriented in different directions.
- Dew, ice, and fog precipitation also occurs in cities and can be of significance to the water budget, especially for certain surfaces. These forms of precipitation may also be relevant to applications such as plant diseases, insect activity, road safety, and finding supplementary sources for water resources.

## 2.4.5  Solar Radiation

Solar radiation data are very useful inputs for several climate variables, such as atmospheric stability; daytime cloud activity; turbulence statistics; the fluxes of momentum, heat, and water vapor; determination of mixing height; and pollutant dispersion and models. The data can also be used to represent daylight levels in buildings and pedestrian comfort. Adding solar radiation measurements to an automatic station is very simple, relatively inexpensive, and highly recommended.

Solar radiation measurement sites are often located in rural or remote locations specifically to avoid the aerosol and gaseous pollutants of cities that "contaminate" their records. For stations located in built-up areas, only incoming solar (global) radiation is likely to be measured; neither incoming long wave nor any fluxes with outgoing components are monitored. All short- and long-wave fluxes are affected by the spatial properties of the atmosphere and the surface of cities, and the same is true for the net all-wave radiation balance that effectively drives the urban energy balance.

The placement of solar radiation sensors on the top of high building is a widely used and preferable practice that avoids horizon obstructions.

The principal exposure requirement for monitoring direct solar radiation is freedom from obstructions to the solar beam at all times and seasons of the year. Furthermore, the site should be chosen so that the incidence of fog, smoke, and airborne pollution is as typical as possible for the surrounding area.

On the one hand, incoming solar radiation is a fundamental forcing variable of urban climate, so its measurement has a high priority in the establishment of an

**Table 2.1** Meteorological radiation instruments used in a network

| Instrument | Parameter to be measured |
|---|---|
| Pyrheliometer | Direct solar radiation |
| Pyranometer | (a) Global radiation |
| | (b) Sky radiation |
| | (c) Reflected solar radiation |
| Spectral pyranometer | Global radiation in broadband spectral ranges |
| Net pyranometer | Net short-wave radiation |
| Pyrgeometer | (a) Upward long-wave radiation (downward looking) |
| | (b) Downward long-wave radiation (upward looking) |
| Net pyrradiometer | Net long-wave radiation |
| Net radiometer | Total radiation balance |

urban automatic station, where its assessment is quite simple and relatively inexpensive. The instruments for measuring incoming radiation fluxes are listed in Table 2.1. The following recommendations should be observed when pyranometers and other incoming flux sensors might be installed:

• Mount the sensor at some height.
• Choose a site free of vibration, such as a stable platform or the roof of a tall building (often ideal).
• Make sure that the site is free from any obstructions (buildings, hills, towers, trees).
• Avoid excessive reflection from very light-colored walls above the local horizon.
• In polluted environments, upper domes should be cleaned on a daily basis.

On the other hand, the reflection of solar radiation, the emission and reflection of long-wave radiation from the underlying surface, and the net results of short-, long- and all-wave radiant fluxes are seldom and poorly observed in urban environments. Difficulties in measuring include finding a representative area of urban surface. In cases of standard exposure, the sensor height is 2 m over a short grass surface. Over an urban area, greater height is necessary to be representative for the given LCZ. Consider a radiometer at a height of 20 m (at the top of a 10 m mast mounted on a 10 m- high building) in a densely developed district (such as LCZ 1, LCZ 2, LCZ4, or LCZ 5). The radiometer faces downward toward the surface. In this case, the 90 % source area has a diameter of 120 m at ground level, which might be sufficient to detect surface structures involving roofs, walls, roads, and ground surface that are in the sun or in shade. It is generally recommended that downward-facing radiometers be placed at least height of $2z_H$, and preferably higher. The radiative properties of the immediate surroundings of the radiation mast should be representative of the urban district of interest.

The backs of inverted sensors are exposed to solar heating, which can be prevented by shielding and insulation. In finding a suitable height for a measuring mast, the maintenance and cleaning of instrument domes should be taken into account to avoid any difficulties.

## 2.5   Measurement Programs in Urban Environments

The meteorological elements to be measured, in order of priority, are:

- Air temperature
- Surface temperature (natural, artificial)
- Soil temperature (optionally)
- Air humidity
- Wind speed and direction, mean wind profile
- Precipitation
- Radiation, incoming fluxes, outgoing and net fluxes, sunshine duration
- Visibility, meteorological optical range (MOR)
- Evaporation (optionally)
- Soil moisture (optionally)
- Atmospheric pressure (if there is no synoptic station in the vicinity)
- Cloud cover
- Present weather.

Operational measurement uncertainty requirements and instrument performance for standard near surface meteorological measurements are summarized in WMO Guidance, 2008. Among the technical parameters, it is highly advisable to study the accuracy requirements for the following parameters before planning a new meteorological site: range, reported resolution, mode of measurement/observation, required measurement uncertainty, sensor time constant, output averaging time, and achievable measurement uncertainty. In the case of meteorological elements, the technical requirements recommended for urban climate meteorological observations are listed in Table 2.2. In any case, a description of measurement techniques, sensors characteristics, sampling procedures, calibration, and maintenance requirements are published in the user's guide or manual of each sensor to be set up. Figure 2.10 depicts a well-equipped urban meteorological and air quality station.

## 2.6   Site Description for METADATA

Station metadata should contain the following aspects of instrument exposure:

(a) height of the instruments above the surface,
(b) type of sheltering and degree of ventilation for temperature and humidity,
(c) degree of interference from other instruments or objects (masts, ventilators),
(d) microscale and toposcale surroundings of the instrument, in particular

    (i) the state of the enclosure's surface, influencing temperature and humidity,
    (ii) nearby major obstacles (buildings, fences, trees) and their size,
    (iii) the degree of horizon obstruction for radiation observations,

**Table 2.2** Accuracy requirements for surface meteorological measurements

| Variable | Range | Reported resolu-tion | Mode of meas./obs. | Required measurement uncertainty | Sensor time constant | Output avera-ging time | Achievable measurement uncertainty | Remarks |
|---|---|---|---|---|---|---|---|---|
| **1.Temperature**<br>1.1 Air temperature | -80-+60°C | 0.1K | I | 0.3K for =-40°C<br><br>0.3K for =-40°C | 20 s | 1 min | 0.2 K | Achievable uncertainty and effective time constant may be affected by the design of thermometer solar radiation screen. Time constant depends on the airflow over the sensor. |
| 1.2 Extremes of air temperature | -80-+60°C | 0.1K | I | 0.3K for =-40°C<br>0.3K for =-40°C | 20 s | 1 min | 0.2 K | |
| **2. Humidity**<br>2.1 Dew-point temperature | -80-+35°C | 0.1K | I | 0.1K | 20 s | 1 min | 0.5 K | Wet-bulb temperature (psychrometer) |
| 2.2 Relative humidity | 0-100% | 1% | I | 1% | 20 s | 1 min | 0.2 K | If measured directly and in combination with air temperature (dry bulb). Large errors are possible due to aspiration and cleanliness problems. |
| | | | | | | Solisl state and others | | |
| | | | | | 40 s | 1 min | 3% | Solid state sensors may show significant temperature and humidity dependence. |
| **3. Atmospheric pressure**<br>3.1 Pressure | 500-1080 hPa | 0.1 hPa | I | 0.1 hPa<br><br>0.2 hPa | 20 s | 1 min | 0.3hPa<br><br>0.2hPa | Both station pressure and MSL pressure. Measurements uncertainty seriously affected by dynamic pressure due to wind if no precautions are taken. Inadequate temperature compensation of the transducer may affect the measurement uncertainty significantly. |
| 3.2 Tendency | Not specified | 0.1 hPa | I | | | | | Difference between instantaneous values |

| Variable | Range | Reported resolu-tion | Mode of meas./obs. | Required measurement uncertainty | Sensor time constant | Output avera-ging time | Achievable measurement uncertainty | Remarks |
|---|---|---|---|---|---|---|---|---|
| **4. Clouds**<br>4.1 Clouds amount | 0/8-8/8 | 1/8 | I | 1/8 | n/a | | 2/8 | Period (30 s) clustering algorithms may be used to estimate low cloud amount automatically. |
| 4.2 Height of cloud base | 0 m-30 km | 10 m | I | 10 m for≤100 m<br>10% for>100 m | n/a | | ~10 m | Achievable measurement uncertainty undetermined because no clear definition exists for instrumentally measured cloud base height (e.g. based on penetration depth or significant discontinuity in the extinction profile).<br><br>Significant bias during precipitation. |
| **5. Wind**<br>5.1. Speed | 0-75 m s⁻¹ | 0.5 m s⁻¹ | A | 0.5 m s⁻¹<br>for ≤ 5 m s⁻¹ | Distance constant 2-5 m | 0 and/or 1 10 min | 0.5 ms⁻¹ | Average over 2 and/or 10 minutes. Non-linear devices. Care needed in design of averaging process. Distance constant is usually expressed as response length. Averages computed over Cartesian components (see WMO Guide 2008 Part III, Chapter 2, section 2.6 |
| 5.2. Direction | 0-360° | 1° | A | 5° | | | | Highest 3 s average should be recorded. |
| 5.3. Gusts | | | | | | | | |
| **6.Precipitation**<br>6.1 Amount (daily) | | | | | | | The larger of 5% or 0.1 mm | Quantity based on daily amounts. Measured uncertainty depends on aerodynamic collection efficiency of gauges and evaporation losses in heated gauges. |
| 6.2 Depth of snow | 0-500 mm | 0.1 mm | T | 0.1mm for ≤ 5mm | n/a | n/a | | Average depth over an area representative of the observing site |
| | 0-25 m | 1 cm | A | 2% for >5 mm<br><br>0 cm for ≤ 5mm<br>5% for >20 cm | | | | |

(continued)

**Table 2.2** (continued)

| Variable | Range | Reported resolu-tion | Mode of meas./obs. | Required measurement uncertainty | Sensor time constant | Output avera-ging time | Achievable measurement uncertainty | Remarks |
|---|---|---|---|---|---|---|---|---|
| **7. Radiation** | | | | 0.1 h | | | The larger of 0.1 h or 2% | Radiant exposure expressed as daily sums (amount) of (net) radiation. |
| 7.1 Sunshine duration (daily) | 0-24 h | 60s | T | 0.4 MJ m$^{-2}$ for ≤ 8 MJ m$^{-2}$ | 20s | n/a | 0.4 MJ m$^{-2}$ for ≤ 8 MJ m$^{-2}$ | |
| 7.2 Net radiation, radiant exposure (daily) | Not specified | 1 J m$^{-2}$ | T | 5% for >8 MJ m$^{-2}$ | 20s | n/a | 5% for >8 MJ m$^{-2}$ | |
| **8. Visibility** | | | | 50 m for ≤ 600 m | | | | Achievable measurement uncertainty may depend on the cause of obscuration. Quantity to be averaged: extinction coefficient (see WMO Guide 2008, Part III, Chapter 2, section 2.6). Preference for averaging logarithmic values. |
| 8.1 Meteorological Optical Range (MOR) | 10 m-100 km | 1 m | I | 10% for > 600 m-≤1500 | <30s | 1 and 10 min | The larger of 20 m or 20% | |
| | | | | 20% for > 1500 m | | | | |
| **9. Evaporation** | 0-100 mm | 0.1 mm | T | 0.1 mm for ≤ 5 mm | n/a | | | |
| 9.1 Amount of pan evaporation | | | | 2% for > 5 mm | | | | |

Source: WMO Guide (2008)

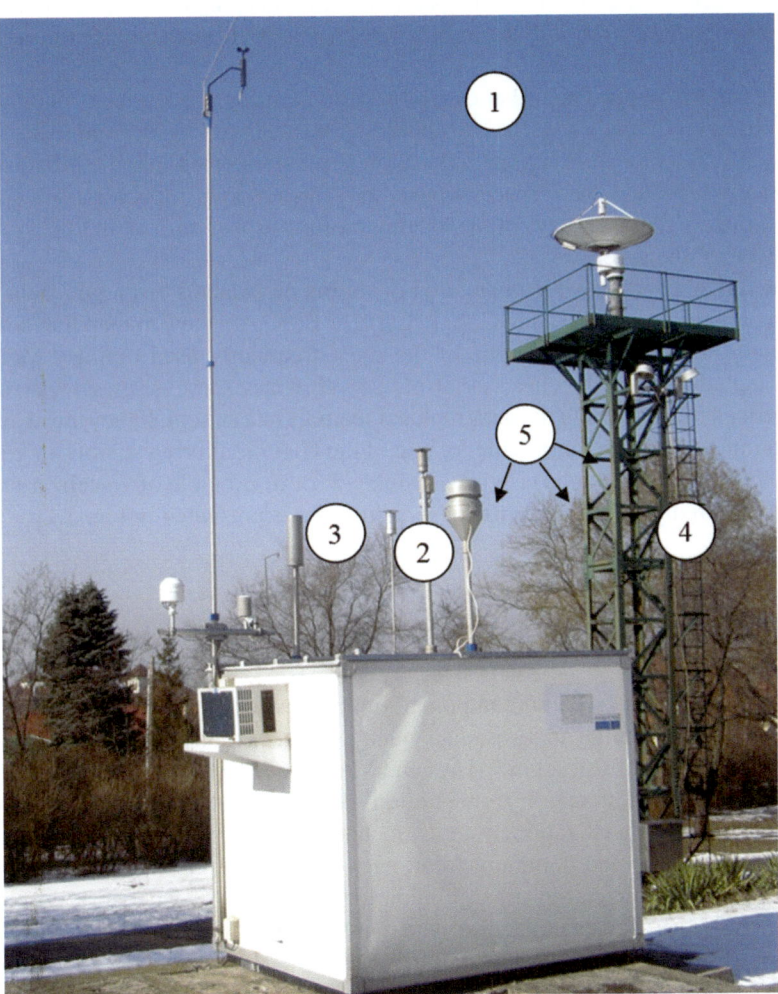

**Fig. 2.10** Example of an urban monitoring station; meteorological sensors: wind (*1*), temperature (*2*), global solar radiation (*3*), precipitation (*4*), air quality (*5* – gas and aerosol samplers)

(iv) surrounding terrain roughness and major vegetation, influencing the wind,
 (v) all toposcale terrain features, such as small slopes, pavements, or water surfaces,
(vi) Major mesoscale terrain features, such as coasts, mountains, or urbanization.

The survey of each site should be reviewed periodically, as environmental circumstances can change over a period of time. A systematic yearly visual check is recommended: If some aspects of the environment have changed, a new site description document should be included in the metadata file. A complete update of the site should be undertaken at least every five years.

The natural relief of the landscape can be disregarded if it is sufficiently distant (>1 km). A method of judging if the relief is representative of the surrounding area is to consider whether a move of the station by 500 m changes the LCZ obtained. If the answer is no, the relief is a natural characteristic of the area and is not taken into account.

One general requirement that cannot be kept at many urban sites is the distance from obstacles, namely that the site should be located well away from trees, buildings, walls, or other obstructions. Instead, it is recommended that the urban station be centered in an open space where the surrounding aspect ratio ($z_{H/W}$) is approximately representative of the locality (see the aspect ratio of an LCZ from the datasheets of the given area).

The full and accurate documentation of station metadata is essential for the evaluation of measurements. Using Google Earth or ESRI ArcView to map the locations of meteorological stations is one of the most frequently used options. Metadata could include maps, sketches, aerial photos, compass surveys, or screens with a fisheye lens for describing the geographical features of a station, if they are available. An example of a documentary file for one of the UHI monitoring stations in Warsaw (at the city centre – Twarda) is given in Table 2.3. An example of visualization files of the surroundings of the Twarda observation site is presented in Fig. 2.11.

## 2.7  Data Transmission and Data Management

Communication, an essential component of any network, consists of the data flow from the sensor to initial analysis, data management, data display, and usage, jointly termed the "cyberinfrastructure" (Hart and Martinez 2006). This infrastructure consists of computer systems, instrumentation, data acquisition, data storage systems and repositories, visualization systems, management services, and technicians, all linked by software and communication networks (Estrin et al. 2003; Brunt et al. 2007). The communication urban climatological monitoring network consists of four main segments: data collection, data management, data display, and data usage.

The majority of weather installations work on a "star" network, relaying information back to the central host server over the Internet via a wired Ethernet connection.

**Table 2.3** Example of a documentary file for the urban station shown on Fig. 2.10

| Number of station | II |
|---|---|
| Station name | UW |
| Address | Warszawa, ul. Twarda 51/55 |
| Geographical coordinates | 52°13'42,7 N, 20°59'37,8 E |
| Observed elements | T, RH, prec, UV, Kglob, Kref, DD/FF |
| Period of observation | 2001–2012 |
| Time resolution | 10 min |
| Function of surrounded area | Research services/residential |
| Settlement intensity | Very dense, multi floor |
| Number of floors | 6–10 |
| Horizon limitation (%) | 65 |
| Ground surface | Artificial (partially clay) |
| Ground water depth | Not applicable |
| Sewage system | Yes |

Urban areas are particularly well placed to utilize wireless technology, as there is an increasing number of municipal wireless access points in urban areas, allowing almost complete coverage in most towns and cities. Hence, with the appropriate permissions granted, these existing municipal wireless networks (open access or subscription wireless access points) can be utilized to relay data from sensors to the host server.

Recently developments in the miniaturization of electronics have produced advances in communications and computing power, with environmental sensors becoming more innovative, reliable, compact, and inexpensive as a result. These advances provide increased potential for urban networks of meteorological sensors, which may now be more numerous and densely spaced, with vastly improved temporal collection and rapid data transmission (Muller et al. 2013). The new generation of atmospheric observation networks will permit new insights into urban atmospheric processes.

The options available for powering sensor networks depend on the location of the sensors, the specific power requirements, and the nature of equipment involved. All short- and long-wave fluxes are affected by the special properties of the atmosphere and the surfaces of cities, and the same is true for the net all-wave radiation balance that effectively drives the urban energy balance (Oke 1988). All of the instruments of radiation measurements, their calibration, the data correction, and most of the field methods are the same for urban environments as for open country sites.

The calibration of equipment and instruments during intercomparison periods is essential to ensure the quality of the data. Sensor networks frequently contain low-cost, nonstandard sensors, and as such all equipment needs to be tested against a traceable "standard" instrument. Ideally, equipment should be calibrated at a national standards and calibration lab, ensuring the reliability of results and allowing for comparisons with other equipment calibrated to the same standard.

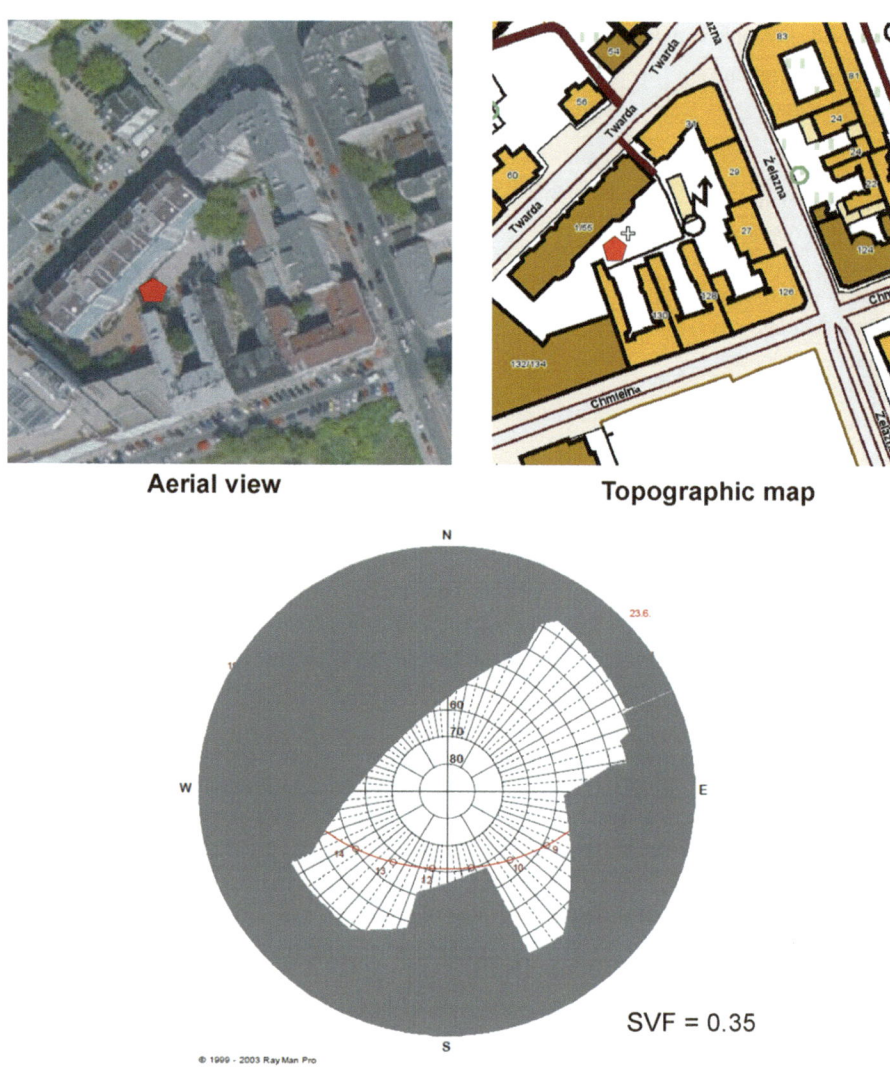

**Aerial view**                                                      **Topographic map**

SVF = 0.35

© 1999 - 2003 RayMan Pro

**Horizon limitation**

**Fig. 2.11** Example of visualization documentary materials for the surroundings of an urban station located at Twarda Street in downtown Warsaw

Documented quality assurance and quality control (QA/QC) procedures must be used in order to provide end-users with high-quality data. Consideration also needs to be given as to where the data are initially stored and processed (including procedures for ensuring that servers are secure and backed-up), archived long-term, and accessed by end-users.

Communication via informal methods such as websites is important for providing information to a variety of stakeholders. Many urban sensor networks have websites through which data can be visualized and downloaded on request.

Calibration processes in a laboratory imply that, under the same circumstances, parallel measurements result between sensors and reference tools. Sensor calibration should help meet the requirements described above in discussing the data quality issues facing institutions that are considering operating an automated surface weather station network. Calibration and maintenance processes should be declared in quality assurance documents for the network.

Data from urban observations should be recorded a well-defined file format, where the header of file contains the station name, observing period, observing element, and units and frequency of observations. The end-data users (researchers, general public, schools) are able to use these records for their own purposes.

## 2.8 The Central-European Urban Heat Island Atlas

The Central-European Urban Heat Island Atlas (UHI Atlas) is a tool for a presentation and exploration of different factors influencing urban heat island phenomena in the Central European area. It can be used to limit the temperature increase in cities by establishing proper short-term and long-term mitigation, risk prevention and management activities (Komac and Ciglič 2014; Ciglič and Komac 2015).

Different influencing factors are presented, such as the altitude, vegetation, land use/cover, and settlement density.

The general goal of this chapter is to present the structure and characteristics of the atlas.

### 2.8.1 UHI Atlas and its database

The database for atlas of urban heat island consists of different data:

- Digital elevation model (DEM),
- Normalized difference vegetation index (NDVI),
- Land surface temperature data (LST),
- Data on land use (Corine Land Cover and Urban Atlas data),
- Night scene image,
- Air temperature at 2 m,
- The data collected from project partners.

The database was elaborated in GIS environment using ArcGIS Desktop and published online using ArcGIS Server programme.

The atlas (Figs. 2.12 and 2.13) is published on-line at: http://gismo.zrc-sazu.si/flexviewers/UHIAtlas or http://zalozba.zrc-sazu.si/p/1352.

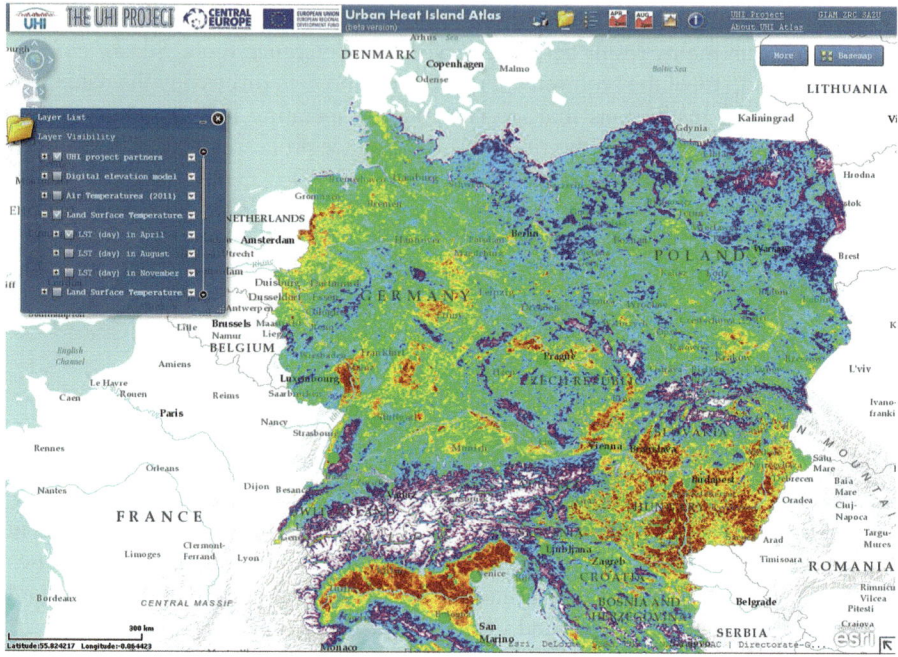

**Fig. 2.12** Print screen of UHI atlas

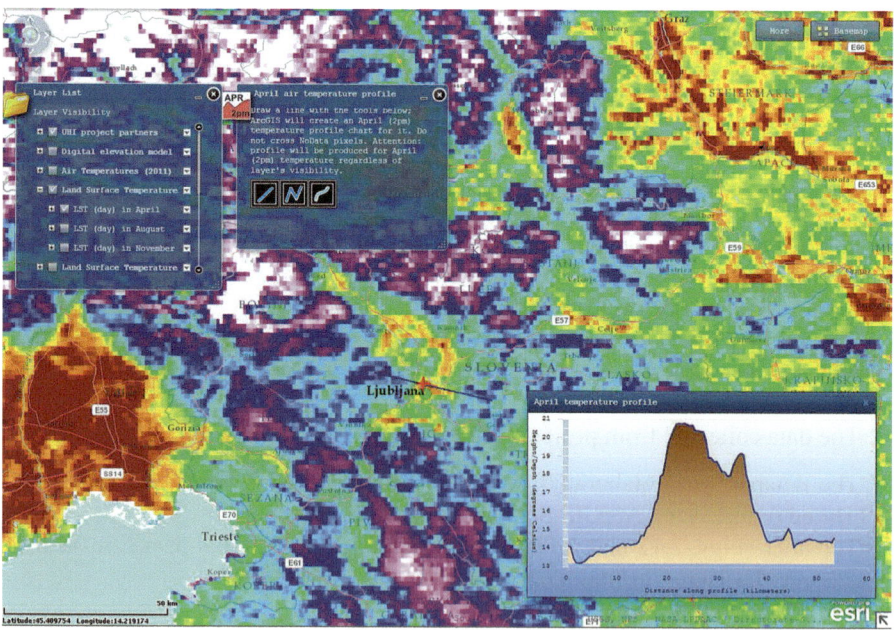

**Fig. 2.13** Print screen of UHI atlas. Its user friendly interface enables users to select between different layers, and to make profiles across April (as below) and August temperatures in Central Europe

## 2.8.2  High Resolution (<0.5 km) Raster Data and Vector Data

### 2.8.2.1  Corine Land Cover

**Format**: Available in raster data (100 m) and vector data
**Temporal coverage**: 2006
**Spatial coverage in UHI atlas**: Central Europe
**Short description of Corine Land Cover**

CORINE (Coordination of information on the environment) provides land use data layers. Two kinds of satellite imagery were used for the CORINE project:

– French SPOT-4 & 5 (60 km swath width, 20 m pixels; VIS, NIR and SWIR bands), and
– Indian IRS P6 (141 km swath width, 23 m pixels; VIS, NIR and SWIR bands).

**Where to find the data layers**: http://www.eea.europa.eu

### 2.8.2.2  Land Cover/Land Use for Cities Included in Project (Urban Atlas)

**Format**: vector data (1 : 10.000)
**Temporal coverage**: 2005–2007
**Spatial coverage in UHI atlas**: Budapest, Vienna, Ljubljana, Prague, Stuttgart, Venice, Warszawa
**Short description of Urban Atlas**

The European Urban Atlas provides reliable, inter-comparable, high-resolution land use maps for 305 Large Urban Zones and their surroundings (more than 100.000 inhabitants as defined by the Urban Audit) for the reference year 2006.

**Where to find the data layers**: http://www.eea.europa.eu

**Format:** raster data (120 m, resampled to 60 m)
**Temporal coverage**: different examples for spring, summer and autumn seasons
**Spatial coverage**: Budapest, Vienna, Ljubljana, Prague, Stuttgart, Venice, Warszawa
**Short description of Landsat images**

Landsat images of reflectance can show us differences among different areas (urban areas, green areas etc.). With some methods and field work measurements it is possible to define land surface temperature and air.

**Where to find it**: http://earthexplorer.usgs.gov/or other image browsers

### 2.8.2.3  Digital Elevation Data SRTM

**Format**: raster data (3 arc sec)
**Temporal coverage**: 2008

**Spatial coverage in UHI atlas**: Central Europe
**Short description of Digital elevation data SRTM**

The SRTM (Shuttle radar topography mission; Jarvis et al. 2008) digital elevation data, produced by NASA originally, has a resolution of 90 m at the equator and is available for over 80 % of the globe.

**Where to find the data layers**: http://srtm.csi.cgiar.org

## 2.9 Raster Data with Low Resolution (≥0.5 km)

### 2.9.1 MODIS NDVI Images

**Format:** raster data (1 km)
**Temporal coverage**: different examples for spring, summer and autumn (16-day average)
**Spatial coverage in UHI atlas**: Central Europe
**Short Description of MODIS NDVI Images**

Some of MODIS (Moderate resolution imaging spectroradiometer) products provide information on vegetation. One of them is NDVI (Normalized difference vegetation index). Vegetation indices are used for global monitoring of vegetation conditions and are used in products displaying land cover and land cover changes.

MODIS NDVI data (product codes MYD13A2 and MOD13A2) are provided every 16 days at 1-km spatial resolution.

**Where to find the data layers**: http://earthexplorer.usgs.gov

### 2.9.2 MODIS LST Images

**Format:** raster data (1 km)
**Temporal coverage**: different examples for spring, summer and autumn (8-day average)
**Spatial coverage in UHI atlas**: Central Europe
**Short Description of MODIS LST Images**

The MODIS global Land Surface Temperature (LST) and Emissivity 8-day data are composed of the daily 1-km LST product (MOD11A1&MYD11A1) as the average values of clear-sky LSTs during an 8-day period.

MOD11A2 & MYD11A2 are comprised of daytime and nighttime LSTs, quality assessment, observation times, view angles, bits of clear sky days and nights, and emissivities estimated from land cover types.

**Where to find the data layers**: http://earthexplorer.usgs.gov

## 2.9.3   VIIRS Night Scene Images

**Format:** raster data (750 m)
**Temporal coverage**: 2012
**Spatial coverage in UHI atlas**: Central Europe
**Short Description of VIIRS**

VIIRS (Visible Infrared Imaging Radiometer Suite), a scanning radiometer, collects visible and infrared imagery and radiometric measurements of the land, atmosphere, cryosphere, and oceans. Its data is used to measure cloud and aerosol properties, ocean color, sea and land surface temperature, ice motion and temperature, fires, and Earth's albedo.

**Where to find the data layers**: http://earthobservatory.nasa.gov
**Credit**: NASA's Earth Observatory, NASA

## 2.9.4   Air Temperature (2 m Above the Ground)

**Format:** raster data (0.0056°)
**Temporal coverage**: selected periods for various seasons in 2011
**Spatial coverage  in UHI atlas:** Central Europe
**Short Description of Air Temperature Data**

Air temperature was calculated on the basis of MODIS Land Surface Temperature (LST) by dr. Klemen Zakšek and dr. Krištof Oštir from Research Centre of the Slovenian Academy of Sciences and Arts (ZRC SAZU). Calculation method is presented in the paper Estimaton of dailiy mean air temperature from MODIS LST in Alpine areas written by Colombi et al. (2007).

**Where to find the data layers**: UHI atlas (http://zalozba.zrc-sazu.si/p/1352)

## 2.10   The Data from the Partners

The data provided by the partners of the UHI project consist of different data layers and present several aspects of urban heat island phenomena and urban heat island influencing factors.

**The Case of Bologna and Modena**
– Meterological stations in the Bologna and Modena area,
– Air quality monitoring system in the Bologna and Modena area,
– Buildings in the municipality of Bologna and Modena,
– Border of the municipalities Bologna and Modena,
– Pilot area in Modena

**The Case of Stuttgart**
- Physiological equivalent temperature (PET) for different areas in Stuttgart,

**The Case of Warszaw**
- Border of the municipality of Warsaw,
- IGSO measurement points,
- UHI index,
- Universal Thermal Climate Index for Warsaw and various health resorts,
- Global solar radiation on the ground level (Mazovian Lowland),
- Reflected solar radiaton (Mazovian Lowland),
- Air temperature (Mazovian Lowland),
- Wind velocity (Mazovian Lowland),
- Subjective Temperature Index (STI) (Mazovian Lowland).

**The Case of Prague**
- Various climate data for precipitation and temperature.

**The Case of Budapest**
- Various climate data for precipitation, temperature, and global radiation.

For each UHI Project partner there is a location marked on the map and important links provided.


## 2.11   Summary

In accordance the aims of the authors, this Manual for UHI assessment helps in selecting measurement locations as well as in identifying sampling infrastructures to be developed in the single urban areas. First of all, it is very important to determine the purpose of the station clearly: (i) to monitor the greatest impact of the city; (ii) to collect data for a more representative or typical district; (iii) to characterize a particular site, where climate problems are perceived to be or where future development is planned. In choosing a location and site for urban stations, there are three scales of interest in urban area studies: macro, local, and micro on horizontal scales. Horizontal and vertical representativeness of the stations should be also specified. After that, the next essential step in selecting an urban station site is to evaluate the physical nature of the urban surroundings and urban terrain (structure, cover, fabric, metabolism).

It is also very important to measure not only UHI components and indicators, i.e. urban and rural temperatures. For a better understanding of UHI phenomena, the measurements of solar radiation, air humidity, wind speed, and precipitation can be very useful. A wide range of technical parameters (such as time resolution, accuracy, range, uncertainty, and calibration requirements) of these meteorological measurements must be also mentioned in the development of UHI observation techniques and evaluation processes.

For a given urban zone, the complexity of the urban environment sets special requirements for siting observation equipment to provide representative values that are little affected by nearby buildings or pollution sources. The interpretation of atmospheric conditions between measuring sites requires detailed information about surface characteristics and the use of urban scale numerical models. For a proper selection of appropriate sites for UHI monitoring we can use also geographic information system. UHI atlas is one example for presentation of urban heat island phenomena and influencing factors.

# References

Balázs, B., Unger, J., Gál, T., Sümeghy, Z., Geiger, J., & Szegedi, S. (2009). Simulation of the mean urban heat island using 2D surface parameters: Empirical modelling, verification and extension. *Meteorological Applications, 16*(3), 275–287.
Brunt J., Benson, B., Vande Castle, J., Henshaw, D., & Porter, J. (2007). *LTER network cyberinfrastructure strategic plan*, Version 4. Retrieved October 28, 2011. Available at http://intranet2.lternet.edu/sites/intranet2.lternet.edu/files/documents/LTER_History/Planning_Documents/LTER_CI_Strategic_Plan_.pdf
Ciglič, R., & Komac, B. (2015). *Central-European urban heat island atlas*. Ljubljana: ZRC SAZU. http://gismo.zrc-sazu.si/flexviewers/UHIAtlas/.
Colombi, A., Pepe, M., & Rampini, A. (2007). *Estimation of daily mean air temperature from MODIS LST in Alpine areas*. Rotterdam: New Developments and Challenges in Remote Sensing.
Estrin, D., Michener, W., & Bonito, G. (2003). *Environmental cyberinfrastructure needs for distributed sensor networks: A report from a National Science Foundation sponsored workshop*. La Jolla: Scripps.
Halitsky, J. (1963). Gas diffusion near buildings. *Transactions – American Society of Heating, Refrigerating and Air-Conditioning Engineers, 69*, 464–485.
Harman, I. N. (2003). *The energy balance of urban areas*. PhD thesis. University of Reading.
Hart, J. K., & Martinez, K. (2006). Environmental sensor networks: A revolution in the earth system science? *Earth-Science Reviews, 78*, 177–191.
Jarvis, A., Reuter, H. I., Nelson, A., Guevara, E. (2008). Hole-filled seamless SRTM data V4. *International Centre for Tropical Agriculture (CIAT)*. Available from http://srtm.csi.cgiar.org.
Komac, B., & Ciglič, R. (2014). Urban heat island atlas: A web tool for the determination and mitigation of urban heat island effects. *Geographia Polonica, 87*, 595–595. Warszawa.
Muller, C. L., Chapman, L., Grimmond, C. S. B., Young, D. T., & Cai, X. (2013). Sensors and the city: A review of urban meteorological networks. *International Journal of Climatology, 33*(7), 1585–1600.

Oke, T. R. (1988). Street design and urban canopy layer climate. *Energy and Buildings, 11,* 103–113.

Stewart, I. D., & Oke, T. (2009). A new classification system for urban climate sites. *Bulletin of the American Meteorological Society, 90,* 922–923.

Stewart, I. D., & Oke, T. (2012). "Local Climate Zones" for urban temperature studies. *Bulletin of the American Meteorological Society, 93,* 1879–1900.

Unger, J., Savić, S., & Gál, T., (2011). Modelling of the annual mean urban heat island pattern for planning of representative urban climate station network, *Advances in Meteorology,* Paper 398613. 9 p.

WMO. (2008). *Guide to meteorological instruments and methods of observation,* 7th edn. WMO-No. 8.

# Chapter 3
# Methodologies for UHI Analysis

## Urban Heat Island Phenomenon and Related Mitigation Measures in Central Europe

**Ardeshir Mahdavi, Kristina Kiesel, and Milena Vuckovic**

**Abstract**  A central strand of research work in the realm of urban physics aims at a better understanding of the variance in microclimatic conditions due to factors such as building agglomeration density, anthropogenic heat production, traffic intensity, presence and extent of green areas and bodies of water. The characteristics and evolution of the urban microclimate is not only relevant to people's experience of outdoor thermal conditions in the cities. Higher air temperatures also exacerbate discomfort caused by the overheating of indoor spaces and increases cooling energy expenditures. It can be argued that the solid understanding of the temporal and spatial variance of urban microclimate represents a prerequisite for the reliable assessment of the thermal performance of buildings (energy requirements, indoor thermal conditions). In this context, the present treatment entails a three-fold contribution. First, the existence and extent of the UHI phenomena are documented for a number of Central-European cities. Second, a comprehensive assessment of the effectiveness of UHI mitigation measures in these cities is described that is conducted using advanced numeric modelling instruments. Third, a systematic framework is proposed to identify a number of variables of the urban environment that are hypothesized to influence UHI and the urban microclimate variance. These variables pertain to both geometric (morphological) and semantic (material-related) urban features.

**Keywords**  Urban climate • Urban heat island • Mitigation measures • Simulation • Evaluation

A. Mahdavi (✉) • K. Kiesel • M. Vuckovic
Department of Building Physics and Building Ecology, Vienna University of Technology,
Vienna, Austria
e-mail: bpi@tuwien.ac.at

F. Musco (ed.), *Counteracting Urban Heat Island Effects in a Global Climate Change Scenario*, DOI 10.1007/978-3-319-10425-6_3

## 3.1   Introduction

The characteristics of the urban microclimate are of critical importance with regard to inhabitants' health and well-being (thermal comfort, heat stress, mortality rates) as well as energy and environmental issues (Akbari 2005; Harlan and Ruddell 2011). In the last few years, the general awareness concerning the urban microclimate has been steadily rising. However, given the fact that world-wide an increasing number of people live in cities, further research and planning efforts are needed to better understand and address the effects of urban microclimate, its variance, and its development. Given the complexity of the urban fabric, it is widely recognized that heat storage in urban areas will be higher when compared to unbuilt areas (Grimmond and Oke 1999; Piringer et al. 2002). Generally speaking, the undesired thermal circumstances in the urban environment are caused in part by certain properties of the materials used for construction of buildings, pavements, and roads, the urban layout and structure including topography, morphology, density, and open space configuration, as well as processes and activities such as transportation and industry (Unger 2004; Grimmond 2007; Alexandri 2007; Kleerekoper et al. 2012; Shishegar 2013). These factors can affect, amongst other things, the way solar radiation is absorbed by urban surfaces and the way air masses flow through the urban fabric. Empirical observations in many cities around the world point to significantly higher urban temperatures than the surrounding rural environment. This circumstance is referred to as the urban heat island (UHI) phenomenon (see, for example, Voogt 2002; Arnfeld 2003; Blazejczyk et al. 2006; Oke 1981; Gaffin et al. 2008). Together with climate change, this phenomenon can be crucial to the way we view urban areas as living environments.

Recently, a number of research efforts have been initiated to better understand the very specifics of the UHI phenomenon (see, for example, Arnfeld 2003; Blazejczyk et al. 2006). Some of related foci of these efforts are to describe the characteristics and patterns of UHI (Voogt 2002; Hart and Sailor 2007). Empirical observations have shown that the UHI phenomenon shows different characteristics during different seasons (Gaffin et al. 2008) and that it is pronounced differently during the night and the day (Oke 1981). Furthermore, the intensity of urban heat islands is believed to rise proportionally to the size and population of the urban area (Oke 1972). More recently, Gaffin et al. (2008) performed a detailed spatial study of New York City's current UHI and concluded that summer and fall periods were generally the strongest UHI seasons, consistent with seasonal wind speed changes in the area. A simple quantitative indicator of urban heat island phenomenon is the UHI intensity. The UHI intensity is defined as the difference between urban and rural air temperature (Oke 1972).

Generally, heat island intensities are quantified in the range of 1–3 K, but under certain atmospheric and surface conditions can be as high as 12 K (Voogt 2002). Material properties of urban surfaces (Grimmond et al. 1991; Akbari et al. 2001) as well as evapotranspiration, and anthropogenic heat emission (Taha 1997) can result in higher urban temperatures. To address the implications of the UHI phenomenon, cities (both governmental bodies and affected stakeholders) must implement well-conceived, comprehensive, and collective actions with a high potential to positively influence urban climate and remedy the negative phenomena associated with the urban heat islands.

In this context, the present contribution reports on the results of data analyses and modelling efforts undertaken to investigates the extent of urban heat island phenomena and the potential of relevant mitigation measures in the Central European region (Mahdavi et al. 2013). Thereby, a large set of data was collected and analysed concerning the extent of the UHI effect in multiple cities in Central Europe. Furthermore, to develop and demonstrate approaches toward supporting the process of design and evaluation of UHI mitigation measures, the potential of numerical (simulation-based) urban microclimate analysis models were explored.

As numerical modelling poses certain challenges not only in view of time and computational resources but also model validation and calibration issues, the potential of alternative (or complementary) empirically-based modelling options were investigated. To develop such alternative models, certain features of the urban environment are hypothesized to influence UHI and the urban microclimate variance. The related variables, which pertain to both geometric (morphological) and semantic (material-related) urban features are captured within a systematic framework.

The statistical relationships between the values of such variables and the extent of microclimatic variance provide the basis for simple empirically-based models. These models can be directly used to predict the impact of mitigation measures or indirectly applied to gauge the performance of detailed numerical models of the urban microclimate.

## 3.2   The Urban Heat Island in Central Europe

Metropolitan areas worldwide vary in their spatial configuration. This is typically manifested in the diversity of the respective microclimatic conditions. The present contribution focuses on documenting this diversity in terms of the frequency, magnitude, and time-dependent (diurnal and nocturnal) UHI intensity distribution (during a reference week) and the long-term development of urban and rural temperatures in seven Central-European cities, namely Budapest, Ljubljana, Modena, Padua, Prague, Stuttgart, Vienna, and Warsaw (see Tables 3.1 and 3.2). The magnitude of the UHI effect can be expressed in terms of Urban Heat Island intensity (UHI). This term denotes the temperature difference (in K) between simultaneously measured urban and rural temperatures. The aim was to identify and evaluate the extent of the UHI effect and its variance in the broader geographical context of the participating cities.

As already mentioned, UHI intensity in observed urban areas was derived for a reference summer week (with high air temperature and relatively low wind velocity) selected by each participating city independently. The collected information included hourly data on air temperature, wind speed, and precipitation from two representative weather stations (one urban and one rural).

**Table 3.1** General information about the participating cities

| City | Area [km²] | Population [millions] | Latitude | Longitude | Altitude [m] |
|---|---|---|---|---|---|
| Budapest | 525 | 1.74 | 47° 30' N | 19° 3' E | 90–529 |
| Ljubljana | 275 | 0.28 | 46° 3' N | 14° 30' E | 261–794 |
| Modena | 183 | 0.18 | 44° 39' N | 10° 55' E | 34 |
| Padua | 93 | 0.21 | 45° 25' N | 11° 52' E | 8–21 |
| Prague | 496 | 1.26 | 50° 5' N | 14° 25' E | 177–399 |
| Stuttgart | 207 | 0.60 | 48° 46' N | 9° 10' E | 207–548 |
| Vienna | 415 | 1.73 | 48° 12' N | 16° 22' E | 151–543 |
| Warsaw | 517 | 1.70 | 52° 13' N | 21° 00' E | 76–122 |

**Table 3.2** Information about the urban topology of the participating cities

| City | Topology |
|---|---|
| Vienna | Vienna is located in north-eastern Austria, at the eastern most extension of the Alps in the Vienna Basin. |
| Stuttgart | Stuttgart's center lies in a Keuper sink and is surrounded by hills. Stuttgart is spread across several hills, valleys, and parks. |
| Padua | Padua is located at Bacchiglione River, 40 km west of Venice and 29 km southeast of Vicenza. The Brenta River, which once ran through the city, still touches the northern districts. To the city's south west lie the Euganaean Hills. |
| Budapest | The Danube River divides Budapest into two parts. On the left bank the Buda is located, with over 20 hills within the territory of the capital, and on the right bank the flat area of Pest is located with its massive housing, as well as commercial and industrial areas. |
| Prague | Prague is situated on the Vltava river in the center of the Bohemian Basin. |
| Modena | Modena is bounded by the two rivers Secchia and Panaro, both affluent of the Po River. The Apennines ranges begin some 10 km from the city, to the south. |
| Warsaw | Warsaw is located some 260 km from the Baltic Sea and 300 km from the Carpathian Mountains. Furthermore, Warsaw is located in the heartland of the Masovian Plain. |
| Ljubljana | Ljubljana is located in the Ljubljana Basin between the Alps and the Karst Plateau. |

To obtain a long-term impression of the urban and rural temperature development, mean annual (urban and rural) temperatures and UHI values were derived for a period of up to 30 years, namely from 1980 to 2011 (Modena, Prague, Stuttgart, Warsaw), from 1994 to 2011 (Vienna, Padua), from 2000 to 2011 (Budapest).

Table 3.3 provides an overview of the time periods used for both the short-term and the long-term analyses.

## 3.3 Short-Term Analyses of the Observations

Figure 3.1 shows the cumulative frequency distribution of UHI values for the participating cities for the aforementioned summer reference week. Figures 3.2 and 3.3 show for a reference summer day (representing the reference week) the hourly values of urban temperature and the hourly UHI values respectively.

**Table 3.3**  Overview for the data sets used for the analysis

|  | Reference week | Long-term climate data | |
|---|---|---|---|
|  |  | Urban station | Rural station |
| Budapest | 20–26.8.2011 | 2000–2011 | 2000–2011 |
| Ljubljana | 20–26.8.2011 | 1980–2011 | 1980–2011 |
| Modena | 20–26.8.2011 | 1980–2010 | 1980–2009 |
| Padua | 18–24.8.2011 | 1994–2011 | 1994–2011 |
| Prague | 8–14.7.2010 | 1980–2011 | 1980–2011 |
| Stuttgart | 20–26.8.2011 | 1981–2011 | 1980–2011 |
| Vienna | 20–26.7.2011 | 1994–2011 | 1994–2011 |
| Warsaw | 9–15.6.2008 | 1980–2011 | 1980–2011 |

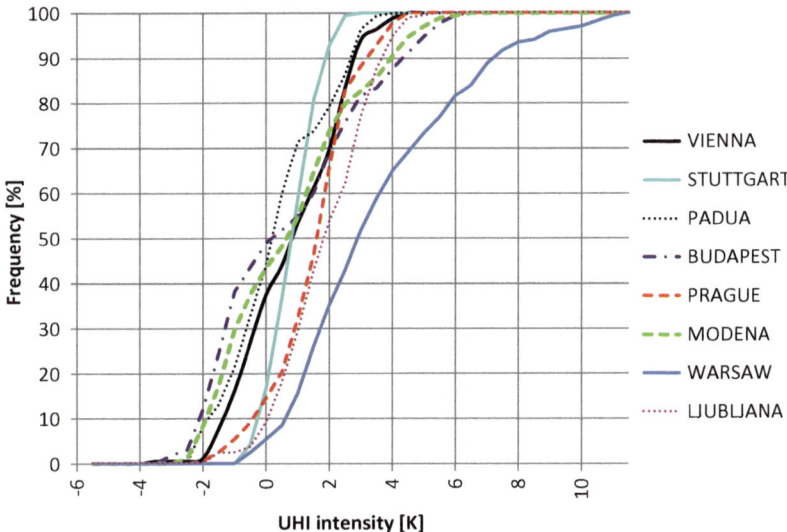

**Fig. 3.1**  Cumulative frequency distribution of UHI intensity for a one week summer period

The reference week data clearly demonstrate the existence and significant magnitude of the UHI effect in participating cities, especially during the night hours
(Fig. 3.3). However, the time-dependent UHI patterns vary considerably across the
participating cities. In Warsaw, for example, UHI intensity level ranges from around
1 K during daytime to almost 7 K during the night, while in Stuttgart levels are
rather steady, ranging from 1 K to 2 K. The UHI pattern differences are also visible
in the cumulative frequency distribution curves of Fig. 3.1. In this Figure, a shift to
the right denotes a larger UHI magnitude.

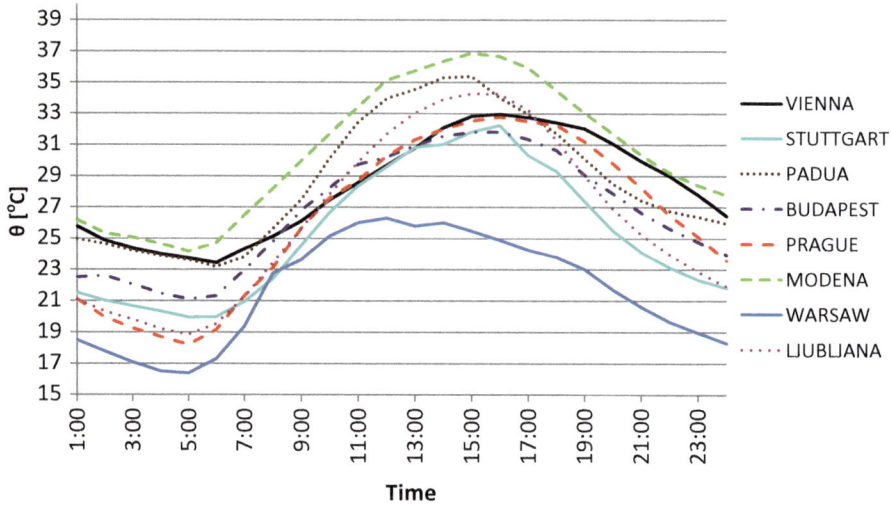

**Fig. 3.2** Mean hourly urban temperature for a reference summer day

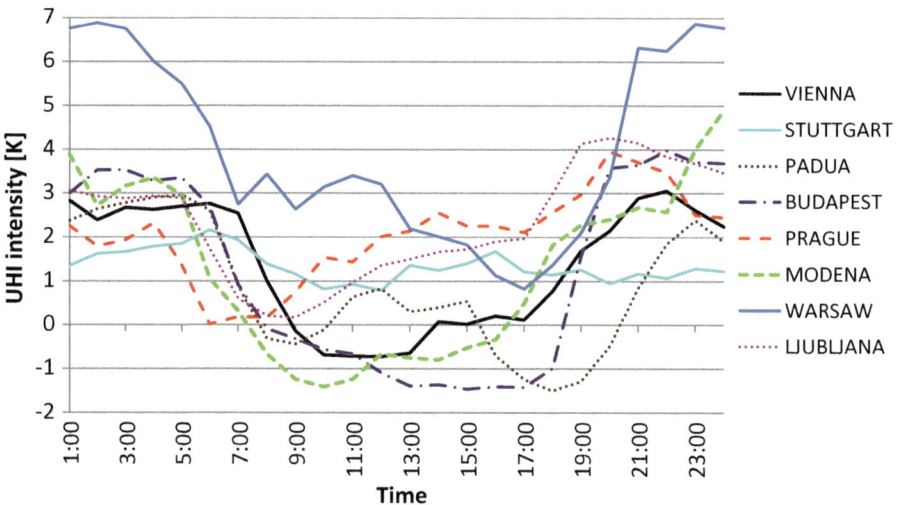

**Fig. 3.3** Mean hourly UHI intensity distribution for a reference summer day

## 3.4  Long-Term Analyses of the Observations

Figures 3.4 and 3.5 show the (mean annual) urban and rural temperatures respectively over a period of 30 years. Figures 3.6 and 3.7 show the long-term UHI intensity trend over the same period. The historical temperature records suggest an upward trend concerning both urban and rural temperatures (see Figs. 3.4 and 3.5). Consistent with regional and global temperature trends, a steady increase in rural temperatures of up to about 2.5 K can be observed in all selected cities with the exception of Budapest. This might be due to the small sample of data set obtained, as this particular weather station was installed in the year of 2000. In the same 30-years period, the mean annual urban temperature rose somewhere between 1 K (Stuttgart) and 3 K (Warsaw). A number of factors may have contributed to this trend, namely increase in population, energy use, anthropogenic heat production, and physical changes in the urban environment (e.g., more high-rise buildings, increase in impervious surfaces). It should be noted that, while both rural and urban temperatures have been increasing, the value of the UHI intensity has been rather steady. Our data suggest increasing UHI intensity trends in Warsaw and Ljubljana, whereas a slight decrease can be discerned from Stuttgart and Prague data (Figs. 3.6 and 3.7).

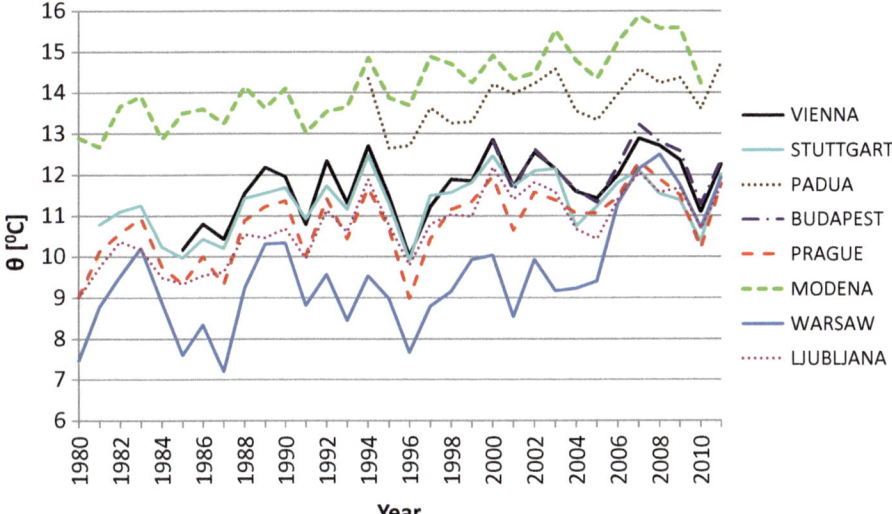

**Fig. 3.4** Development of (mean annual) urban temperatures over a period of 30 years

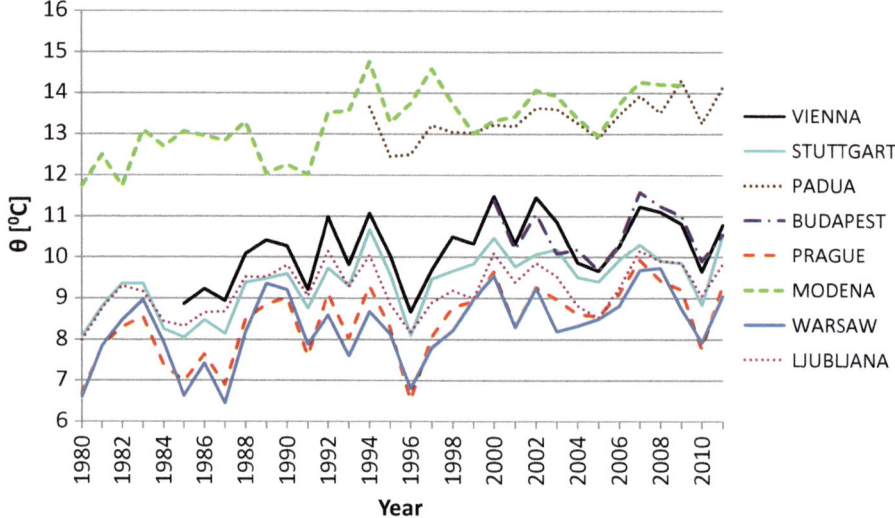

**Fig. 3.5** Development of (mean annual) rural temperatures over a period of 30 years

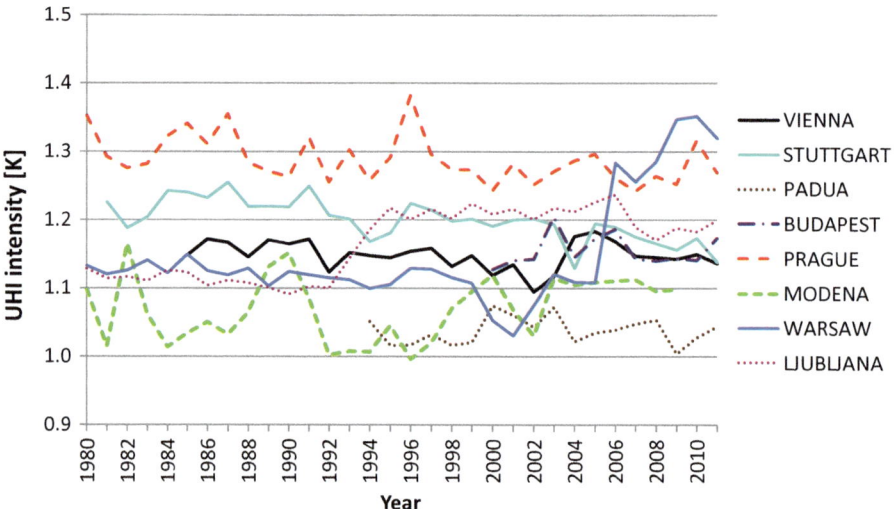

**Fig. 3.6** Long-term development of the UHI intensity over a period of 30 years

## 3.5 Modelling Efforts

Urban microclimate is considered to be a cumulative effect of several circumstances, including small-scale processes such as combustion process of vehicles and meso-scale interactions such as atmospheric forces. To properly model and analyse

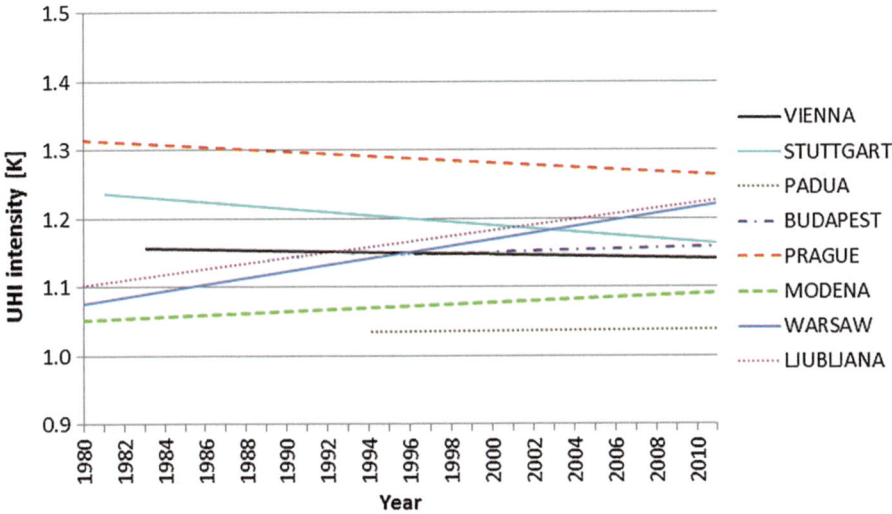

**Fig. 3.7** Long-term UHI intensity trend over a period of 30 years

these effects, computational tools must consider them concurrently and in an integrated fashion. Hence detailed and complex simulation approaches are needed that have the potential to incorporate multiple aspects including hygro-thermal processes and related human comfort issues. The resulting improved predictive performance of proper computational tools would thus provide valuable feedback to planers and decision makers in confronting the UHI phenomenon.

An increasing number of tools are becoming available for microclimatic modelling of urban areas (Mirzaei and Haghighat 2010). Some tools are rather limited in terms of the range of pertinent variables they consider. Other, more detailed tools display limitations in terms of domain size and resolution. Nonetheless, numerical models still present a valuable resource for the assessment of complex thermal processes in the urban field. Within the context of this contribution, we focus on a state of art CFD-based numeric simulation environment ENVI-met (Huttner and Bruse 2009). This tool was selected as it has the capability to simulate the urban micro-climate while considering a relatively comprehensive range of factors (building shapes, vegetation, different surface properties). The high-resolution output generated by this tool includes air, soil, and surface temperature, air and soil humidity, wind speed and direction, short wave and long wave radiation fluxes, and other important microclimatic information.

Project partners undertook an extensive modelling effort including the following steps. First, a specific area within each city was selected. The idea was to select areas that are either targeted for the implementation of mitigation measures or represent likely candidates for such measures ("pilot action areas"). Second, these areas were specified in detail with regard to required model input information (i.e., geometric and semantic properties). Third, the existing microclimatic circumstances

**Table 3.4** Summary of envisioned mitigation measures

|            | Scenario 1                           | Scenario 2                  | Scenario 3  | Scenario 4          | Scenario 5 |
|------------|--------------------------------------|-----------------------------|-------------|---------------------|------------|
| **Budapest**   | Green area + Water bodies        | Trees                       | –           | –                   | –          |
| **Ljubljana**  | Green area                       | Water bodies                | –           | –                   | –          |
| **Modena**     | Green area                       | Cool walls                  | Green roofs | Pervious ground     | Cool roofs |
| **Padua**      | Green area + Trees               | Cool pavements              | Cool roofs  | S1 + Cool pavements | –          |
| **Prague**     | New urban develoment             | Green roofs                 | –           | –                   | –          |
| **Stuttgart**  | Green area                       | Trees                       | Water bodies| –                   | –          |
| **Vienna**     | Trees                            | Green roofs                 | Combined    | –                   | –          |
| **Warsaw**     | Green area + Trees + Green roofs | S1 + Pervious pavements     | –           | –                   | –          |

for these areas (base case) were modelled using the aforementioned simulation environment. Fourth, candidate mitigation measures were defined for each of these areas (see Table 3.4 for an overview). Fifth, the envisioned mitigation measures were virtually implemented in the simulation environment and corresponding output was generated. Sixth, the base case conditions were compared with the predicted post-mitigation circumstances to provide a quantitative basis for the evaluation of the effectiveness of the envisioned mitigation measures.

To illustrate the kinds of information and analyses that can be obtained from the modelling process, relevant results are provided below for three cities, namely Vienna, Padua, and Warsaw. Toward this end, Figs. 3.8, 3.9, and 3.10 show the mean hourly temperature in the course of a reference summer day in Vienna, Padua, and Warsaw for the base case and three mitigation scenarios. Figures 3.11, 3.12, and 3.13 show the corresponding temperature differences between the base case and the applicable mitigation scenarios in the course of a reference summer day. These results point to the potential of various mitigation measures to reduce air temperature levels in hot summer days in the selected cities. As it could be expected, different mitigation measures display different levels of impact. For example, in case of the targeted area in Vienna, green roofs do not appear to noticeably influence the air temperature in the urban canyon. Trees, on the other hand, do impact the air temperature. The combination of these two measures proved in this case to be most effective. With regard to the temporal pattern of the effects, it can be noted that the difference in air temperature is more pronounced during evening and night hours.

To further investigate the temporal nature of UHI intensity values (and their sensitivity to various mitigation measures), we introduced the concepts of Cumulative Temperature Increase (CTI) and Cumulative Temperature Decrease (CTD). CTI and CTD are computed as the cumulative sum of all positive and negative values respectively in the course of a reference day:

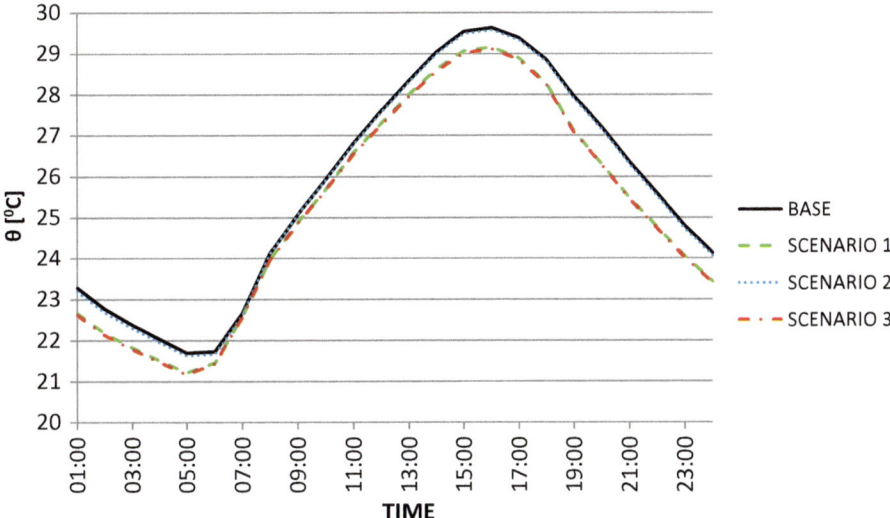

**Fig. 3.8** Mean hourly temperature in the course of a reference summer day in Vienna for the base case and three mitigation scenarios

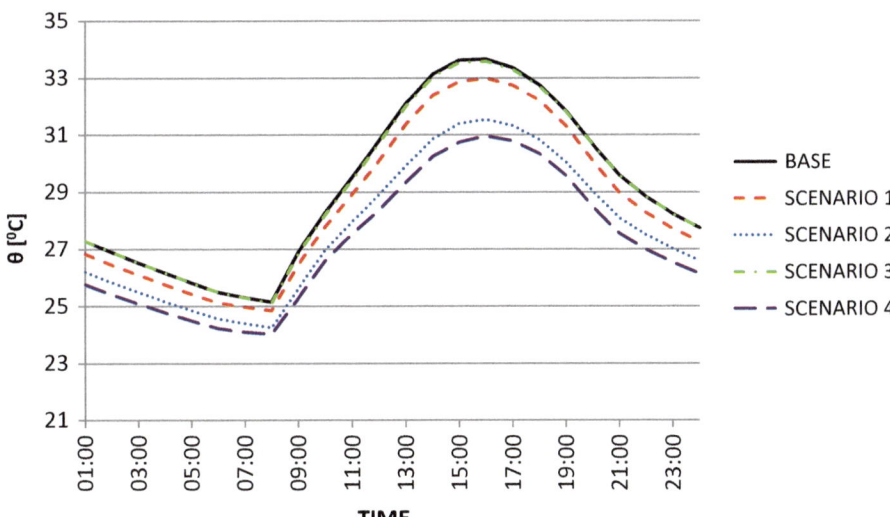

**Fig. 3.9** Mean hourly temperature in the course of a reference summer day in Padua for the base case and four mitigation scenarios

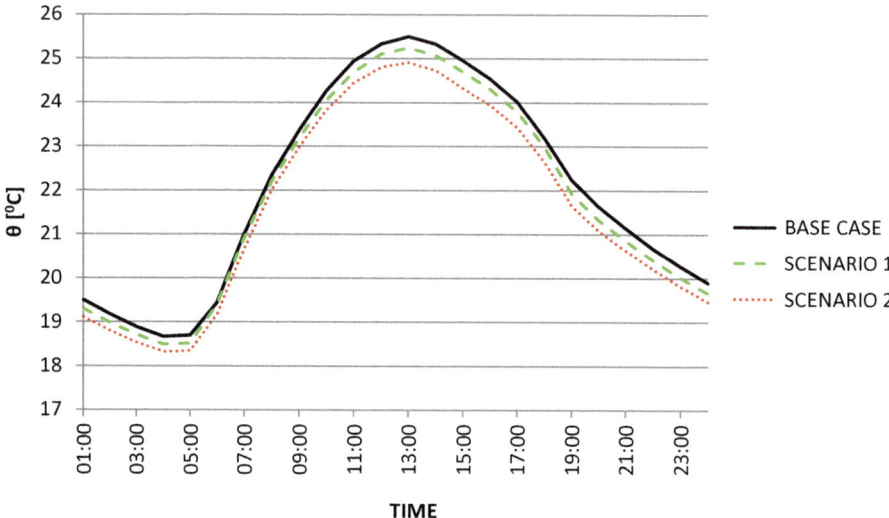

**Fig. 3.10** Mean hourly temperature in the course of a reference summer day in Warsaw for the base case and two mitigation scenarios

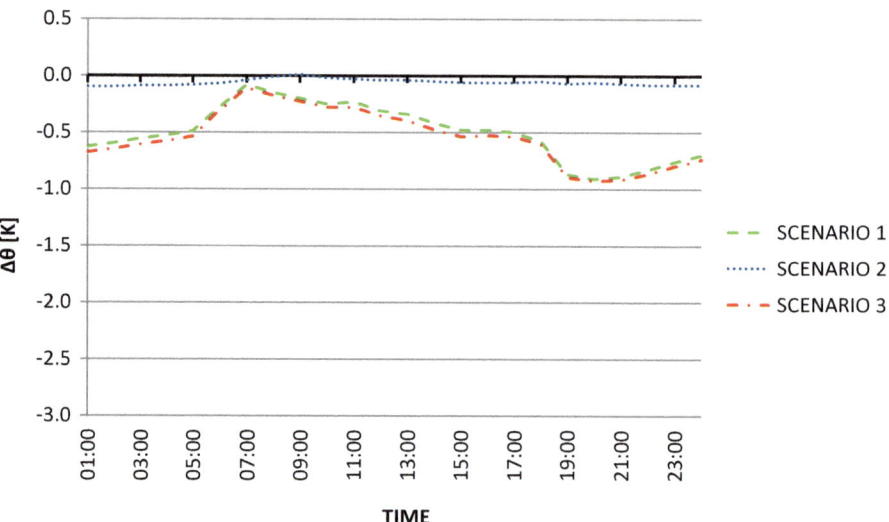

**Fig. 3.11** Temperature difference between the base case and three mitigation scenarios in the course of a reference summer day in Vienna

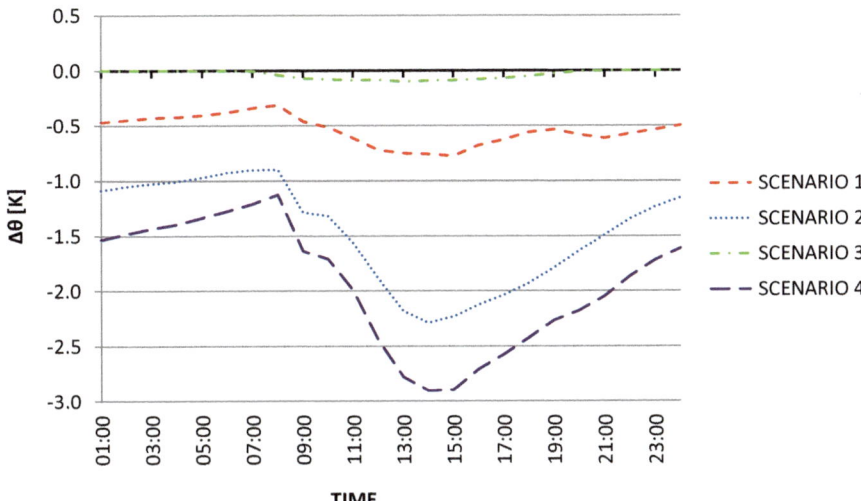

**Fig. 3.12** Temperature difference between the base case and four mitigation scenarios in the course of a reference summer day in Padua

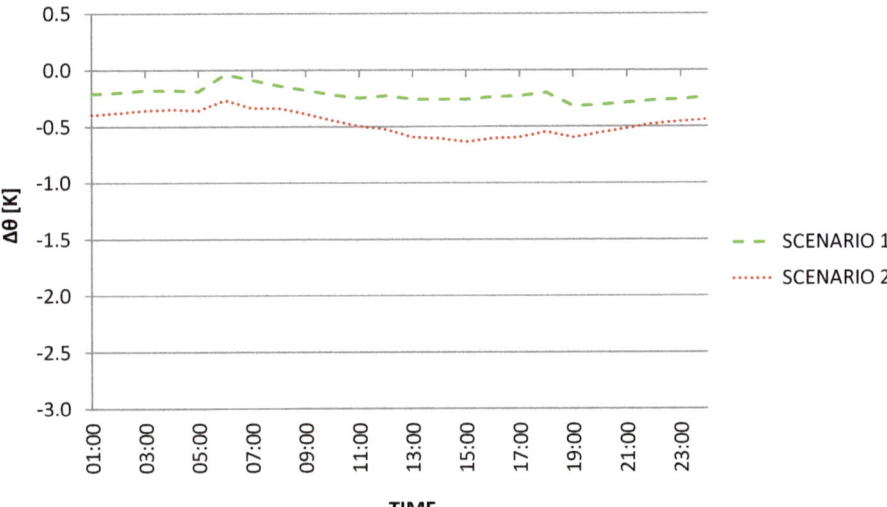

**Fig. 3.13** Temperature difference between the base case and two mitigation scenarios in the course of a reference summer day in Warsaw

**Table 3.5** Summary of predicted CTD and CTI values for summer and winter seasons for all mitigation measures in all cities

| City | Mitigation measures | | Summer | | Winter | |
|------|---------------------|---|--------|-----|--------|-----|
| | | | CTD | CTI | CTD | CTI |
| Budapest | S1 | New urban development + Green area + Water bodies | 16 | 7 | 18 | 0 |
| | S2 | Trees | 2 | 0 | 1 | 0 |
| Ljubljana | S1 | Green area | 1 | 0 | 2 | 0 |
| | S2 | Water bodies | 1 | 0 | 0 | 0 |
| Modena | S1 | Green area | 21 | 0 | N/A | N/A |
| | S2 | Cool walls | 0 | 4 | 0 | 1 |
| | S3 | Green roofs | 6 | 0 | 1 | 0 |
| | S4 | Pervious ground | 5 | 0 | N/A | N/A |
| | S5 | Cool roofs | 2 | 0 | 1 | 0 |
| Padua | S1 | Green area + Trees | 13 | 0 | 1 | 0 |
| | S2 | Cool pavements | 35 | 0 | 6 | 0 |
| | S3 | Cool roofs | 1 | 0 | 0 | 2 |
| | S5 | S1 + Cool pavements | 47 | 0 | 6 | 0 |
| Prague | S1 | New urban development | 0 | 21 | 0 | 31 |
| | S2 | Green roofs | 0 | 26 | 0 | 30 |
| Stuttgart | S1 | Green area | 7 | 0 | 99 | 0 |
| | S2 | Trees | 7 | 1 | 0 | 6 |
| | S3 | Water bodies | 0 | 0 | 11 | 0 |
| Vienna | S1 | Trees | 12 | 0 | 1 | 0 |
| | S2 | Green roofs | 1 | 0 | 0 | 0 |
| | S3 | Combined | 13 | 0 | 1 | 0 |
| Warsaw | S1 | Green area + Trees + Green roofs | 5 | 0 | 0 | 0 |
| | S2 | S1 + Pervious pavements | 11 | 0 | 1 | 0 |

$$CTI = \sum_{i=1}^{24} \left( \theta_{B,i} - \theta_{S,i} \right) \text{forallintervalswhen} \theta_{B,i} < \theta_{S,i} \qquad (3.1)$$

$$CTD = \sum_{i=1}^{24} \left( \theta_{B,i} - \theta_{S,i} \right) \text{forallintervalswhen} \theta_{B,i} > \theta_{S,i} \qquad (3.2)$$

A summary of the results (predicted CTD and CTI values for summer and winter seasons for all mitigation measures in all cities) is provided in Table 3.5.

These results illustrate the potentially significant utility of the modelling tools and approaches for decision making processes pertaining to the proper choice of UHI mitigation measures. However, as with other areas of applied numerical modelling, certain important challenges must be addressed. One issue is related to the rather extensive time and computational resources that are necessary for proper

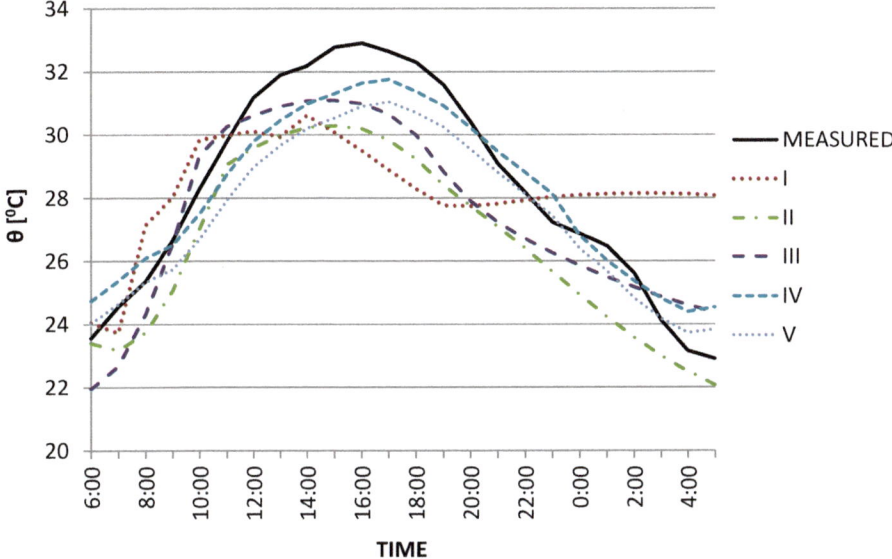

**Fig. 3.14** Comparison of the measured mean hourly temperature in the course of a summer day (22nd July 2010) in Vienna with simulation results conducted with various simulation setting and forcing parameter options (cases I to V)

deployment of complex simulation tools. A second challenge pertains to the issue of model reliability: Even highly detailed and mathematically consistent simulation tools may yield erroneous results given incomplete (or inaccurate) input data.

The high level of domain complexity involved in urban microclimate modelling aggravates this model reliability problem. To exemplify this circumstance, consider the simple case of predicting the air temperature in a specific location in the city of Vienna. In this case, the actual air temperature measurements at this location were compared with simulated results, whereby different tool versions and settings as well as different forcing parameters were considered. The results (see Fig. 3.14) suggest that predictions based on computation may significantly deviate from actual measurements, thus undermining the practical usability of modelling tools. An important approach to address the model reliability issues focuses on model calibration potential (Maleki et al. 2014).

In addition to numerical modelling, we explored the potential of simple empirically-based relationships between fundamental features of the urban setting (morphology, materials) and basic microclimatic variables. Such relationships would not only provide efficient means for rough estimations of the effectiveness of mitigation measures, but would also provide a basic plausibility check for the results of numerical computation. Hence, within the framework of the project, we developed a systematic framework toward definition and derivation of fundamental variables of a selected urban area. These variables are hypothesized to influence UHI and the urban microclimate variance. They pertain to both geometric (morphological) and semantic (material-related) urban features and are captured within a formal and systematic framework.

## 3.6    A Systematic Framework for the Representation of Urban Variables

Within the UHI project, a systematic framework was developed (Mahdavi et al. 2013) to assess – for a specific urban location, hereafter referred to as urban unit of observation (U2O) – the urban heat island phenomenon, to specify potential mitigation measures, and to evaluate such measures via adequate empirically-based calculation methods. The framework involves the following steps:

  (i) Definition of "Urban Units of Observation" (U2O): These are properly bounded areas within an urban setting selected as the target and beneficiary of candidate UHI mitigation measures;
 (ii) Description of the status quo of U2O in terms of a structured set of geometric and physical properties;
(iii) Specification of the existing extent of UHI in terms of proper indicators;
(iv) Specification of the candidate mitigation measures in terms of projected changes to the geometric and/or physical properties captured in step ii above;
 (v) Prediction of the effect of mitigation measures using empirically-based calculation methods;
(vi) Expression of the mitigation measures' impact in terms of predicted changes in the extent of UHI.

**Table 3.6** Variables to capture the geometric properties of an U2O

| Geometric properties | Definition |
| --- | --- |
| Sky view factor | Fraction of sky hemisphere visible from ground level |
| Aspect ratio | Mean height-to-width ratio of street canyons |
| Built area fraction | The ratio of building plan area to total ground area |
| Unbuilt area fraction | The ratio of unbuilt plan area to total ground area |
| Impervious surface fraction | The ratio of unbuilt impervious plan area to total ground area |
| Pervious surface fraction | The ratio of unbuilt impervious surface area to total ground area |
| Equivalent building height | The ratio of built volume (above terrain) to total ground area |
| Built surface fraction | The ratio of total built surface area to total built area |
| Wall surface fraction | The total area of vertical surfaces (walls) |
| Roof surface fraction | The total area of horizontal surfaces (roofs) |
| Effective mean compactness | The ratio of built volume (above terrain) to total surface area (built and unbuilt) |
| Mean sea level | Average height above sea level |

**Table 3.7** Variables to capture the surface and material properties of an U2O

| Surface/material properties | Definition |
|---|---|
| Reflectance/albedo | Fraction of reflected direct and diffuse shortwave radiation |
| Emissivity | Ability of a surface to emit energy by radiation (longwave) |
| Thermal conductivity | Property of a material's ability to conduct heat, given separately for impervious and pervious materials |
| Specific heat capacity | Amount of heat required to change a body's temperature by a given amount, given separately for impervious and pervious materials |
| Density | Mass contained per unit volume, given separately for impervious and pervious materials |
| Anthropogenic heat output | Heat flux density from fuel combustion and human activity (traffic, industry, heating and cooling of buildings, etc.) |

In this framework, the notion of U2O is applied to systematically address the local variation of the urban climate throughout a city. A spatial dimension (diameter) of approximately 400–1000 m has been targeted for U2O.

As the urban microclimate is believed to be influenced by different urban morphologies, structures, and material properties, a set of related variables were identified and included in our framework (Tables 3.6 and 3.7) based on past research (Nowak 2002; Piringer et al. 2002; Burian et al. 2005; Ali-Toudert and Mayer 2006) and our own investigations (Mahdavi et al. 2013; Kiesel et al. 2013).

The geometric properties are meant to capture the urban morphology of an U2O. The physical properties describe mainly the thermal characteristics of urban surfaces. These properties are often considered as fundamental factors in view of the heat balance of urban systems (Rosenfeld et al. 1995).

To derive the specific values of the U2O variables for the selected urban areas, we used data provided by the city of Vienna in a form of a Digital Elevation Model (DEM). The DEM consisted of a terrain and a surface model, including building footprints in form of closed polygons associated with building height data (which indicates the height of the building eaves). QGIS (Quantum GIS 2013), an open source Geographic Information System, was used to visualize, manage, and analyse the data. A specific set of algorithms was developed (Glawischnig et al. 2014) and further used for the quantitative analysis of the microclimatic attributes.

To exemplify and illustrate the application of the aforementioned algorithms and procedures, Fig. 3.15 depicts computed values of a selected set of geometric and semantic variables for four locations across Vienna. The selected locations include both low-density suburban and high-density urban typologies in Vienna (Table 3.8). Given the specific arrangement of the respective scales in this representation (descending versus ascending order of the scale numbers), it can support the recognition of distinct differences between the selected locations.

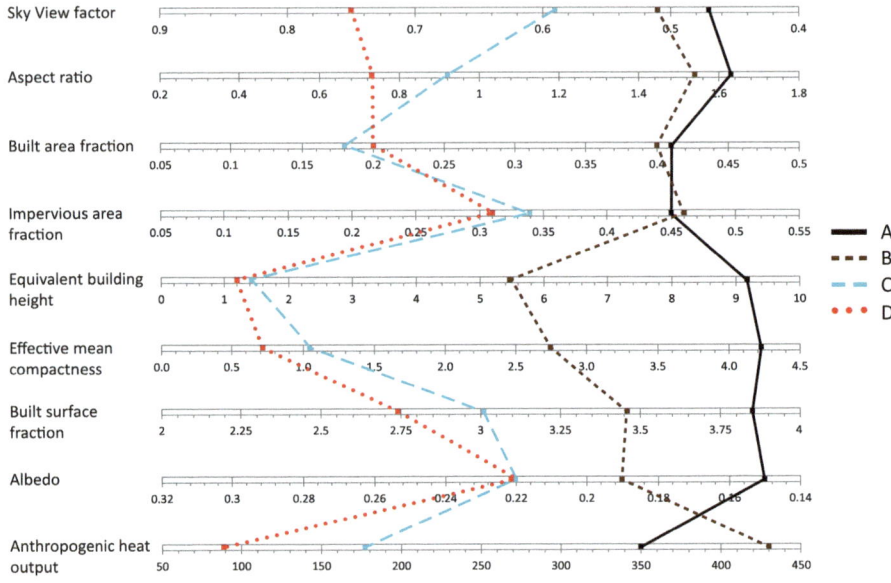

**Fig. 3.15** The computed values of a number of U2O variables for the urban areas around the aforementioned five weather station locations

**Table 3.8** Information regarding the selected locations in the city of Vienna

|   | Name | Type | Elevation |
|---|------|------|-----------|
| A | Innere Stadt | Urban (city center) | 177 m |
| B | Gaudenzdorf | Urban | 179 m |
| C | Hohe Warte | Urban (peripheral) | 198 m |
| D | Donaufeld | Suburban | 161 m |

A clear shift to the left in Fig. 3.15 denotes a more suburban character, while the shift to the right denotes a more urban character.

Once U2Os and their respective variables are derived, the existence and extent of the correlations between urban microclimate variance and the U2O variables are explored. These statistically significant correlations could provide a useful basis toward developing empirically-based predictive models. Such models could support, amongst other things, decision making processes with regard to the selection of appropriate mitigation measures that are intended to address the UHI phenomena.

Ongoing work in this area involves the collection of information on urban microclimate variance and the analysis of its hypothesised relationship to U2O variables. While certain U2O variables (impervious surface fraction, anthropogenic heat emission) are hypothesised to positively correlate with indicators of UHI phenom-

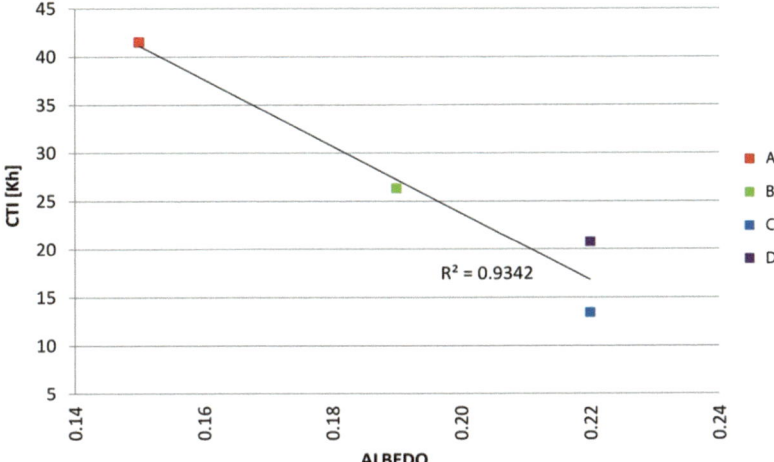

**Fig. 3.16** Correlation between the albedo and CTI values for summer period for selected areas (A to D) in Vienna

enon, others (sky view factor, albedo) are more likely to display a negative correlation. While the state of progress in this area has not reached the point to warrant solid and generally valid relationships, initial findings appear to be promising. For example, Fig. 3.16 illustrates the relationship between urban surface albedo (determined for four location across Vienna) and the measured summer-time CTI in those areas.

## 3.7 Conclusion

We presented the results of EU-supported project concerned with the extent of the UHI phenomena in a number of Central European cities. The objectives of this project are to provide a common understanding of the UHI effects and to conceive and evaluate appropriate mitigation and adaptation measures. Short-term and long-term data with regard to urban and rural temperatures demonstrate the existence and significant magnitude of the UHI effect in a number of Central European cities. Furthermore observations based on hourly data display distinguished patterns implying larger UHI intensities during the night hours. To address the need for effective means of evaluating and mitigating UHI effects a comprehensive modelling effort was undertaken. Thereby, the ramifications of potential mitigation measures in selected areas of the participating cities were investigated using advance numeric modelling tools and techniques. Moreover, a systematic framework was

developed and tested, which proposed and tested a number of geometric (morphological) and semantic (material-related) variables of the urban environment. These
variables are hypothesized to influence UHI and the urban microclimate variance.
Currently, the suggest link is being explored and statistically analysed. This work is
expected to not only provide empirical data for the validation of numeric models,
but also to support the formulation of simplified approaches toward estimation of
mitigation measures effectiveness in view of UHI phenomena.

**Acknowledgements** This project was funded in part within the framework of the EU-Project
"Development and application of mitigation and adaptation strategies and measures for counteracting
the global Urban Heat Island phenomenon" (Central Europe Program, No 3CE292P3).

# References

Akbari, H. (2005). *Energy saving potentials and air quality benefits of urban heat island mitiga-
tion*. Berkeley: Lawrence Berkeley National Laboratory.
Akbari, H., Pomerantz, M., & Taha, H. (2001). Cool surfaces and shade trees to reduce energy use
and improve air quality in urban areas. *Solar Energy, 70*(3), 295–310.
Alexandri, E. (2007). *Green cities of tomorrow?* Paper presented at the Sustainable Construction,
Materials and Practices, SB07, Portugal.
Ali-Toudert, F., & Mayer, H. (2006). Numerical study on the effects of aspect ratio and orientation
of an urban street canyon on outdoor thermal comfort in hot and dry climate. *Buildings and
Environment, 41*, 94–108.
Arnfeld, A. J. (2003). Two decades of urban cli-mate research: A review of turbulence, exchanges of
energy and water, and the urban heat island. *International Journal of Climatology, 23*(1), 1–26.
Blazejczyk, K., Bakowska, M., Wieclaw, M. (2006). *Urban heat island in large and small cities*.
Paper presented at the 6th international conference on urban climate, Göteborg, Sweden.
Burian, S., Suk Han, W., & Brown, M. J. (2005). *Morphological analysis using 3D building data-
bases: Oklahoma City*. Oklahoma: Los Alamos National Laboratory.
Gaffin, S. R., Rosenzweig, C., Khanbilvardi, R., Parshall, L., Mahani, S., Glickman, H., Goldberg,
R., Blake, R., Slosberg, R. B., & Hillel, D. (2008). Variations in New York City's urban heat
island strength over time and space. *Theoretical and Applied Climatology, 94*, 1–11.
Glawischnig, S., Kiesel, K., Mahdavi, A. (2014, May). *Feasibility analysis of open-government
data for the automated calculation of the micro-climatic attributes of urban units of observa-
tion in the city of Vienna*. Paper presented at the 2nd ICAUD International Conference in
Architecture and Urban Design, Epoka University, Tirana, Albania.

Grimmond, C. S. B. (2007). Urbanization and global environmental change: Local effects of urban warming. *Cities and Global Environmental Change, 173*(1), 83–88.

Grimmond, C. S. B., & Oke, T. R. (1999). Heat storage in urban areas: Local-scale observations and evaluation of a simple model. *Journal of Applied Meteorology, 38*(7), 922–940.

Grimmond, C. S. B., Cleugh, H., & Oke, T. R. (1991). An objective urban heat storage model and its comparison with other schemes. *Atmospheric Environment, 25B*(3), 311–326.

Harlan, S. L., & Ruddell, D. M. (2011). Climate change and health in cities: Impacts of heat and air pollution and potential co-benefits from mitigation and adaptation. *Current Opinion in Environmental Sustainability, 3*(3), 126–134.

Hart, M. & Sailor, D.J. (2007, September). *Assessing causes in spatial variability in urban heat island magnitude.* Paper presented at the Seventh Symposium on the Urban Environment, San Diego, CA

Huttner, S. & Bruse, M. (2009, June–July). *Numerical modelling of the urban climate – A preview on ENVI-met 4.0.* Paper presented at the 7th International Conference on Urban Climate ICUC-7, Yokohama, Japan.

Kiesel, K., Vuckovic, M., Mahdavi, A. (2013). *Representation of weather conditions in building performance simulation: A case study of microclimatic variance in Central Europe.* Paper presented at the 13th International Conference of the International Building Performance Simulation Association, IBPSA, France.

Kleerekoper, L., van Esch, M., & Salcedo, T. B. (2012). How to make a city climate-proof, addressing the urban heat island effect. *Resources, Conservation and Recycling, 64*, 30–38.

Mahdavi, A., Kiesel, K., Vuckovic, M. (2013, April). *A framework for the evaluation of urban heat island mitigation measures.* Paper presented at the SB13 Munich Conference, Germany.

Maleki, A., Kiesel, K., Vuckovic, M., & Mahdavi, A. (2014). Empirical and computational issues of microclimate simulation. Information and Communication Technology. *Lecture Notes in Computer Science, 8407*, 78–85.

Mirzaei, P. A., & Haghighat, F. (2010). Approaches to study urban heat island – Abilities and limitations. *Building and Environment, 45*(10), 2192–2201.

Nowak, D. J. (2002). *The effects of urban trees on air quality.* Syracuse: USDA Forest Service.

Oke, T. R. (1972). City size and the urban heat island. *Atmospheric Environment, 7*(8), 769–779.

Oke, T. R. (1981). Canyon geometry and the nocturnal urban heat island comparison of scale model and field observations. *Journal of Climatology, 1*, 237–254.

Piringer, M., Grimmond, C. S. B., Joffre, S. M., Mestayer, P., Middleton, D. R., Rotach, M. W., Baklanov, A., De Ridder, K., Ferreira, J., Guilloteau, E., Karppinen, A., Martilli, A., Masson, V., & Tombrou, M. (2002). Investigating the surface energy balance in urban areas – Recent advances and future needs. water. *Air and Soil Pollution: Focus, 2*(5–6), 1–16.

Quantum GIS. (2013). http://www.qgis.org/en/site/

Rosenfeld, A. H., Akbari, H., Bretz, S., Fishman, B. L., Kurn, D. M., Sailor, D. J., & Taha, H. (1995). Mitigation of urban heat islands: Materials, utility programs, updates. *Energy and Buildings, 22*(3), 255–265.

Shishegar, N. (2013). Street design and urban microclimate: Analysing the effects of street geometry and orientation on airflow and solar access in urban canyons. *Journal of Clean Energy Technologies, 1*(1), 52–56.

Taha, H. (1997). Urban climates and heat islands: Albedo, evapotranspiration, and anthropogenic heat. *Energy and Buildings, 25*(2), 99–103.

Unger, J. (2004). Intra-urban relationship between surface geometry and urban heat island: Review and new approach. *Climate Research, 27*, 253–264.

Voogt, J. A. (2002). Urban heat island. *Encyclopedia of Global Environmental Change, 3*, 660–666.

# Chapter 4
# Relevance of Thermal Indices
# for the Assessment of the Urban Heat Island

**Andreas Matzarakis, Letizia Martinelli, and Christine Ketterer**

**Abstract** Urban areas, with their specific characteristics, modify the atmosphere and produce their own meso- and micro climate. The major aspect of this chapter is the discussion of methods for the quantification and assessment of the urban micro-climate and the most known and world-wide studied phenomenon, the Urban Heat Island (UHI). Four urban measurement stations and one rural measurement station are used to quantify the temporal and spatial climatic characteristics in Stuttgart, Germany. For the quantification of the urban micro-climate and the UHI human thermal, comfort indices were applied. These indices, namely Physiologically Equivalent Temperature and the Universal Thermal Climate Index, are used to describe the integral effect of urban thermal atmosphere, based on the energy exchange of the human body. These indices, following the concept of equivalent temperature, are applied to quantify the integral effect of air temperature, air humidity, wind and radiation fluxes, expressed as mean radiant temperature.

**Keywords** Thermal Comfort • Physiologically Equivalent Temperature • Human Biometeorology

A. Matzarakis (✉)
Albert-Ludwigs-University of Freiburg, Werthmannstr. 10, D-79085, Freiburg, Germany

Research Center Human Biometeorology, German Meteorological Service, Stefan-Meier-Str. 10, D-70104, Freiburg, Germany
e-mail: andreas.matzarakis@dwd.de

L. Martinelli
Chair of Environmental Meteorology, Albert-Ludwigs-University, Freiburg, Germany

Department of Planning, Design and Technology of Architecture, Sapienza University, Via Flaminia 72, 00196 Rome, Italy

C. Ketterer
Albert-Ludwigs-University of Freiburg, Werthmannstr. 10, D-79085, Freiburg, Germany

iMA Richter & Roeckle, Eisenbahnstrasse 43, 79098, Freiburg, Germany

F. Musco (ed.), *Counteracting Urban Heat Island Effects in a Global Climate Change Scenario*, DOI 10.1007/978-3-319-10425-6_4

## 4.1   Introduction

Urban areas, with their artificial materials and specific morphology, act as an obstacle to the atmosphere, altering energy-balance, the chemical composition as well as the wind field (Landsberg 1981; Oke 1982; Oke and Cleugh 1987; Helbig et al. 1999). The urban heat island (UHI), describing the urban-rural surface and air temperature differences, is the most prominent and world-wide studied phenomenon of urban climate (eg Böhm and Gabl 1978; Katsoulis and Theoharatos 1985; Kuttler et al. 1996; Runnalls and Oke 2000; Johansson and Emmanuel 2006).

   In fact, the intensity of the urban heat island depends on eg land-use, building ratio, population density and vegetation (eg Landsberg 1981; Oke 1981; 1982; 1988). Before establishing mitigation and adaptation measures counteracting the urban heat island, city planners and officials need to comprehend the spatial and temporal dimensions of the meteorological and climatological conditions in a city (Matzarakis et al. 2008; Ketterer and Matzarakis 2014a, b). As city dwellers are the main target of city planners, the integral effect of air temperature, air humidity, wind speed and radiation fluxes on humans in a city has to be quantified and assessed (Eliasson 2000; Ketterer and Matzarakis 2014a, b). Hence, modern human-biometeorological methods for quantification of the spatial and temporal distribution of the UHI as well as to assess mitigation and adaptation measures for improving outdoor meteorological conditions have to be applied (Kuttler 2011, 2012; Matzarakis 2013).

   Urban planners require information about the human biometeorological conditions in terms of frequencies (eg number of days or hours per year or season), as well as the quantification of temperature differences between different planning scenarios Fröhlich and Matzarakis (2011, 2013). The quantification of heat stress and its reduction by planning measures is a big challenge, especially in the light of climate change (Matzarakis and Endler 2010). Due to climate change, the mean air temperature is expected to increase and also heat waves are assumed to became more frequent, more intense and longer lasting (Matzarakis and Amelung 2008; Meehl and Tebaldi 2004; Schär et al. 2004; Muthers and Matzarakis 2010; Matzarakis and Nastos 2011). Thus, there is a demand for the assessment and quantification of adaptation measures improving the urban climate, ie street morphology, different types of vegetation (Hwang et al. 2011; Ketterer et al. 2013; Lin et al. 2012; Matzarakis 2001, 2006, 2007, 2010). This approach is twofold: the analysis and description of single places for urban planning measures and the construction of maps for the detection of areas with frequent heat stress (Svensson et al. 2003).

   The aim is to show and describe methods based on long term measured data and their analysis for a comprehensive quantification of urban-rural differences and possible strategies for adaptation and mitigation in urban areas, focused on micro scale conditions.

## 4.2 Methods and Data

The application of thermal indices based on the human energy balance gives detailed information on the effect of complex thermal environments on humans (Höppe 1999). It is related to the close relationship between the human thermoregulatory mechanism and the human circulatory system. The human body does not have any selective sensors for the perception of individual climatic parameters.

Thermoreceptors can register the temperature of the skin and blood flow passing the hypothalamus and response thermoregulatorily (Höppe 1984, 1993, 1999). These temperatures, however, are influenced by the integrated effect of all climatic parameters, which are in some kind of interrelation, ie affect each other (VDI 1998; Höppe 1999).

Commonly used thermal indices, based on the human energy balance, are Predicted Mean VotePMV (Fanger 1972), Physiologically Equivalent Temperature PET (Mayer and Höppe 1987; Höppe 1999; Matzarakis et al. 1999), Standard Effective Temperature SET* (Gagge et al. 1986) or Outdoor Standard Effective Temperature Out_SET* (Spagnolo and Dear 2003), Perceived Temperature pT (Staiger et al. 2012) and Universal Thermal Climate Index UTCI (Jendritzky et al. 2012). These thermal indices require the same meteorological input parameters: air temperature, air humidity, wind speed, short and long wave radiation fluxes. These input parameters have a temporal and spatial variability, which have a huge influence on thermal indices. Wind speed and mean radiant temperature have the highest variability and are modified by surroundings and obstacles in complex urban areas. Thus, it is particularly important to calculate correctly these parameters and to perform the measurements with high quality and exactness (ie including artificial ventilation and radiation shield for air temperature measurements).

The basis for these thermal indices is the energy balance equation for the human body:

$$M + W + R + C + E_D + E_{Re} + E_{Sw} + S = 0 \tag{1}$$

where, M represents the metabolic rate (internal energy production), W the physical work output, R the net-radiation of the body, C the convective heat flow, $E_D$ the latent heat flow to evaporate water diffusing through the skin (imperceptible perspiration), $E_{Re}$ the sum of heat flows for heating and humidifying the inspired air, $E_{Sw}$ the heat flow due to evaporation of sweat, and S the storage heat flow for heating or cooling the body mass.

The individual terms in this equation have positive signs if they result in an energy gain for the body and negative signs in the case of an energy loss (M is always positive, W, $E_D$ and $E_{sw}$ are always negative). The unit of all heat flows is in Watt (Höppe 1999).

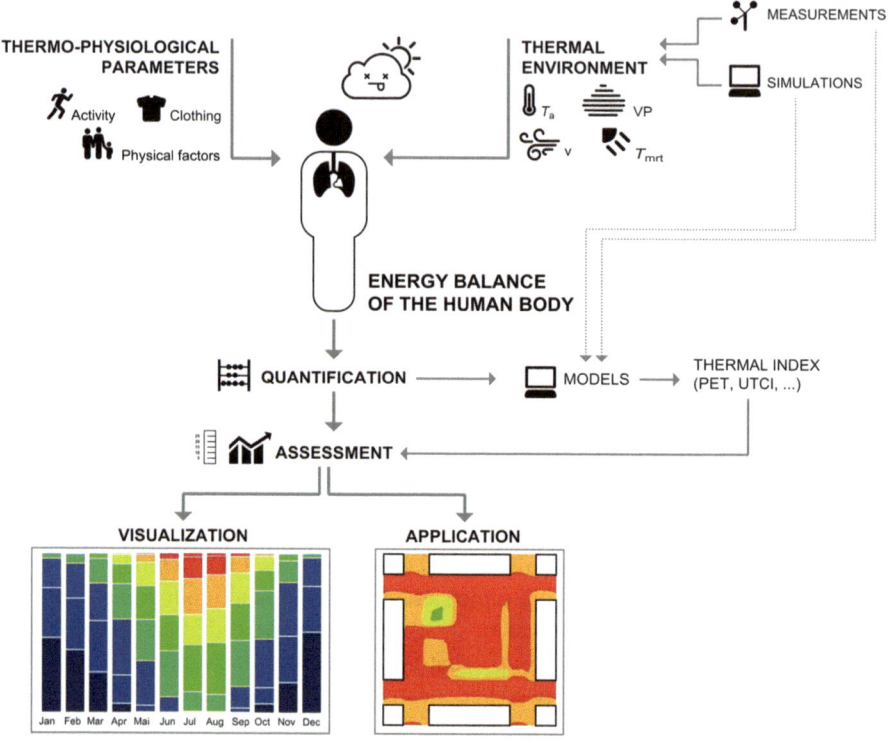

**Fig. 4.1** Flowchart of the human-biometeorological assessment of the thermal environment

The individual heat flows in Eq. 1, are controlled by the following meteorological parameters (VDI 1998; Höppe 1999):

- air temperature: $C$, $E_{Re}$
- air humidity: $E_D$, $E_{Re}$, $E_{Sw}$
- wind velocity: $C$, $E_{Sw}$
- radiant temperature: $R$

The thermo-physiological parameters heat resistance of clothing (clo) and activity of humans (in W) are required in addition (Fig. 4.1).

From the energy balance, which is primarily designed for the calculation of a thermal index like PET, are the indices which enable the user to predict "real values" of thermal quantities of the body, ie skin temperature, core temperature, sweat rate or skin wetness.

For this purpose, it is necessary to take into account all basic thermoregulatory processes, like the constriction or dilation of peripheral blood vessels and the physiological sweat rate (Höppe 1993; 1999).

An example of thermophysiological heat-balance model is the Munich Energy Balance Model for Individuals (MEMI) (Höppe 1993), which is the basis for the calculation of the Physiologically Equivalent Temperature, PET.

## 4.2.1   Physiologically Equivalent Temperature

The Physiologically Equivalent Temperature (PET) is the equivalent temperature at a given place (outdoors or indoors) to the air temperature in a typical indoor setting with core and skin temperatures equal to those under the conditions being assessed. Thereby, the heat balance of the human body with a work metabolism 80 W (light activity, added to basic metabolism) and a heat resistance of clothing 0.9 clo) is maintained (Höppe 1999).

The following assumptions are made for the indoor reference climate:

- mean radiant temperature equals air temperature ($Tmrt = Ta$).
- air velocity (wind speed) is fixed at $v = 0.1$ m/s.
- water vapor pressure is set to 12 hPa (approximately, equivalent to a relative humidity of 50 % at $Ta = 20$ °C).

The procedure for the calculation of PET contains the following steps:

1. calculation of the thermal conditions of the body with MEMI for a given combination of meteorological parameters.
2. insertion of the calculated values for mean skin temperature and core temperature into the model MEMI and computation the energy balance equation system for the air temperature Ta (with $v = 0.1$ m/s, $VP = 12$ hPa and $Tmrt = Ta$).
3. the resulting air temperature is equivalent to PET.

PET offers the advantage of a widely known unit (degrees Celsius), which makes results more comprehensible to regional or urban planners, who are not necessarily very familiar with the modern human-biometeorological methods (Matzarakis et al. 1999). The assessment classes of PET (Table 4.1) are valid only for the assumed values of internal heat production (80 W) and thermal resistance of the clothing (0.9) (Matzarakis and Mayer 1997).

The meteorological input parameters have to be measured or transferred to the average height of a standing person's gravity center, 1.1 m above ground (Matzarakis et al. 2009). These meteorological parameters can be measured or calculated by numerical models. PET can be calculated by models like RayMan (Matzarakis et al. 2007, 2010).

## 4.2.2   Universal Thermal Climate Index

The Universal Thermal Climate Index UTCI (Jendritzky et al. 2012) is defined as the air temperature (Ta) of the reference condition causing the same model response as the actual condition. Thus, UTCI represents the air temperature, which would produce, under reference conditions, the same thermal strain as in the actual thermal

**Table 4.1** Ranges of the physiological equivalent temperature (PET) for different grades of thermal perception by human beings and physiological stress on human beings, internal heat production: 80 W, heat transfer resistance of the clothing: 0.9 clo (according to Matzarakis and Mayer 1996, 1997)

| PET | Thermal perception | Grade of physiological stress |
| --- | --- | --- |
| | very cold | extreme cold stress |
| 4 °C | --------------------- | ------------------------- |
| | cold | strong cold stress |
| 8 °C | --------------------- | ------------------------- |
| | cool | moderate cold stress |
| 13 °C | --------------------- | ------------------------- |
| | slightly cool | slight cold stress |
| 18 °C | --------------------- | ------------------------- |
| | Comfortable | no thermal stress |
| 23 °C | --------------------- | ------------------------- |
| | slightly warm | slight heat stress |
| 29 °C | ---------------------------- | ------------------------------------------- |
| | warm | moderate heat stress |
| 35 °C | --------------------- | ------------------------- |
| | hot | strong heat stress |
| 41 °C | --------------------- | ------------------------- |
| | very hot | extreme heat stress |

environment. Both meteorological and non-meteorological (metabolic rate and thermal resistance of clothing) reference conditions were defined:

- wind speed (v) of 0.5 m/s at 10 m height (approximately 0.3 m/s in 1.1 m),
- mean radiant temperature (Tmrt) equal to air temperature,
- vapor pressure (VP) that represent relative humidity of 50 %, at high air temperatures (>29 °C) the reference air humidity is defined as 20 hPa.
- representative activity to be that of a person walking with a speed of 4 km/h (1.1 m/s). This provides a metabolic rate of 2.3 MET (135 W/m²).

The adjustment of clothing insulation is a powerful behavioral response to changing atmospheric conditions. Thereby, the conception behind UTCI was to consider seasonal clothing adaptation habits of Europeans based on available data from field surveys, in order to obtain a realistic representation of this behavioral action.

The categorization of UTCI is based on physiological response of an organism at actual environmental conditions depending on the responses for the reference conditions and thermal load (ie heat or cold stress) (Table 4.2). UTCI values between 18 and 26 °C may comply closely with the definition of the "thermal comfort zone" supplied in the Glossary of Terms for Thermal Physiology (International Union of Physiological Sciences – Thermal Commission 2003) as: "The range of ambient temperatures, associated with specified mean radiant temperature, humidity, and air movement, within which a human in specified clothing expresses indifference to the thermal environment for an indefinite period".

**Table 4.2**   UTCI equivalent temperature categorized in terms of thermal stress

| UTCI (°C) range | Stress category | Physiological responses |
|---|---|---|
| above +46 | extreme heat stress | Increase in $Tre$ time gradient |
| | | Steep decrease in total net heat loss |
| | | Averaged sweat rate >650 g/h, steep increase |
| +38 to +46 | very strong heat stress | Core to skin temperature gradient < 1 K (at 30 min) |
| | | Increase in $Tre$ at 30 min |
| +32 to +38 | strong heat stress | Dynamic Thermal Sensation ($DTS$) at 120 min > +2 |
| | | Averaged sweat rate > 200 g/h |
| | | Increase in $Tre$ at 120 min |
| | | Latent heat loss >40 W at 30 min |
| | | Instantaneous change in skin temperature > 0 K/min |
| +26 to +32 | moderate heat stress | Change of slopes in sweat rate, Tre and skin temperature: mean ($Tskm$), face ($Tskfc$), hand ($Tskhn$) |
| | | Occurrence of sweating at 30 min |
| | | Steep increase in skin wetness |
| +9 to +26 | no thermal stress | Averaged sweat rate > 100 g/h |
| | | $DTS$ at 120 min < 1 |
| | | $DTS$ between −0.5 and +0.5 (averaged value) |
| | | Latent heat loss >40 W, averaged over time |
| | | Plateau in $Tre$ time gradient |
| +9 to 0 | slight cold stress | $DTS$ at 120 min < −1 |
| | | Local minimum of $Tskhn$ (use gloves) |
| 0 to −13 | moderate cold stress | $DTS$ at 120 min < −2 |
| | | Skin blood flow at 120 min lower than at 30 min (vasoconstriction) |
| | | Averaged $Tskfc$ < 15 °C (pain) |
| | | Decrease in $Tskhn$ |
| | | $Tre$ time gradient < 0 K/h |
| | | 30 min face skin temperature < 15 °C (pain) |
| | | $Tmsk$ time gradient < −1 K/h (for reference) |
| −13 to −27 | strong cold stress | Averaged $Tskfc$ < 7 °C (numbness) |
| | | $Tre$ time gradient < −0.1 K/h |
| | | $Tre$ decreases from 30 to 120 min |
| | | Increase in core to skin temperature gradient |
| −27 to −40 | very strong cold stress | 120 min $Tskfc$ < 0 °C (frostbite) |
| | | Steeper decrease in $Tre$ |
| | | 30 min $Tskfc$ < 7 °C (numbness) |
| | | Occurrence of shivering |
| | | $Tre$ time gradient < −0.2 K/h |
| | | Averaged $Tskfc$ < 0 °C (frostbite) |
| | | 120 min $Tskfc$ < −5 °C (high risk of frostbite) |
| below −40 | extreme cold stress | $Tre$ time gradient < −0.3 K/h |
| | | 30 min $Tskfc$ < 0 °C (frostbite) |

(Blazejczyk et al. 2010)

Following abbreviations are used: rectal temperature Tre (°C), mean skin temperature Tskm (°C), face skin temperature Tskfc (°C), sweat production Mskdot (g/min), heat generated by shivering Shiv (W), skin wittedness wettA (%) of body area, skin blood flow VblSk (%) of basal value, Dynamic Thermal Sensation DTS

## 4.3 Exemplary Results

Thermal biometeorological conditions are described using the thermal indices PET and UTCI for Stuttgart. Thereby, the results are focused on the comparison of thermal indices and air temperature.

Figure 4.2 shows the air temperature conditions for the period 2000 to 2011 based on hourly data, where all the hourly data of the examined period is shown. A comparison is made between mean conditions and the specific year 2003, which experienced an extraordinary hot summer. The two upper graphs show the pattern of air temperature for the long period and the year 2003. It can be noted that, during summer 2003, air temperature exceed very often the 30 °C level. This heat wave in

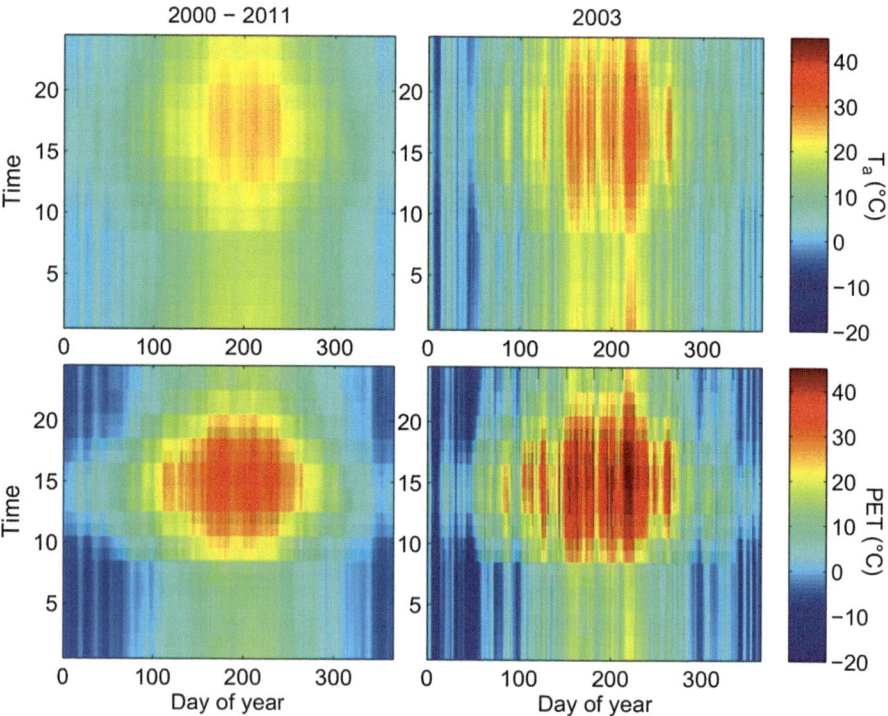

**Fig. 4.2** Ta (*upper graphs*) and PET (*lower graphs*) for the period 2001–2011 and 2003 for station Schwabenzentrum in Stuttgart

**Table 4.3**  Location and altitude of the measurement stations

| Measurement station | Lat (N) | Long (E) | Altitude (asl) | Feature |
|---|---|---|---|---|
| Neckartal (N) | 48:47 | 09:13 | 224 m | River valley, urban |
| Schwabenzentrum (S) | 48:46 | 09:10 | 250 m | City center, on top of a 25 m high Building |
| Schnarrenberg (Sb) | 48:50 | 09:12 | 314 m | top of a SW-exposed hill |
| Hohenheim (H) | 48:42 | 09:02 | 405 m | Suburb |
| Echterdingen (E) | 48:41 | 09:14 | 371 m | Airport |

2003 had many negative effects on human health, which results in high mortality and morbidity rate in Western Europe (Koppe et al. 2003; Le Tertre et al. 2006). The daytime PET values are higher than Ta during the day. During the night, PET is lower than air temperature. During the winter, the values of air temperature are mostly higher than the values of PET, due to the effect of wind, low irradiation and air humidity.

Another possibility to describe thermal comfort conditions and urban rural differences can be the long terms analysis for different stations. In Stuttgart are located five stations and they have all been selected for the analysis (Table 4.3).

Beanplots, developed by (Kampstra 2008), display the density curve of the data together with median or mean, percentiles or standard deviation. To obtain the typical bean shape, the density curve is mirrored along the central y-axis. Fig. 4.3 depicts beanplots for the described stations in Stuttgart (Neckartal, Schnarrenberg, Schwabenzentrum, Hohenheim and Echterdingen) for the period 2000–2011.

The plots show the seasonal pattern for winter, spring, summer and autumn for Ta (upper graphs), PET (middle) and UTCI (lower graph). The differences between the stations for Ta are small.

In particular, the daily minimum values are higher for the urban stations, compared to suburban site Hohenheim or the rural reference station Echterdingen. This fact can be explained by the heat storage in the city, but also by the differences in altitude, that reach more than 150 m.

The density distribution of PET is governed by radiation fluxes from spring to autumn, which is also the most impacting meteorological variable. In winter, the density distribution of PET is similar to the one of air temperature. Air temperature is the factor that influences PET most during winter.

The minimum values of UTCI (Fig. 4.3 lower panel) are lower than PET. This can be explained by adapting clothing model of UTCI during summer, but also by different assessment scale.

In some cases, esp. for application in planning or for the detection of extreme events for human health, issues conditions can be analyzed in terms of thresholds of air temperature or any thermal index. Table 4.4 shows the amount of days per year for different levels of maximum air temperature (Tamax > 30 and > 35), minimum air temperature (Tamin > 18, > 20 and > 23), the conditions for the maximum value of PET (PETmax > 30, > 35 and > 41) according to the assessment classes

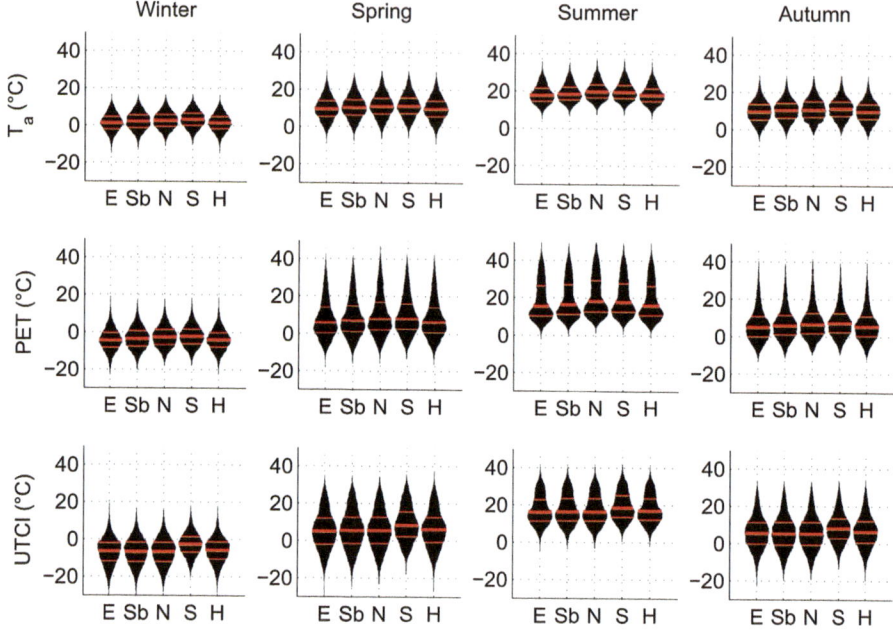

**Fig. 4.3** Bean plots for air temperature (*upper graphs*), PET (*middle graphs*) and UTCI (*lower graphs*) for the Stations Echterdingen (E), Schnarrenberg (Sb), Neckartal (N), Schwabenzentrum (S) and Hohenheim (H) for the period 2000–2011

**Table 4.4** Number of days on which defined thresholds are exceeded

|  | Echterdingen | Schnarrenberg | Neckartal | Schwabenzentrum | Hohenheim |
|---|---|---|---|---|---|
| $T_{amax} > 30$ | 8 | 13 | 10 | 14 | 7 |
| $T_{amax} > 35$ | 1 | 2 | 1 | 2 | 1 |
| $T_{amin} > 18$ | 3 | 15 | 11 | 20 | 6 |
| $T_{amin} > 20$ | 0 | 4 | 2 | 6 | 1 |
| $PET_{max} > 30$ | 93 | 85 | 96 | 99 | 92 |
| $PET_{max} > 35$ | 55 | 57 | 58 | 61 | 54 |
| $PET_{max} > 41$ | 16 | 26 | 19 | 23 | 15 |
| $PETmin > 18$ | 0 | 0 | 0 | 0 | 0 |

(Table 3.7.2.1) and also the PETmin for the levels of > 18. It can be seen that there is a difference in the conditions for maxima over Ta > 30 and only 1 or 2 days for Ta > 35 °C. In addition, it can be seen that Tamin conditions over > 18 and > 20 occur mostly in the urban area. Maximum values of PET are represented by a similar picture; however, the PET is generally higher than air temperature, while extreme heat stress (PET > 41 °C) can occur in more than 20 days per year in the city stations. The daily minimum value of PET never exceeds the threshold of 18 °C in Stuttgart (the frequency is below 0.1 %).

**Table 4.5** Number of hours per year on which defined thresholds were exceeded in the period 2000–2010

|  | Echterdingen | Schnarrenberg | Neckartal | Schwabenzentrum | Hohenheim |
|---|---|---|---|---|---|
| Daytime (10–16) |  |  |  |  |  |
| $T_a > 30$ | 40.2 | 53.6 | 70.5 | 76.5 | 37.5 |
| $T_a > 35$ | 2.3 | 4.6 | 7.5 | 7.2 | 2.3 |
| $T_a > 41$ | 0.0 | 0.0 | 0.0 | 0.0 | 0.0 |
| PET > 30 | 534.5 | 560.3 | 539.3 | 587.4 | 522.5 |
| PET > 35 | 252.9 | 271.2 | 298.0 | 289.9 | 244.4 |
| PET > 41 | 54.9 | 64.4 | 97.7 | 75.1 | 51.6 |
| Nighttime (22–6) |  |  |  |  |  |
| $T_a > 18$ | 11.9 | 44.6 | 52.5 | 121.0 | 27.1 |
| $T_a > 20$ | 4.3 | 17.5 | 21.5 | 61.5 | 10.1 |
| $T_a > 23$ | 0.4 | 3.9 | 4.4 | 21.5 | 2.1 |
| PET > 18 | 188.5 | 305.2 | 292.8 | 516.6 | 277.4 |
| PET > 20 | 75.0 | 149.4 | 154.0 | 307.5 | 130.5 |
| PET > 23 | 12.4 | 36.3 | 40.5 | 115.0 | 29.6 |

Beyond the information given by the amount of days that manifest specific events, the amount of hours can describe more precisely the occurrence of specific conditions, thus providing more valuable information. Table 4.5 depicts the total amount of hours for the thresholds for the examined stations in Stuttgart urban and rural area. It can be clearly observed that the amount of hours for Ta and PET is quite higher in the urban areas than in the rural one, esp. for Ta > 35 and PET > 41 during day time (10 to 16 LST). At nighttime (22 to 6 LST) the picture is a clearer indication that urban areas are quite hotter than rural areas.

## 4.4 Discussion and Conclusion

There is an increased demand for the quantification and the assessment of the Urban Heat Island (UHI). Most of the studies deal with the comparison of different station in urban and rural areas based on different temporal resolutions and many studies report about differences in air temperature about maxima of more than 10 K and for mean conditions about 2–3 K (Ketterer and Matzarakis 2014a, b). In fact, cities are warmer than rural areas and the formation of the UHI depends mostly on energetic aspects of the urban structures. Most influencing factors are the limited horizon, the storage of heat of urban fabrics and anthropogenic heat emissions (Landsberg 1981; Oke 1982). There are several aspects to quantify UHI beside the energetic aspects (heating and cooling) and air pollution. Recent studies focus on the reduction of UHI based on the modification of urban structures, using reflective materials or increasing evaporative cooling, however, these studies are mostly focusing only one some aspects of the energy budget of the urban areas and concentrate only on air temperature reduction. This aim is achieved using models with different spatial resolutions and designing exemplary measurements for case studies. All these results and information are demanded and can have a direct application. However, it has to be mentioned that results of thermal indices (PET or UTCI), compared to studies based only on air temperature, produce different outcomes. For example, increasing albedo of urban surfaces can decrease air temperature, but at the same time it increases the values of PET, due to increased shortwave reflection. This is a relevant topic not only during summer, but throughout the all year, because it affects the use of the urban spaces, where people want to spend time outdoors.

Finally, easy understandable graphs and figure are a possibility for better communication between different users and disciplines.

# References

Błażejczyk, K., Broede, P., Fiala, D., Havenith, H., Holmér, I., Jendritzky, G., Kampmann, B., & Kunert, A. (2010). Principles of the new universal thermal climate index (UTCI) and its application to bioclimatic research in European scale. *Miscellanea Geographica, 14*, 91–102.

Böhm, R., & Gabl, K. (1978). Die Wärmeinsel einer Großstadt in Abhängigkeit von verschiedenen meteorologischen Parametern. *Archiv für Meteorologie, Geophysik und Bioklimatologie Serie B, 26*, 219–237.

Eliasson, I. (2000). The use of climate knowledge in urban planning. *Landscape and Urban Planning, 48*, 31–44.

Fanger, P. O. (1972). *Thermal comfort: Analysis and applications in environmental engineering.* New York: McGraw-Hill.

Fröhlich, D., & Matzarakis, A. (2011). Hitzestress und Stadtplanung – Am Beispiel des Platz der alten Synagoge in Freiburg im Breisgau. *Gefahrstoffe – Reinhaltung der Luft, 71*, 333–338.

Fröhlich, D., & Matzarakis, A. (2013). Thermal bioclimate and urban planning in Freiburg – Examples based on urban spaces. *Theoretical and Applied Climatology, 111*, 547–558.

Gagge, A.P., Fobelets, A., & Berglund, L. (1986). A standard predictive index of human response to the thermal environment. *ASHRAE Transactions, 92*, 709–731.

Helbig, A., Baumüller, J., & Kerschgens, M. J. (Eds.). (1999). *Stadtklima und Luftreinhaltung.* Berlin/Heidelberg: VDI-Buch. Springer.

Höppe, P. (1984). Die Energiebilanz des Menschen, Wiss. Mitt. Meteorol. Inst. Univ. München No. 49

Höppe, P. (1992). Ein neues Verfahren zur Bestimmung der mittleren Strahlungstemperatur im Freien. *Wetter und Leben, 44*, 147–151.

Höppe, P. (1993). Heatbalancemodelling. *Experientia, 49*, 741–746.

Höppe, P. (1999). The physiological equivalent temperature – a universal index for the biometeorological assessment of the thermal environment. *International Journal of Biometeorology, 43,* 71–75.

Hwang, R. L., Matzarakis, A., & Lin, T. P. (2011). Seasonal effect of urban street shading on long-term outdoor thermal comfort. *Building and Environment, 46,* 863–870.

International Union of Physiological Sciences – Thermal Commission. (2003). Glossary of terms for thermal physiology. *Journal of Thermal Biology, 28,* 75–106.

Jendritzky, G., de Dear, R., & Havenith, G. (2012). UTCI – Why another thermal index? *International Journal of Biometeorology, 56,* 421–428.

Johansson, E., & Emmanuel, R. (2006). The influence of urban design on outdoor thermal comfort in the hot, humid city of Colombo, Sri Lanka. *International Journal of Biometeorology, 51,* 119–133.

Kampstra, P. (2008). Beanplot: A boxplot alternative for visual comparison of distributions. *Journal of Statistical Software, Code Snippets, 28,* 1–9.

Katsoulis, B. D., & Theoharatos, G. A. (1985). Indications of the urban heat island in Athens, Greece. *Journal of Climate and Applied Meteorology, 24,* 1296–1302.

Ketterer, C., & Matzarakis, A. (2014a). Human-biometeorological assessment of adaptation and mitigation measures for replanning in Stuttgart, Germany. *Landscape and Urban Planning, 112,* 78–88.

Ketterer, C., & Matzarakis, A. (2014b). *Human-biometeorological quantification of the urban heat island in a city with complex topography – the case of Stuttgart.* Germany: Urban Climate.

Ketterer, C., Ghasemi, I., Bertram, A., Reuter, U., Rinke, R., Kapp, R., & Matzarakis, A. (2013). Veränderung des thermischen Bioklimas durch stadtplanerische Umgestaltungen – Beispiel Stuttgart-West. *Gefahrstoffe-Reinhaltung der Luft, 73,* 323–329.

Koppe, C., Jendritzky, G., & Pfaff, G. (2003). *Die Auswirkungen der Hitzewelle 2003 auf die Gesundheit.* Klimastatusbericht, 152–162. Available at: http://www.dwd.de/bvbw/generator/DWDWWW/Content/Oeffentlichkeit/KU/KU2/KU22/klimastatusbericht/einzelne__berichte/ksb2003__pdf/09__2003,templateId=raw,property=publicationFile.pdf/09_2003.pdf.

Kuttler, W. (2011). Climate change in urban areas. Part 2, Measures. *Environmental Sciences Europe, 23,* 21.

Kuttler, W. (2012). Climate Change on the Urban Scale – Effects and Counter-Measures in Central Europe, Human and Social Dimensions of Climate Change, Prof. Netra Chhetri (Ed.), InTech, DOI: 10.5772/50867. Available from: http://www.intechopen.com/books/human-and-social-dimensions-ofclimate-change/climate-change-on-the-urban-scale-effects-and-counter-measures-in-central-europe.

Kuttler, W., Barlag, A.-B., & Robmann, F. (1996). Study of the thermal structure of a town in a narrow valley. *Atmospheric Environment, 30,* 365–378.

Landsberg, H. E. (1981). *The urban climate.* London: The academic press.

Le Tertre, A., Lefranc, A., Eilstein, D., Declercq, C., Medina, S., Blanchard, M., Chardon, B., Fabre, P., Filleul, L., Jusot, J.-F., Pascal, L., Prouvost, H., Cassadou, S., & Ledrans, M. (2006). Impact of the 2003 heatwave on all-cause mortality in 9 French cities. *Epidemiology (Cambridge, Mass.), 17,* 75–79.

Lin, T. P., Tsai, K. T., Hwang, R. L., & Matzarakis, A. (2012). Quantification of the effect of thermal indices and sky view factor on park attendance. *Landscape and Urban Planning, 107,* 137–146.

Matzarakis, A. (2001). Die thermischeKomponente des Stadtklimas, Wiss. Ber. Meteorol. Inst. Universität Freiburg No. 6

Matzarakis, A. (2006). Weather and climate related information for tourism. *Tourism and Hospitality Planning & Development, 3,* 99–115.

Matzarakis, A. (2007). Climate, human comfort and tourism. In B. Amelung, K. Blazejczyk, A. Matzarakis (Eds.), *Climate change and tourism: Assessment and coping strategies* (pp. 139–154). Freiburg - Warsow - Maastricht: Own publishing

Matzarakis, A. (2010). Climate change: Temporal and spatial dimension ofad aptation possibilities at regional and localscale. In C. Schott (Ed.), *Tourism and the implications of climate change:*

*issues and actions* (Bridging tourism theory and practice, Vol. 3, pp. 237–259). Bingley: Emerald Group Publishing.

Matzarakis, A. (2013). Stadtklima vor dem Hintergrund des Klimawandels. *Gefahrstoffe – Reinhaltung der Luft, 73*, 115–118.

Matzarakis, A., & Amelung, B. (2008). Physiologically equivalent temperature as indicator for impacts of climate change on thermal comfort of humans. In M. C. Thomson et al. (Eds.), *Seasonal forecasts, climatic change and human health* (Advances in global change research, Vol. 30, pp. 161–172). Dordrecht: Springer-Sciences and Business Media.

Matzarakis, A., & Endler, C. (2010). Climate change and thermal bioclimate in cities: Impacts and options for adaptation in Freiburg, Germany. *International Journal of Biometeorology, 54*, 479–483.

Matzarakis, A., & Mayer, H. (1996). Another kind of environmental stress: Thermal stress. WHO collaborating centre for air quality management and air pollution control. *Newletters, 18*, 7–10.

Matzarakis, A., & Mayer, H. (1997). Heat stress in Greece. *International Journal of Biometeorology, 41*, 34–39.

Matzarakis, A., & Nastos, P. (2011). Human-biometeorological assessment of heat waves in Athens. *Theoretical and Applied Climatology, 105*, 99–106.

Matzarakis, A., Mayer, H., & Iziomon, M. G. (1999). Applications of a universal thermal index: physiological equivalent temperature. *International Journal of Biometeorology, 43*, 76–84.

Matzarakis, A., Rutz, F., & Mayer, H. (2007). Modelling radiation fluxes in simple and complex environments – application of the RayMan model. *International Journal of Biometeorology, 51*, 323–334.

Matzarakis, A., Röckle, R., Richter, C.-J., Höfl, H.-C., Steinicke, W., Streifeneder, M., & Mayer, H. (2008). Planungsrelevante Bewertung des Stadtklimas – Am Beispiel von Freiburg im Breisgau. *Gefahrstoffe - Reinhaltung der Luft, 68*, 334–340.

Matzarakis, A., De Rocco, M., & Najjar, G. (2009). Thermal bioclimate in Strasburg – The 2003 heat wave. *Theoretical and Applied Climatology, 98*, 209–220.

Matzarakis, A., Rutz, F., & Mayer, H. (2010). Modelling radiation fluxes in simple and complex environments – basics of the RayMan model. *International Journal of Biometeorology, 54*, 131–139.

Mayer, H., & Höppe, P. R. (1987). Thermal comfort of man in different urban environments. *Theoretical and Applied Climatology, 38*, 43–49.

Meehl, G. A., & Tebaldi, C. (2004). More intense, more frequent, and longer lasting heat waves in the 21st century. *Science, 305*, 994–997.

Muthers, S., & Matzarakis, A. (2010). Use of Beanplots in climatology and biometeorology – A comparison with boxplots. *Meteorologische Zeitschrift, 19*, 639–642.

Oke, T. R. (1981). Canyon geometry and thenocturnal urban heat island: Compari-son of scale model and field observations. *Journal of Climatology, 1*(3), 237–254. http://dx.doi.org/10.1002/joc.3370010304

Oke, T. R. (1982). The energetic basis of the urban heat island. *Quar-terly Journal of the Royal Meteorological Society, 108*(455), 1–24. http://dx.doi.org/10.1002/qj.49710845502

Oke, T. R. (1988). Street design and urban canopy layer climate. *Energy and Buildings, 11*(1–3), 103–113. http://dx.doi.org/10.1016/0378-7788(88)90026-6

Oke, T. R., & Cleugh, H. A. (1987). Urban heat storage derived as energy balance residuals. *Boundary-Layer Meteorology, 39*, 233–245.

Runnalls, K. E., & Oke, T. R. (2000). Dynamics and controls of the near-surface heat island of Vancouver, British Columbia. *Physical Geography, 21*, 283–304.

Schär, C., Vidale, P. L., Lüthi, D., Frei, C., Häberli, C., Liniger, M. A., & Appenzeller, C. (2004). The role of increasing temperature variability in European summer heat waves. *Nature, 427*, 332–336.

Spagnolo, J., & de Dear, R. A. (2003). Field study of thermal comfort in outdoor and semi-outdoor environments in subtropical Sydney Australia. *Building and Environment, 38*, 721–738.

Staiger, H., Laschewski, G., & Grätz, A. (2012). The perceived temperature – a versatile index for the assessment of the human thermal environment. *Part A: scientific basics. International Journal of Biometeorology, 56*, 165–176.

Svensson, M., Thorsson, S., & Lindqvist, S. (2003). A geographical information system model for creating bioclimatic maps – examples from a high, mid-latitude city. *International Journal of Biometeorology, 47*, 102–112.

VDI. (1998). Methods for the human-biometerological assessment of climate and air hygiene forurban and regional planning. Part I: Climate, VDI guideline 3787. Part 2. Beuth, Berlin.

# Chapter 5
# Decision Support Systems for Urban Planning

**Davide Fava, Graziella Guaragno, and Claudia Dall'Olio**

**Abstract** Decision Support System (DSS) is an interactive software-based tool that has been realized in the framework of the project "Development and application of mitigation and adaptation strategies for counteracting the global Urban Heat Islands phenomenon (UHI)" implemented through CE programme 2007–2013 co-financed by the ERDF. DSS is a simplified database management tool that allow the use of the project deliverables uploaded on the web sites with a user-friendly approach.

Users can access to the needed project's output through knowledge's needs oriented pathway.

**Keywords** Decision Support System • UHI • Mitigation actions • Urban planning • Climate change

## 5.1 Introduction

The construction of Decision Support System (DSS) has been realized in the framework of the Work Package 6 (WP6), in particular with the action 6.1.

The main activity of the WP6 was the UHIs simulation of future alternative scenarios related to the development of the urban areas chosen for the eight pilot actions.

In addition, WP6 was intended to define and realize a set of support actions for fostering the implementation of urban & spatial planning strategies in each involved region.

In particular, it was foreseen a progressive integration of mitigation and adaptation strategies in current urban planning tools.

D. Fava
Democenter-Sipe Foundation and Emilia Romagna Region,
Via Pietro Vivarelli, 2, 41125 Modena, Italy
e-mail: d.fava@democentersipe.it

G. Guaragno (✉) • C. Dall'Olio
Territorial Planning and Mountain Development Service, Emilia Romagna Region, Italy
e-mail: GGuaragno@regione.emilia-romagna.it; CDallOlio@regione.emilia-romagna.it

© The Author(s) 2016
F. Musco (ed.), *Counteracting Urban Heat Island Effects in a Global Climate Change Scenario*, DOI 10.1007/978-3-319-10425-6_5

Furthermore, DSS is designed to address policies in actions of mitigation and adaptation in the framework of the Urban Heat Island phenomenon.

## 5.2  Development of the UHI Project's DSS

During the project implementation the partners agreed to choose the hypothesis described below:

A simplified database management tool that allow the use of the project deliverables uploaded on the web sites with a user-friendly approach.

In this case users can access to the needed project's output through knowledge's needs oriented pathway.

To manage the urban development it is compulsory to manage a multitude of purposes and address many different goals, often conflicting, to satisfy the needs of different stakeholders. This poses considerable challenges to policy makers and urban planner. The need for enhanced urban plan decision support systems is evident in the same complexity of the UHI phenomenon.

In the following paragraphs is described the operation of the DSS chosen within the project on UHI.

## 5.3  UHI's Decision Support System

### 5.3.1  Structure

The UHI DSS is composed by a structure of html, javascript and php files that allows, via communication with a database, the end users to remotely access the project deliverables produced by UHI partners. In fact, considering the type of files and its structure, the software is a website, that can be installed and run on any platform that provides services http and MySQL.

The logical structure of the files is likely (Fig. 5.1):

From the home page (index.html) you can access to:

- the logical framework of the project, which is a presentation of the UHI project; with a specific links to the official website of the project ;
- the Decision Support System, which allows, through a logical structure, obtaining a specific report based on the user's requests;;
- the Consultation Tool, which is a grid with specific questions about the results of the project deliverables produced by the partners; this tool is useful for users who know in advance what kind of information they are looking for.

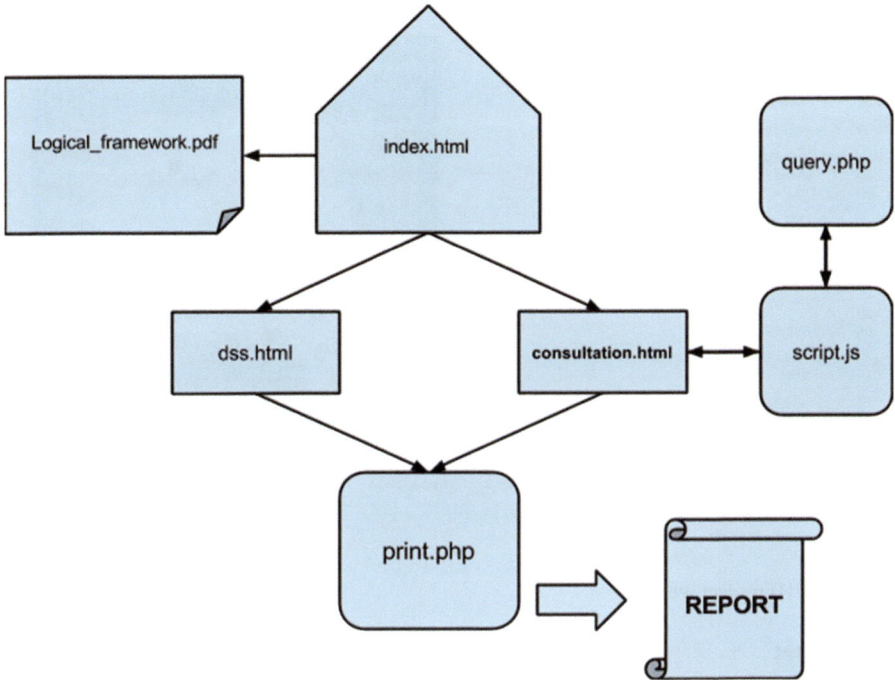

**Fig. 5.1**  UHI DSS logical structure of the files

The two php files (query.php and print.php) shall only be used to query the database that stores the data that is composing the report: in this case, query.php is part of the communication AJAX of the page consultation.html (through script.js); while print.php dynamically creates a page report based on the choices you make, whether they come from the DSS or from the Consultation Tool.

The database named db_dss_uhi is composed of five separate tables:

- intervention_type: stores data regarding the various types of intervention carried out in the pilot actions;
- legislation: containing the legislative framework for each region involved in the project UHI *;
- pilot_actions: the table with a detailed description of all pilot projects related to UHI project *;
- skills: the list (provided by partners) of the experts who have contributed to the UHI project with a short description of them and their personal contacts;
- software: a list of tools used by partners with a short description;

**Fig. 5.2** UHI project website

**Fig. 5.3** First page of the DSS

## 5.3.2  The Interface

This paragraph shows the real functioning of the DSS as was finally built with the help of the below pictures.

The link of the DSS can be found on the homepage of the UHI website (Fig. 5.2).

By clicking on the link highlighted above, the first page of the DSS appears like this (Fig. 5.3):

**Fig. 5.4** Link to the project logical framework

As the user can immediately see, from this page you can enter other three links to three different tools:

- Logical framework (magnifying glass)
- Consultation tool (open book)
- Decision Support System (right arrow)

### 5.3.3  Project Logical Framework

Clicking on the left icon below, the user can have an overview of the entire UHI project (Fig. 5.4).

As shown in the picture below in this page the user can find: a brief description of the project; the partnership; the framework analysis; the establishment of a UHI monitoring network; mitigation and adaptation strategies; pilot and capitalization actions for limiting UHI's effects (Fig. 5.5).

### 5.3.4  The Consultation Tool

The interface of the Consultation Tool consists of 7 menus, multiple choice or drop-down menu, in which user is asked to select some areas of interest (Fig. 5.6).

- Climate long term perspective, where the user can see through a map the following data:

  – change in annual mean temperature;

- heat wave frequency;
- changes in annual near-surface temperature.

- The Urban Heat Island phenomenon, where the user can select a city name in the drop-down menu between the various European cities that have been studied during the project:

**URBAN HEAT ISLAND Central Europe PROJECT**
**Development and application of mitigation and adaptation strategies and measures for counteracting the global Urban Heat Islands phenomenon (UHI)**

**The UHI project**

The UHI project, starting from a deep **analysis of the phenomenon** (*WP3*) is designed to both develop mitigation and risk prevention, and management strategies.

The general objective of the project is to establish a **Transnational attention**, as well as policies and practical actions, for the prevention, adaptation and mitigation of the natural and man-made risks arising from the Urban Heat Island phenomenon (*WP2 & WP4*).

In particular, **mitigation strategies** consist in the adoption of urban and land planning models that prevent the establishment of UHI, while **adaptation strategies** aim at reducing the impact of phenomena related to UHI, such as summer bioclimatic discomfort (*WP5*). The innovative strategy of the UHI project is to interact two disciplines that traditionally don't communicate each other: **meteoclimatology** and **urban planning.** Through this interaction will be implemented particular strategies to guide the choices of development and urban renewal.

**The UHI partnership**

**FRAMEWORK ANALYSIS (*WP3*)**

WP3 considers two main scientific aspects: the **characteristics of UHI phenomenon** both in terms of causes and effects on environment and population, and its relationships with **climate change trends.**
Activities are focused on CE area, including an analysis of already existent UHIs, as well as a study of those situations that could constitute a potential for an increase of UHIs.
Additionally a list of existing **rules and legislation** toward UHI phenomenon in CE regions are prepared.

**Fig. 5.5** Logical framework of the project

Dear user, this tool DECISION SUPPORT SYSTEM (DSS) is part of the UHI Project implemented through the CENTRAL EUROPE Programme co-financed by the ERDF. UHI Project aims at developing mitigation and risk prevention and management strategies concerning the urban heat island (UHI) phenomenon.

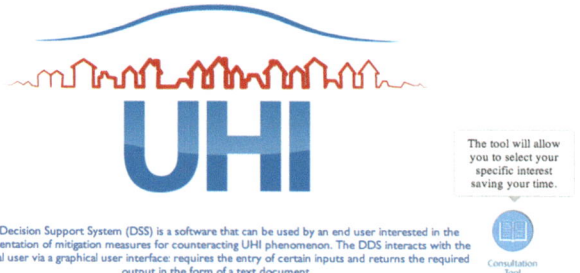

The tool will allow you to select your specific interest saving your time.

This Decision Support System (DSS) is a software that can be used by an end user interested in the implementation of mitigation measures for counteracting UHI phenomenon. The DDS interacts with the external user via a graphical user interface: requires the entry of certain inputs and returns the required output in the form of a text document.

Consultation Tool

**Fig. 5.6** Link to the consultation tool

- Budapest;
- Modena;
- Warsaw;
- Padua;
- Stuttgart;
- Ljubljana;
- Wien;
- Prague.

• Spatial and urban planning, where the user can, through the menu, select the following items of the interested area:

- local rules and regulation;
- incentives, financing, regulatory action;
- E.U. rules and regulation.

• Mitigation strategies, where there is the possibility to choose the mitigation action you want to deepen and find out its benefits on UHI. This sub section is divided into three items:

- Buildings

  • Cool roofs;
  • Green roofs
  • Green facades;
  • Facade surface and construction selection/retrofit;
  • Geometry of Urban Canyon.

- Pavements

  • Cool pavements;
  • Pervious pavements.

- Green areas

  • Planting trees within the urban canyon;
  • Park, green areas.

• Pilot areas, where the user can choose one of the eight pilot actions have been implemented that he wants to deepen:

- Padua/Venice –Italy;
- Modena/Bologna – Italy;
- Wien – Austria;
- Stuttgart – Germany;
- Warsaw – Poland; Prague –
- Czech Republic; Ljubljana – Slovenia;
- Budapest – Hungary.

• Skills, where the user can choose the skill which is interested in:

- Meteoclimatology;
- Biometeoclimatology;

  - Urban planning;
  - Health;
  - Municipality;
  - Innovation;
  - Engineering;
  - Building skill;
  - Environment;
  - Communication.

- Simulation tools, where the user can, simulating the effect of a mitigation action in the urban area object of the pilot, choose the aspect that he is interested in deepening.

  - Global scale

    - Weather forecast;
    - Climate of your region;
    - Future climatic scenarios;
    - Weather risks for your region due to climate change.

  - Urban/district scale

    - Urban micrometeorology;
    - Influence of land use on urban climate;
    - Effectcs of street characteristics (orientation, width, buildings height, etc.) on urban climate;
    - Effects of urban topography on climate;
    - Influence of land use on urban climate;
    - Influence of different kind of roofs on urban climate;
    - Influence of different kind of pavements on urban climate;

  - Human perception of climate;

    - Knowing how land use influences the human state of wellness.

  - Building

    - Influence of different kind of roof on building;
    - Effects on micrometeorological variables due to different materials and properties of the building;
    - Influence of good/bad adopted measures on human health/wellness at building scale.

The following pictures show how are graphically displayed the contents above outlined (Figs. 5.7, 5.8 and 5.9).

Once required the information requested by the user, the software searches in the documents produced in the WPs of the project the material/documentation or deliverable (re-organized in a logical framework); this content has been revised in order to obtain a final compact document with all the information selected. The results are

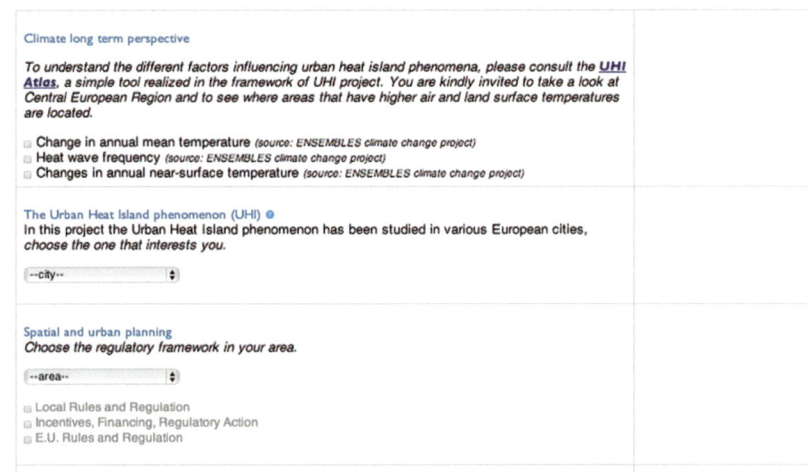

Fig. 5.7 The first three menus of the consultation tool

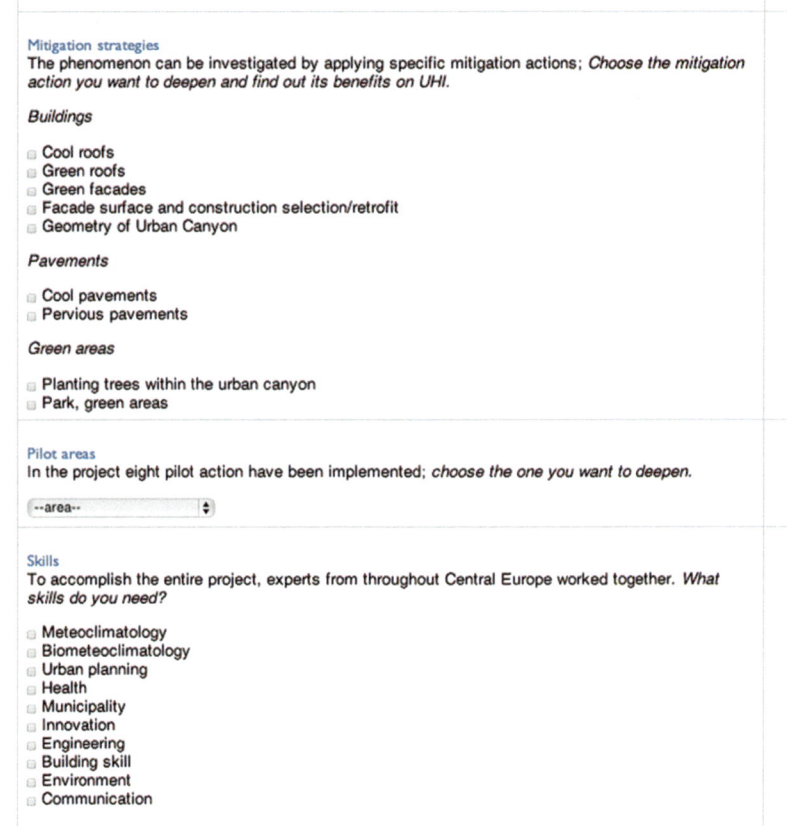

Fig. 5.8 The subsequent three menus

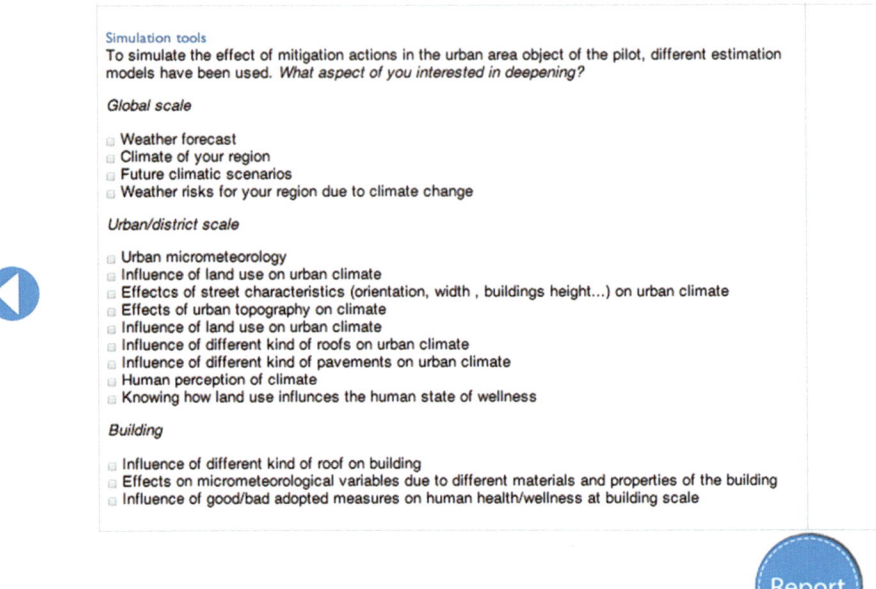

**Fig. 5.9** The simulation tool

**Fig. 5.10** Link to the UHI's DSS

**Outline your pilot study**

ision process, underlining the logical interaction between the different potential choices. In this case you'll be able to explore the project
approach designing an healthy and sustainable city.

**Fig. 5.11** Interactive map

finally displayed in html pages for instant opening for a quick reference, then the user can print it as a pdf.

## 5.3.5   *The DSS*

**By clicking the icon in the middle of the screen, user is led within the DSS.**
By clicking on the different menu items the user can simulate the approach to his pilot study based on the data we gathered in our 8 pilot tests (Fig. 5.10).
This section has 5 menus:

• Location: an interactive Central Europe Map allows the user to select the region of interest, as shown the picture below (Fig. 5.11)
• Scale of intervention: the user can choose between two option, building or urban;
• Typology of intervention: the choice here is between:

 – Menu item A)

   • Existing structure;
   • New construction.

 – Menu item B)

- Facedes;

  - Roofs;
  - Surface lots;
  - Urban structure;
  - Urban green.

- Economic assessment: here the user can choose the action he would like to undertake between:

  – Cool roofs;
  – Green roofs
  – Green facades;
  – Facade surface and construction selection/retrofit;
  – Geometry of Urban Canyon.
  – Cool pavements;
  – Pervious pavements.
  – Planting trees within the urban canyon;
  – Park, green areas.

After entering a few data, the software proceeds a quick calculation of the action selected.

- Skills, where the user can choose the skill which is interested in:

  – Meteoclimatology;
  – Biometeoclimatology;
  – Urban planning;
  – Health;
  – Municipality;
  – Innovation;
  – Engineering;
  – Building skill;
  – Environment;
  – Communication;

The selections made in the menus help to match with the material presented above, relating them in a logical way.

This tool aims at presenting the same data of the Consultation Tool but in a more organic way: a combination of a raw data and documents selected shape a report and finally a model.

Indeed the choices are collected and reported in a final html page as a report.

## 5.4  DSS Input

The inputs of the DSS are all data stored in the database that contribute for processing the final information. In our case inputs are established based on the work developed in the previous Work Packages: WP3, WP4, and WP5.

From the above-cited WPs here follows a synthesis of the most important aspects necessary to set up the DSS.

### 5.4.1  WP3

WP3 was divided in 2 main actions:

- Action 3.1 was basically a review action in which one has collected different studies about the Urban Heat Island phenomenon. Action 3.1 was split in subactions. This analysis focused on anthropogenic causes that generate the UHI phenomenon and the survey techniques used to study it.

The analysis take into account both (1) technical-scientific issues and (2) urban planning and land use regulation.

1. Technical-scientific issues

   - 1.a. reviewing knowledge (causes and related factors: anthropogenic causes that generate UHI phenomenon i.e. peculiar urban and building characteristics, particular industrial activities, etc.; the patterning of UHI phenomenon; the measures adopted to fight the intensification of UHI; the survey techniques used to study the phenomenon). Many relationships were found in literature to assess the UHI phenomenon based on different kinds of data. Namely we have a collection of indices to establish the increase of temperature and the maximal increase of temperature based on:

     - Population data;
     - Urban geometric factors of buildings as well as infrastructures;
     - Meteorological data.

   - 1.b. scheduling of existing infrastructures to meteorological and environment data assessment in different project areas. A big range of methods can be adopted in order to study the UHI phenomenon. For each project area different data can be available. Based on this availability or lack of information DSS is able to provide as much as possible of accuracy in the results.

2. Urban planning and land use regulation

   Urban planning and land use regulation assume an important role to regulate the possible actions of adaptation and mitigation. DSS take into account these in order to formulate answers. Of course these are not static and can be supported by the DSS to assess and to plan in a better way the urban development taking into account in particular:

- 2.a. Review of different rules and regulation set up by involved local government (reviewing of the local main rules and regulations to plan the urban development and the land use; incentives and regulatory actions in support of environmental restoration, energy conservation and to fight climate change put in act from the different local authorities).
- 2.b. Review of the main European legislation concerning urban and spatial planning and concerned issues.

• Action 3.2. approaches the problem of UHI vs Climate Change: it aims to study the interaction between UHI and climate change phenomena as well as to understand the influences and correlations between them. Thanks to this Action have been set up indicators establishing relations among urban planning and human activities (main causes of UHI) with climate change trends, estimated on the basis of temperature shifting and other parameters. Regional climate model simulations are able to provide an estimation of the future climate scenarios (temperature, humidity, precipitation, wind speed, cloud cover, etc.) which may serve as outer conditions for the assessment of UHI phenomenon in the CE cities. The simulations can be made with WRF, e.g., for a time slice of 10 years and statistical output on means and standard deviations of the meteorological variables can be supplied. Afterwards a downscaling to regional/town scales is required. Regional climate model uses available boundary conditions provided by existing global climate model. Climate change has the potential to alter shifts in average temperature, cloud cover, wind, speed and precipitation. The strong relationship between wind speed and cloud cover and UHI imply that changes in the magnitude of the UHI effect over the current century will depend at least in part on how cloud cover and wind speed change. Higher temperature might increase evaporation demand and may indirectly create a positive feedback, augmenting the tem-

In summary, tasks of WP3 were to provide the collection of techniques to assess the UHI phenomenon from which a selection have been chosen to be implemented in DSS. Moreover WP3 analyzes the European and local legislation within which decisions can be taken and from which DSS development is dependent.

Another important topic is the connection between UHI and climate change. Climate change will amplify the UHI phenomenon and its consequences for environment and human health have to be carefully taken into account. DSS has to consider then the projections to future scenarios to manage more useful information for administrators and stakeholders.

perature difference between city and rural setting. Though the magnitude of urban–suburban differences may not increase, the population affected by severe pollution episodes may increase as UHI-like conditions become more frequent in outlying suburban locations, presenting additional challenges for policy-makers.

## 5.4.2   WP4

A large work to be implemented in DSS comes also from WP4 that is divided in three main actions.

- Within Action 4.1. a permanent Transnational Network (TN) among experts scientific and institutional involved was set up. TN monitors UHI in CE area; in addition it develops shared and coordinated strategies in urban planning and land using. Task of the Transnational Focus Groups was to manage the knowledge flow between partners and stakeholders and to share competence and knowledge on thematic issues with Local Working Groups.
- Action 4.2. Methodology and areas definition: definition of sensible indicators, sampling procedures and analysis tools are fundamental issues that need to be shared for a common methodology and compare different characteristics of urban areas. A gold standard in assessment of UHIs and in the respective data sampling, accessing and processing has been defined. Cities without a monitoring network suitable for monitoring UHI should take gold standard as a prototype in creating a new monitoring system. Cities with already existing monitoring systems was asked to adapt their systems to this gold standard, to allow a better coverage of phenomenon and to enhance the comparability between different cities.
- Action 4.3. CE UHIs web database and Atlas: the shared web databases implemented through input from existing local partners/institutions allows to monitor the specific situation. Here, the measurements and data obtained and analysed have been designed to describe as well as possible the intensity of phenomenon and its characteristics. Direct surveys conducted by applying both traditional urban biometeorology techniques and remote sensing techniques, allowed to collect many data and information about the micro–macro meteorological conditions. CE Atlas implementation foresees digitalization and geo-referencing of data collected. Creation of a GIS based data processing tool, where all informa-

In summary from WP4 we acquired for DSS the gold standard to define and to monitor UHIs with common and accepted rules. Moreover Act. 4.3 gave rise to the web database from which DSS will acquire information for model running.

tion about detected UHIs of CE area where loaded and put in relation with meteorological and climatic data and trends as well as to spatial planning information.

### 5.4.3  WP5 Mitigation and Adaptation Strategies

Starting from scientific and institutional framework and from assessment tools provided by previous WPs 3, 4, WP5 was focused on approaches to model for long-terms mitigation strategies and short-medium-term adaptation strategies to encounter UHI.

WP5 deals with the following specific questions: Given the results of WPs 3 and 4:

What are the common and differential features of the UHI that effect the regions studied?

What set of mitigation and adaptation measures and options should be considered as potentially effective and subjected to detailed modelling studies?

How could "top-down" (low-resolution) meteorological prediction models and bottom-up (high-resolution) building models be combined to provide an environment modelling for parametric study of the aforementioned mitigation and adaptation measures and strategies?

Having identified a coupled top-down and bottom-up UHI modelling environment, what would be the outcome and implications (recommendations, guidelines) of the parametric modelling studies of alternative mitigation and adaptation measures?

Three Actions are developed in the WP5 framework: Act. 5.1, 5.2, and 5.3.

- Action 5.1. Extent of UHI effects and corresponding potential Mitigation and Adaptation (M&A) measures: Within the framework of this action, the common and differential features of UHI effects in the selected regions have been identified by the corresponding partners. A set of candidate (potentially effective) M&A measures have been collected and reviewed by the interdisciplinary and transnational research team. Thereby, the mitigation strategies provide the definition and application of urban& spatial-planning approaches (e.g. widening of green areas and rows, spread distribution of populated areas preferring short buildings surrounded by gardens, canyon effect) that prevent UHIs emergences. Likewise, relevant construction parameters for buildings (e.g. surfaces characteristics of external building components) have been considered. As to adaptation strategies, the phenomenon of summer bioclimatic discomfort has been addressed by setting up warning and prevention systems. Mitigation strategies, already mentioned in report 3.1.1., can be divided into three main categories, namely buildings, pavements and vegetation. For buildings one can work on directly either on their heat exchange material properties (valid for new and old buildings), or on the building/street geometry in order to modify the intensity of the so

called canyon effect (valid only for new buildings). To the first category belong
the following actions:

– Cool roofs/façades: By applying building materials with high solar reflectance (high albedo) and high infrared emittance (cool materials) one can
  reduce the cooling load of a building, and thus one can limit the heat emitted
  by the use of air conditioning facilities to the building's surroundings. Though
  cool materials can be applied for both roof and façade surfaces, it is usually
  easier and cheaper to apply them to existing roofs, making the method of
  "cool roof" creation more prevalent in daily practice.
– Green roofs: a green roof means a roof covered by a vegetative layer of variable width. Compared to traditional high absorptive roof surfaces, green roofs
  bringing about lower surface temperatures, thus leading to lower cooling
  loads in the building itself. Irrigation of these roofs can reduce the air temperature next to the roof, as an outcome of the evapotranspiration effect of the
  vegetative layer. When applied to a large group of buildings, green roof are
  believed to have a significant effect of urban air temperature.
– Green façades: vegetation can also be used as an overlaying layer for facades,
  either by enabling the growth of climber plants directly on the façade surface
  (thus shading parts of it) or by adding a more substantial bedding "wall"
  which connects to the façade and functions also as an insulating and evapotranspiration layer similar to that of a green roof.
– Street geometry, which is defined by the shape of the buildings and their orientation, might also have an impact on UHI intensity levels by increasing air
  flow rate through the street, thus replacing the warmer air "trapped" between
  buildings. Although changing the geometry of existing streets is limited, large
  urban projects regularly introduce the opportunity to increase the effects of
  advection by a careful design of their geometry.
– Pavements play also an important role in the formation of the UHI phenomenon, since conventional paving materials (mainly concrete and asphalt) tend
  to absorb large amounts of solar radiation during daytime and to release it to
  the cooler surrounding air (cool pavements). Another property of these paving
  materials is their limited permeability to water, which prevents the absorption
  of water in the ground and thus reduces the evaporation potential of the ground
  surface which may help in reducing air temperatures (pervious pavements).

  • Cool pavements are built of materials which have higher albedo (higher
    solar reflectance) values than conventional paving materials, usually
    because of their lighter color.
  • Pervious pavements: pavements of pervious materials which enable the
    draining of water through the porous pavement surface. The water is thus
    absorbed in the subsoil and evaporates when the paving material is heated,
    resulting in lower surface temperatures of the pavement. Pervious pave-
    ments are also helpful for the storm water management, and then for the
    prevention from floods.

- – Vegetation: trees and vegetation reduces ambient air temperature by evapo-transpiration and shading and is therefore expected to help in mitigating UHI intensity levels (Kurn et al. 1994). The common practices within this scope are the planting of trees and vegetation in existing urban fabric (mainly city streets), or the creation or preservation of wider green areas (parks, groves) within the urban fabric.

- • Action 5.2. Establishment of an effective UHI modelling environment. The purpose of this action is to establish a coupled "top-down" (meteorological) and bottom-up (built environment) computational modelling environment. Thereby, low-resolution (large-grid) meteorological models provide data on large-scale UHI effects. This data are subsequently used as boundary conditions for medium-small scale thermal modelling tools of the built environment. For this scope, the potential of transfer functions has been explored. These functions derive from weather station data, high-resolution micro-climatic conditions at immediate proximity of built structures. The available simulation tools, used by the project partners are listed. These models and their coupling are the core of the DSS elaboration.

- • *Action 5.3.* Definition of mitigation and adaptation strategies. Given the above coupled modelling environment, the relative performance (predicted degree of success) for various alternative M&A strategies and measures could be examined and numerically described. A set of strategies are formulated to be applied at national and transnational scales to address the UHI phenomena. Such M&A measures portfolio includes specific urban and spatial planning guidelines as well as risk management recommendations.

## 5.5  Main Acronyms

CE:        Central Europe
DSS:       Decision Support System
LP:        Lead Partner
WP:        Work Package
AF:        Application Form
UHI:       Urban Heat Island
ACT:       Action
DBMS:      Database Management Software
MBMS:      Model Base Management Software
DGMS:      Dialogue Generation Management Software
SMS:       Short Message Service
TN:        Transnational Network
M&A:       Mitigation and Adaptation

# Reference

Kurn, D., Bretz, S., Huang. B., & Akbari, H. (1994). The potential for reducing urban air temperatures and energy consumption through vegetative cooling. In *ACEEE summer study on energy efficiency in buildings*. Pacific Grove: American Council for an Energy Efficient Economy.

# Part II
# Pilot Actions in European Cities

# Chapter 6
# UHI in the Metropolitan Cluster of Bologna-Modena: Mitigation and Adaptation Strategies

**Stefano Zauli Sajani, Stefano Marchesi, Paolo Lauriola, Rodica Tomozeiu, Lucio Botarelli, Giovanni Bonafè, Graziella Guaragno, Federica Fiumi, Michele Zanelli, Lodovico Gherardi, Marcello Capucci, Catia Rizzo, and Filippo Bonazzi**

**Abstract** The pilot action took place in a district of Modena, the Villaggio Artigiano, characterized by the presence of disused small industrial buildings, which is part of a wider redevelopment context and regeneration process.

The innovative mixture of instruments proposed by the Municipality to better re-use the territory and to estimate the environmental restoration achieved with the urban interventions, is a starting point to give the planner flexible and easy to use instruments.

**Keywords** Cluster Bologna-Modena • Villaggio Artigiano • Urban redevelopment • Urban planning • Urban indexes

S. Zauli Sajani (✉) • S. Marchesi • P. Lauriola • R. Tomozeiu • L. Botarelli • G. Bonafè
ARPA Emilia-Romagna, Bologna, Italy
e-mail: szauli@arpa.emr.it

G. Guaragno • F. Fiumi
Territorial Planning and Mountain Development Service, Emilia Romagna Region, Italy

M. Zanelli
Urban Quality and Residencial Policy Department, Emilia Romagna Region, Italy

L. Gherardi
Territorial Cooperation Contact Point, Emilia Romagna Region, Italy

M. Capucci
Urban Planning Department, Modena, Italy

C. Rizzo
Urban Design and Transformation office, Modena, Italy

F. Bonazzi
Territorial Planning and Private Building office, Modena, Italy

© The Author(s) 2016                                                                                       131
F. Musco (ed.), *Counteracting Urban Heat Island Effects in a Global Climate Change Scenario*, DOI 10.1007/978-3-319-10425-6_6

## 6.1 Implementing Solutions for Climate Change in Urban Context

The effect of climate change on urban scale is often seen as a simple projection of a global risk. In fact, the urban environment is certainly characterized by particularly critical in relation to the effects of extreme events alluvial and heat waves, which is closely connected to the "climate change". In reality, however, the local and the global influence each other mutually shaping opportunities and constraints. In this sense, an integrated approach to mitigation and adaptation is the only way to reduce the impact of climate change and to turn a threat into an opportunity for sustainable territorial development, economic and social too.

Measures to reduce heat island effect in this regard are a prime example of an action that fits both the perspective of adaptation to climate change and their mitigation. Reduce hardship bioclimatic in urban involves not only the ability to innovate in terms of use of materials and construction techniques but also to change economic and social structures, ensuring over time the quality of life and the environment.

In this chapter we want to draw a picture of the relevance of the heat island phenomenon in relation to climate change and illustrate the potential of the interventions of urban planning. The urban heat island is certainly one of the best known effects of urbanization on local climate.

## 6.2 The Metropolitan Cluster Of Bologna-Modena

### 6.2.1 Urban and Environmental Framework

Emilia-Romagna Region's territory, in the Padan area, includes the metropolitan area of Bologna and other main conurbations located in the Emilian area and in the coastal area. The first main urban conurbation develops from Bologna along via Emilia (Emilia Street), including the cities of Modena, Reggio Emilia up to Parma, and is characterized by high density settlements with high-intensity exchanges; the second one, distributed along the coastline, concerns the intensely built-up touristic area from Rimini to Cervia.

Emilia-Romagna's location makes it part of two National corridors which respectively connects the Apennine Mountains to the Adriatic Sea and North and South Italy, including: A1 and A14 highways, Piacenza-Rimini railway and a stretch of the high speed railway Milan-Bologna-Rome. Consequently, with a road network of 10.792 Km (which consists of 643 Km of highways and connection roads, 2907 Km of state roads and 7242 Km of provincial roads), the region has a key role for transport integration within National and European contexts (Fig. 6.1).

In the past forty years, a high intensity construction activity has affected the region and this has led to the spreading of settlements and of production and service sector activities. The main cities have lost inhabitants to the hinterland and consequently a sort of "city-area" characterized by a high and widespread urbanization index (153 inhabitant/kmq) has grown along via Emilia.

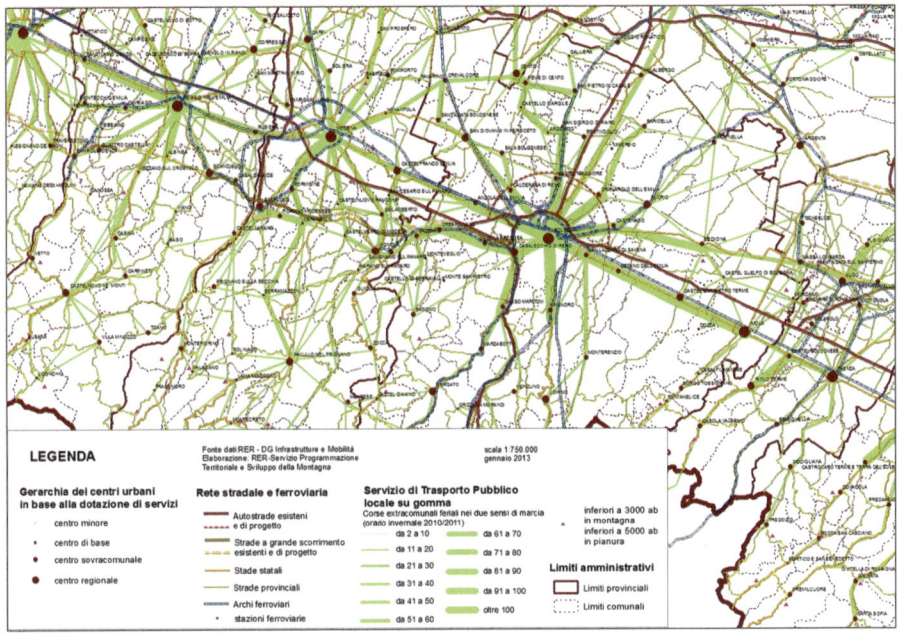

**Fig. 6.1** Hierarchy of urban centres – Road and rail network

In addition to that, during the past twenty years Bologna's metropolitan area has doubled, and the conurbations of central Emilia and along the coastline (where the 50 % of the regional population lives). have been affected by an increase in urbanization and land consumption up to 8–13 % of the total surface (Fig. 6.2).

The pilot area is located in the core of this regional system described above. It is among the more developed areas at the regional and European level as far as its socio-economic development is concerned. The three provinces of Bologna, Modena and Reggio Emilia host 56 % of the regional industries. The local economy is based on the manufacturing sector, especially ceramic, which contributes to make Modena's province a key productive centre within Europe.

The presence of firms, together with service providers and houses, has a negative impact on road conditions and on the quality of live in the cities and it is the cause of polluting emissions reaching emergency levels for increasingly longer periods. The use of renewable energies in this area is also still low.

In the metropolitan cluster of Bologna-Modena, despite the presence of some of the bigger rivers in the region, the high density settlements (which characterize the area) have reduced the space for natural environment. Furthermore, the development of man-made infrastructures has a negative impact on landscape and creates more obstacles to policies aimed at (1) integrating the metropolitan areas of Bologna and Modena; (2) decongesting the central areas (along via Emilia) and (3) mitigating the environmental alterations linked to critical traffic conditions (air and noise pollution), which have negative consequences, especially on the health of children and the elderly; (4) improving the connection infrastructures and public transport.

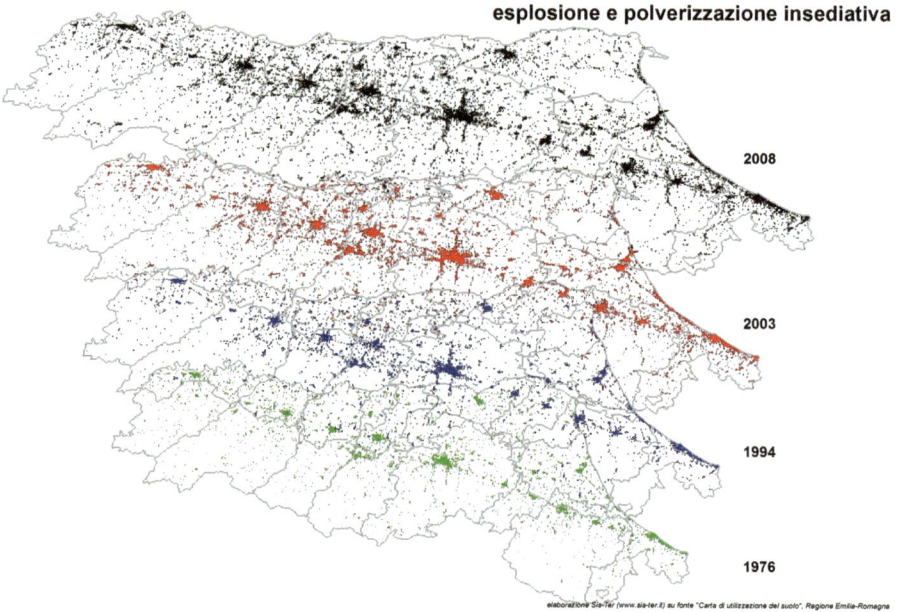

**Fig. 6.2** Urbanization 1976-2008

The regional climate is sub-continental with strong difference between summer and winter: summers are hot and muggy and winters are cold, rainy, foggy, moderately snowy and long. On average, the temperature reaches minimum values at dawn and maximum during mid-afternoon. Since 2001 an absolute maximum of 38,7°C and a minimum of −10°C has been detected.

The urban area of Modena is usually hotter and drier than the rural one. The biggest temperature differences are at night, and go from 2 °to 8°C, especially in the summer.

Furthermore, while variations in humidity tend to be wider at night, they can also be quite considerable in the daytime, during the winter months.

During the last 20 years the climate has undergone a strong change, if compared to the period 1961–1990: the average temperature has increased (+1,1 °C) and so has the maximum temperature (especially in summer, + 2 °C) and changes have been registered with regards to seasonal cycles and the intensity of rainfall.

Issues such as the quality of the air, energy and water cycles, renewable energy and land consumption reduction have been set as priorities by the regional governments.

Appendix A contains rules and regulation set up by Emilia-Romagna local governments and Appendix B contains a summary of incentives, financing and regulatory actions implemented at local levels to facilitate a sustainable land use and to support environmental restoration, energy conservation and reduction of phenomena related to climate change.

**Fig. 6.3** POC

## 6.2.2   Pilot Area Identification Methodology and Description

Emilia-Romagna Region has decided to select the pilot area of the "Villaggio Artigiano" (Craftsman Village) in Modena as the Administration was preparing the Municipal Operational Plan (POC prescribed by the regional planning law) called "Urban Redevelopment of the West Face of Modena", and expressed an interest in experimenting and integrating practical solutions aimed at containing the UHI phenomenon into urban planning (Fig. 6.3).

In this framework, the Municipality of Modena has concluded the approval of a specific Plan of Urban Redevelopment of the Villaggio Artigiano (http://www.comune.modena.it/laboratoriocitta/laboratoriocitta/i-progetti-del-laboratorio/poc-mow).

Given the new environmental context in which the Village is included, the plan entails interventions aimed at fostering high-performance at the overall urban as well as environmental systems.

The pilot action lies in the urban area of Modena, in the western sector of the city; this is an area which might have been considered "almost suburban" until recently, but is now a rather central location of the city. Today, the Villaggio

**Fig. 6.4**  Villaggio Artigiano – Planimetry

Artigiano is an area that is immediately identifiable by its unique triangular shape, framed by two streets coupling to the surrounding urban fabric and by the historical Bologna-Milano railway line (Fig. 6.4).

The idea of the Village was conceived in 1954. Modena's municipal authorities decided to create a "craft area", after the economic crisis of the postwar period, with the aim of boosting the economic recovery.

The Council allocated 15 ha of land to a "village for craftsmen" on the far western outskirts of the city, in the district Madonnina.

Within six years, 74 new companies, their owners, new entrepreneurs (especially workers who had been laid off by large companies, people with specific expertise) found a place and started their trade in the wasteland between the railway and the Via Emilia.

Participation in the project, however, greatly exceeded the initial expectations of the Administration: the two triangular areas divided into 60 lots initially planned were immediately granted, the Village was then extended to the current size (almost 500,000 square meters), with about 200 businesses set up.

The "Villaggio Artigiano" presents a building structure that is rather recognizable even nowadays: perpendicular roads constitute a mesh with traditional orientation, deriving from Roman centuriation and subpartition of rural areas: 4 long streets are oriented from north-east to south-west and other shorter roads, orthogonal to the former, delimit all built lots.

The lots are all rectangular in shape. In the north, where the oldest portion of the Village is located, lot sizes are smaller, whereas those in the more recent south portion are larger.

The elements that make the Villaggio Artigiano a privileged project area today derive mainly from two sets of issues: one of an urban, economic and social nature, related to the ongoing problems of the area, and the other linked to the context of the Villaggio Artigiano and its strong, untapped potential.

In a nutshell, the themes considered in the redevelopment plan are:

- **Identity value** of the Villaggio Artigiano: it is a "piece" of the city history and an example of that "model Modena" that mainly contributed to the economic and social development of the city. For this reason, it is important, from the perspective of identity and business, to promote the renewal of the Village without compromising on its productive nature, and to boost the vitality of the area that seems to derive from the building typology and flexibility, which combines "home & shop" with very particular architectural languages.
- **Economic** and entrepreneurial **value**: seen in retrospect, and with a modern perspective, Villaggio Artigiano was a major help in what today is called the start-up of new businesses. To this one should add, the opportunity for artisans to have their houses built in the vicinity of their workshop, thus reducing significantly their personal and family costs of residence and transport. This particularly applies to the early settlements
- **Urban planning value**: with regards to urban design and urban planning, the Village, located next to the Old Town, but also well connected to the extra-urban transport axes, is a center of gravity as far as the redevelopment of the whole western sector of the city is concerned. This feature will be greatly enhanced by new road connections and new forms of public transport. In fact, the Village is highly affected by the planned diversion of the railway line bounding the area, which will leave a large "diagonal" line of interconnection to the city center for new public spaces and transport services.

These features make the Village an ideal testing ground from the point of view of urban sustainability through the recovery of the existing fabric, ground-saving oriented, and the increase in the functional mix.

One of the main goals of this redevelopment plan is to renew the buildings in the Village, by means of a deep restructuring of the existing edifices, respecting the size and volume relations among them and producing a new estate body, which carries on and updates the typical evolution process of the Villaggio Artigiano (Fig. 6.5).

Therefore, the proposed regulatory actions are aimed at promoting the transformation of the Village, increasing the functional mix among production, which remains prevalent, services and residence, the latter to be rethought in new and experimental ways (home studio, new types of home-workshop, residential complexes with shared facilities etc.). In addition to these possible changes, on the public side, the main object of the redevelopment Plan is to redesign public spaces for meeting and socializing: rethinking and reorganizing the street mobility, creating parking areas and green spaces (using, for example, the large diagonal line on which the Village is grafted), signposting the presence of trade and services as an opportunity to generate significant spaces for urban quality of the neighborhood.

**Fig. 6.5** Villaggio Artigiano – Urban morphology

To date, the administration has launched a series of initiatives aimed at urban regeneration (buildings and public areas), as well as at economic and social improvement, summarized as follows:

– new urban-building rules;
– coordinated project for public space: the Village has its only public spaces in the streets, which have a very small section, are anonymous and not suitable for a non-automotive mobility. Through simulations and sectorial studies various options were examined for the transformation of the road network aimed to interconnect pedestrian, bicycle and automobile paths and to facilitate public accessibility and therefore the settlement of business and services.
– to exploit the dis-used railroad area, to be reinvented as "gateways" to the Village, as a large walk urban connecting two parts of the city historically divided; the Village has indeed a well-defined urban morphology, which makes it easy to identify, but it's also "closed" to the rest of the city (Fig. 6.6).

Taking the modified environmental conditions of the area into consideration, the redevelopment plan aims to promote measures envisaging high-level performances, in order to ensure the environmental, as well as urban, upgrade of the area.

To counteract the alarming impact of UHI phenomenon the municipality of Modena began to consider the main environmental problems, to identify effective methods of construction.

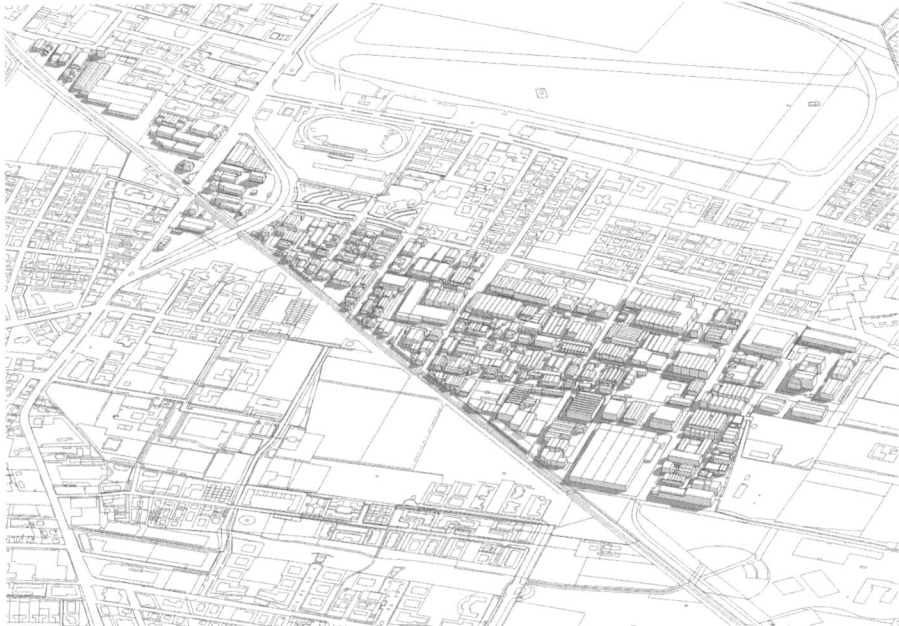

**Fig. 6.6** Villaggio Artigiano – Planimetric view

The Plan aims to identify a synthesis of the main environmental issues related to the area of intervention, in order to derive a calculation method that can show the environmental performance achieved in the redevelopment of the individual lots.

The main environmental issues focalized in the Village are:

– reduction of the Urban Heat Island
– reduction of energy consumption
– reduction of the hydraulic risk

The "Artisan Village Guidelines" summarize the main features of the environmental method prepared by the administration. It is worth stressing that there are various analyses still under way for the final validation of various technical and procedural issues; as a consequence, although the structural characteristics have already been defined, the approach implemented may still be modified or emended.

In any case, the following items were thoroughly considered:

– Appropriate Building Massing:
– Energy Efficiency
– Passive strategies including: highly insulated; massing well arranged for summer radiation and also winter – optimized utilization of daylight,

Also using:

- Buffer zones (such as winter gardens) to harvest passive solar energy and allow natural ventilation under cold/windy conditions
- External solar shading

### 6.2.3 UHI Phenomenon in the Urban Area of Modena and Application of Models to Simulate Mitigation Measures

Analyses on the UHI phenomenon in the urban area of Modena have been performed with a focus on the summer season. We compared data from stations located within the urban area and from stations located in the rural area. The findings showed that minimum temperatures in the urban area were higher than in the rural one. The differences between urban and rural minimum temperatures were generally larger during spring and summer, when they reached values up to 6 °C. The highest intensities of urban heat island effect were found around midnight. On the other hand, the correlation between maximum values of temperature was the opposite: rural temperatures tend to be about 1 °C higher than the urban ones. Relevant positive trends were present in the 30-year time-series of temperature. Long term trend of Heating Degree Days (HDD) and Cooling Degree Days (CDD) were also analysed (Figs. 6.7 and 6.8). These parameters show to what extent the temporal

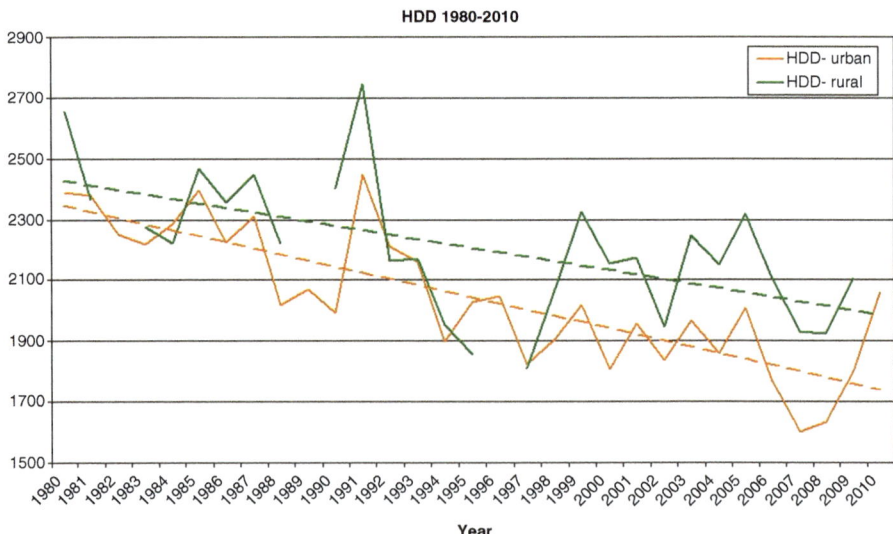

**Fig. 6.7** Long term trend of Heating Degree Days (HDD) for the urban and the rural area

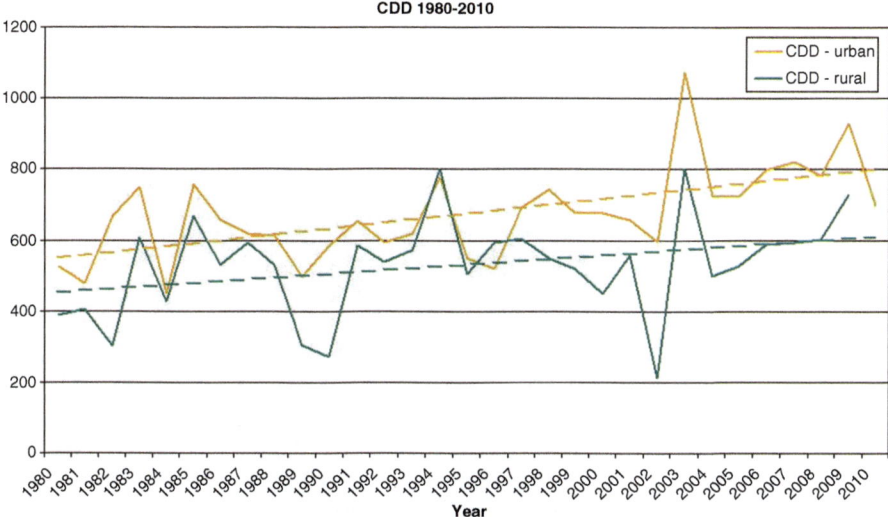

**Fig. 6.8** Long term trend of Cooling Degree Days (CDD) for the urban and the rural area

trend of temperatures is either below (HDD) a predefined bioclimatic thresholds or above (CDD). A markedly decreasing trend was present in HDD time-series, while an opposite, less marked trend, appears for CDD.

Two main simulation tools have been applied to the pilot area in order to estimate the effect of some mitigation measures from a quantitative point of view. These simulations were aimed at assessing the effects of types of mitigation actions, and not the effects of specific interventions.

The first model applied in the pilot area was RayMan, a simulation model able to calculate short- and long-wave radiation fluxes inside a complex urban environment. Output from RayMan model consists in the values of several thermal indices derived from human heat-balance model. RayMan calculates the mean radiant temperature using a simplified radiation balance applied to a person which is exposed to:

- direct solar radiation;
- long wave radiation from ground, building walls and vegetation
- reflected radiation from the same surfaces.

In the present study, RayMan model was applied on a car parking area inside the Villaggio Artigiano. Firstly, the model was run for the actual situation in the domain (reference run) in a typical summer day in August. Then, some changes were introduced in the model domain (scenario runs) and the net effects of the mitigation measures on the thermal field and on the bio-meteorological conditions were estimated. A number of tests were carried out considering various combination of vegetation, type of materials for pavements and facades, height of buildings. From

**Fig. 6.9** Example of 2-m
height field of temperature
as simulated by ENVI-met
model

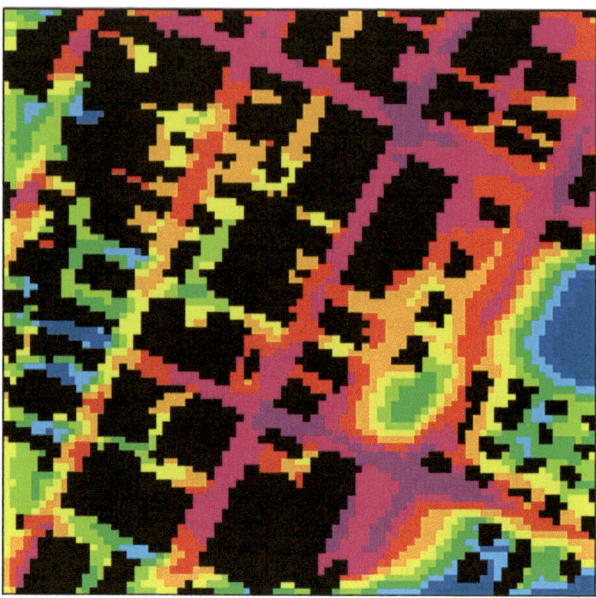

the point of view of thermal comfort, it was quite clear that most effective discomfort reductions were obtained introducing trees in the domain.

The shade from trees produced the largest impact and the mitigation effect was estimated in around 2 °C in the peak hours. A further point worth of notice is that, the pervious surface obtained by replacing the asphalt and/or concrete pavement with grass had a positive impact on the thermal comfort in its turn; however, the absolute value of this effect was much lower than in the scenario where trees were introduced, and the temperature reductions were below 0.5°. Modification of building heights showed rather small differences in the values of bio-climatic indices.

The second model used to simulate the impact of different scenarios was ENVI-met, a three dimensional, non-hydrostatic model of the atmosphere, based on the fundamental laws of fluid-dynamics and thermodynamics. ENVI-met is a much more complex model than Rayman and is able to simulate the tri-dimensional field of the usual meteorological variables taking into account the interaction between atmosphere, urban surfaces and vegetation characterizing the complex urban fabric.

The model domain was a square of 400 m × 400 m, about a half of the whole Villaggio Artigiano (Fig. 6.9). The horizontal resolution was 5 × 5 m (81 × 81 grid points). Vertical resolution was set to 3 m, with the exception of the first model layers, which were split into 4 additional layers with the aim of showing a better representation of the interaction between the atmosphere and the surface elements. The simulation were run for the typical summer conditions for the city of Modena. Various mitigation measures were considered: insertion of green elements (grass and trees), change of the albedo of walls, roofs and pavements, insertion of pervious

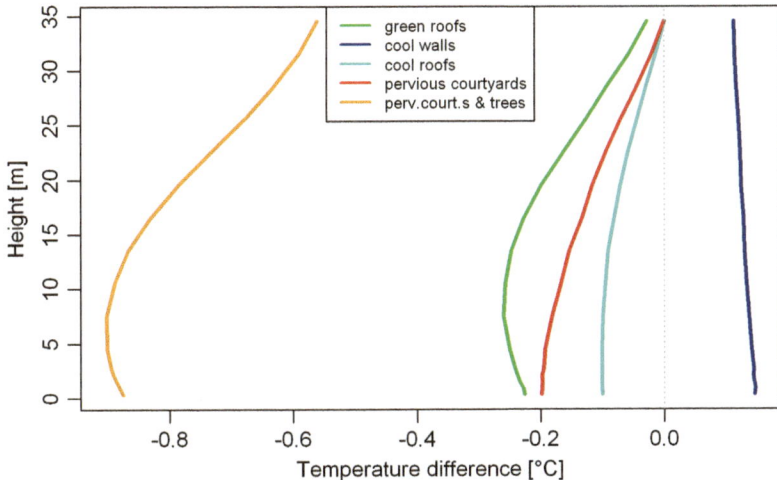

**Fig. 6.10** Vertical profiles of "all day" mean differences between each scenarios, labeled as indicated in the legend, and the control run

surfaces in substitution of asphalt and pavements. The findings showed a well-defined ranking in the impact of mitigation measures (Fig. 6.10).

"Green courtyard with trees" was the most effective mitigation measure. A reduction of temperatures was evident for the entire course of the day. The "green roofs" scenario showed a peak of the cooling effect with respect to the control run at a height of around 10 m, with a slightly smaller impact at ground level. On the contrary, "cool roofs" and "pervious courtyard" scenarios showed a peak of cooling at ground level with a relevant decrease of the impact going upwards. On the other hand, scenario with "cool walls" showed a negative effect in terms of mitigation of urban heat island. The cooling effect induced in this scenario was possibly explained by the unsatisfactory consideration of heat balance in model equation for buildings.

### 6.2.4   *Experimental Environmental Quality Index to Assess the UHI's Mitigation Actions in a Building Lot*

As described in the previews paragraphs, in the Villaggio Artigiano pilot area the following main environmental issues are taken into consideration:

– reduction of the Urban Heat Island
– reduction of energy consumption
– reduction of the hydraulic risk (not strictly connected to UHI phenomenon)

Starting from the experiences carried out by other cities, the Municipality of Modena decided to create a new calculation methodology, to be tested in the redevelopment of the Village.

Several municipalities are already equipped with calculation methodologies, in order to highlight the environmental performance achieved by the redevelopment of urban lots.

These calculation methodologies, also known as "urban indexes", are mainly focused on the urban impact created by the analysed phenomena, and are usually characterized by a simple algorithm.

The main existing urban indexes are the BAF, "Biotope Area Factor", used in Berlin, Malmo and Seattle, and the RIE, "Reduction of the Building's Impact", used in Bolzano and Bologna.

Both indexes calculate the ecological value of an urban lot.

Unfortunately, these indexes have limitations which make their application to "Villaggio Artigiano" impossible.

The Biotope Area Factor considers only the "green" surfaces and is therefore not suitable to assess the positive effect of a "cold roof".

The index of Reduction of the Building's Impact considers only the horizontal surfaces and is therefore not suitable to assess the positive effect of a "green" wall.

The main goal of Modena Municipality was to create a calculation methodology capable to:

- analyse all the surfaces that make up the urban lot, like the courtyard areas, building walls and roof,
- analyse different types of surfaces, like green or cool surfaces,
- stand up to existing indexes in terms of simplicity of data input,
- make appropriate approximations, like urban indexes typically do,
- highlight the environmental performance on the basis of indicators that predict physical tangible phenomena,
- implement procedures and values already defined by laws or municipal regulations.

The three main environmental issues (reduction of the Urban Heat Island, reduction of energy consumption, reduction of the hydraulic risk) have been linked to two indicators that have been used to derive two indices, the index "RATE" of "reduction in the absorption of thermal energy" and the index "HYPER" of "reduction of hydraulic risk".

The creation of the index "RATE" followed four stages:

1. a list of types of surfaces that include the typical materials of the existing context and the materials used today in new projects.

   Altogether 20 types of surfaces has been identified: 7 for the courtyard areas, 6 for the walls and 7 for the roofs. Some of these describe existing interventions, typical of the existing context, while others describe innovative interventions of recent use. The interventions that can be implemented in the courtyard areas are: garden or flower bed, tree or shrub, self-locking pavement, lawn driveway, "cold" asphalt, normal asphalt, gravel pavement.

The interventions that can be implemented in the walls of the building are: green wall with frame on wall, green wall integrated into the wall, ventilated wall with frame on wall, light plaster, dark plaster, wall with exposed brick.

The interventions that can be implemented on the roof of the building are: green roof, "cold" roof, tiled "cold" roof, photovoltaic roof, tiled roof, light flat roof, dark flat roof.

2. identifying a physical quantity, as an indicator, and a calculation methodology.

The indicator chosen to estimate the index "rate" is the thermal energy absorption of each surface of the lot, due to the incident solar radiation.

The absorption of thermal energy of a surface represents the amount of energy that that area is not able to reflect and disperse instantaneously, and therefore represents the amount of thermal energy that will be released for a certain period of time. The absorption of thermal energy is therefore able to highlight the capacity of a material to affect the Urban Heat Island.

In order to calculate the absorption of thermal energy a formula is used, which correlates the technical and physical characteristics of a surface with the incident solar radiation.

The technical and physical features used for the calculation of a generic surface, are: solar reflectance, emissivity and thermal resistance.

The index "RATE" calculates the sum of the absorption of thermal energy of all surfaces that constitute the urban lot.

3. analysing the technical and physical features of every surface in the list and of the existing context, to estimate the indicator. For each of the 20 types of interventions, the respective values of solar reflectance, emissivity and thermal resistance have been estimated. As for walls and roof, different values of thermal resistance have been calculated, depending on the year of construction.

To quantify the incident solar radiation, the values defined by the norm UNI 10349 are used. Municipality of Modena has obtained a value to be applied on the roof and different values to be applied to the walls, depending on their orientation. Regarding courtyard areas, an average value was estimated, taking into account the loss of energy caused by the shading by building volume.

4. adopting the typical/needed approximations of urban indexes.

The Artisan Village Guidelines "Feasibility study of environmental quality indexes to be applied to building lots", made by Municipality of Modena, are attached in Appendix C and the link to the calculation software can be found on the website of the Municipality of Modena:

http://www.comune.modena.it/laboratoriocitta/laboratoriocitta/pubblicazioni-eventi/villaggio-artigiano-di-modena-ovest/esiti-del-progetto

and on the website of the Emilia-Romagna Region:

http://territorio.regione.emilia-romagna.it/programmazione-territoriale

## 6.2.5   Adaptation Strategy to Heat Risk: Assessment
of a Possible Development of the Heat Risk Alert System
Based on the Use of Emergency Ambulance Data

Increased temperatures and extreme heat can lead to a rise in mortality. In EU countries, mortality is estimated to increase by 1–4 % for each one-degree rise in temperature, meaning that heat related mortality could rise by 30 000 deaths per year by the 2030s and 50 000–110 000 deaths per year by the 2080s.

As regard to the National level the Ministry of Health, in cooperation with the Ministry of Civil Protection, an "Early warning national system to prevent heat waves" has been operating since 2004, after the terrible summer of 2003. Furthermore, in 2005 a "National Operational Plan to prevent effects on human health from heat waves" was issued, and in 2006 "Guidelines to prepare monitoring plans in order to respond to heat waves" were provided to assist local authorities:

http://www.salute.gov.it/emergenzaCaldo/paginaInternaMenuEmergenzaCaldo.jsp
   ?id=413&menu=strumentieservizi

In coordination with national plan in Emilia-Romagna, a risk prevention local plan was designed and implemented to reduce the risk of damage and casualties due to summer heat waves.

Every year the plan guidelines are updated by the Emilia-Romagna Regional Government while ARPA (Regional Agency for Environmental Protection) issues forecasts throughout the summer.

The adaptation action then consists of:

–  **Alert system** managed by ARPA, alerting when temperature and humidity level raise above a risky threshold
–  **Emilia-Romagna Regional Government coordination actions** to assist most exposed people groups.

### 6.2.5.1   Alert System

ARPA Emilia-Romagna has been endeavouring to provide forecasting systems of some environmental risk factors to local authorities for several years now. Among such factors, the prediction of heat waves has gained relevance, particularly in relation to climate change projections for the coming decades.

The heat waves forecast service has been operating since 2004. ARPA Emilia-Romagna manages a specific website platform: http://www.arpa.emr.it/index.asp?idlivello=97

The forecast service is active between 15 May and 15 September. The forecast is done with 72 h forewarning.

| | 25% | 30% | 35% | 40% | 45% | 50% | 55% | 60% | 65% | 70% | 75% | 80% | 85% | 90% | 95% | 100% | THOM'S DISCOMFORT INDEX | |
|---|---|---|---|---|---|---|---|---|---|---|---|---|---|---|---|---|---|---|
| 42° | 32 | 32 | 33 | 33 | 34 | 34 | 35 | 35 | 36 | 36 | 37 | 37 | 37 | 38 | 38 | 38 | Up to 21 | No discomfort |
| 41° | 31 | 32 | 32 | 33 | 33 | 34 | 34 | 35 | 35 | 35 | 36 | 36 | 37 | 37 | 37 | 37 | | |
| 40° | 30 | 31 | 31 | 32 | 32 | 33 | 33 | 34 | 34 | 35 | 35 | 35 | 36 | 36 | 36 | 37 | From 21 to 24 | Less than half population feels discomfort |
| 39° | 30 | 30 | 31 | 31 | 32 | 32 | 33 | 33 | 34 | 34 | 34 | 35 | 35 | 35 | 36 | 36 | | |
| 38° | 29 | 30 | 30 | 31 | 31 | 31 | 32 | 32 | 33 | 33 | 34 | 34 | 34 | 35 | 35 | 35 | | |
| 37° | 28 | 29 | 29 | 30 | 30 | 31 | 31 | 32 | 32 | 32 | 33 | 33 | 33 | 34 | 34 | 34 | From 25 to 27 | More than half population feels discomfort |
| 36° | 28 | 28 | 29 | 29 | 30 | 30 | 30 | 31 | 31 | 32 | 32 | 32 | 33 | 33 | 33 | 34 | | |
| 35° | 27 | 27 | 28 | 28 | 29 | 29 | 30 | 30 | 30 | 31 | 31 | 32 | 32 | 32 | 33 | 33 | | |
| 34° | 26 | 27 | 27 | 28 | 28 | 29 | 29 | 29 | 30 | 30 | 30 | 31 | 31 | 31 | 32 | 32 | From 28 to 29 | Most population feels discomfort and deterioration of psychophysical conditions |
| 33° | 26 | 26 | 27 | 27 | 27 | 28 | 28 | 29 | 29 | 29 | 30 | 30 | 30 | 31 | 31 | 31 | | |
| 32° | 25 | 25 | 26 | 26 | 27 | 27 | 27 | 28 | 28 | 29 | 29 | 29 | 30 | 30 | 30 | 30 | | |
| 31° | 24 | 25 | 25 | 26 | 26 | 26 | 27 | 27 | 27 | 28 | 28 | 28 | 29 | 29 | 29 | 30 | | |
| 30° | 24 | 24 | 24 | 25 | 25 | 25 | 26 | 26 | 27 | 27 | 27 | 28 | 28 | 28 | 29 | 29 | | |
| 29° | 23 | 23 | 24 | 24 | 25 | 25 | 25 | 26 | 26 | 26 | 27 | 27 | 27 | 27 | 28 | 28 | From 30 to 32 | The whole population feels an heavy discomfort |
| 28° | 22 | 23 | 23 | 23 | 24 | 24 | 25 | 25 | 25 | 25 | 26 | 26 | 26 | 27 | 27 | 27 | | |
| 27° | 22 | 22 | 22 | 23 | 23 | 23 | 24 | 24 | 24 | 25 | 25 | 25 | 26 | 26 | 26 | 26 | | |
| 26° | 21 | 21 | 22 | 22 | 22 | 23 | 23 | 23 | 24 | 24 | 24 | 25 | 25 | 25 | 25 | 26 | Over 32 | Sanitary emergency due to the the very strong discomfort which may cause heatstrokes |
| 25° | 20 | 21 | 21 | 21 | 22 | 22 | 22 | 23 | 23 | 23 | 23 | 24 | 24 | 24 | 25 | 25 | | |
| 24° | 20 | 20 | 20 | 21 | 21 | 21 | 22 | 22 | 22 | 22 | 23 | 23 | 23 | 24 | 24 | 24 | | |
| 23° | 19 | 19 | 20 | 20 | 20 | 21 | 21 | 21 | 21 | 22 | 22 | 22 | 22 | 23 | 23 | 23 | | |
| 22° | 18 | 19 | 19 | 19 | 19 | 20 | 20 | 20 | 21 | 21 | 21 | 21 | 22 | 22 | 22 | 22 | | |

**Fig. 6.11** Thom's discomfort index table

The Risk Alert alarm is based on the Thom's Discomfort Index (DI). DI is a measure of the reaction of the human body to a combination of heat and humidity (Fig. 6.11).

This index combines the values of humidity and temperature parameters to describe the conditions of physiological discomfort due to heat and humidity. The threshold of bioclimatic discomfort used for the Alert system were identified through a study of mortality conducted in the urban area of Bologna for the years 1989-2003.

– Weak discomfort

Weak discomfort conditions are defined when DI average daily value is 24. Under such conditions, the population feels discomfort but there are no increases in mortality.
– Discomfort

Discomfort conditions are defined when DI average daily value is 25. Under these conditions the weaker sections of the population, and especially the elderly, may experience health effects of various kinds, including headaches, dehydration. Such symptoms may cause fatalities in some extreme cases. The total mortality, natural causes and cardiovascular diseases increase on average by about 15 %, and mortality from respiratory causes up to 50 %.
– Strong discomfort

Strong discomfort conditions are defined when DI daily average value is 26 (the average daily index values never surpass this mark) or when an index level more or equal to 25 persists for 3 or more days. Under these conditions the categories of persons suffering from heat-related illnesses increase. The total mortality, natural causes and cardiovascular ailments, rise by an average of about 30 %. The mortality from respiratory causes raises of about 80 %.

Every day the system automatically alerts all concerned institutions (Healthcare District Services, Civil Protection…) via an email. The email states ALERT or NO

ALERT in the object field, depending on if the DI is higher or equal to 24, or if it is lower than 24. Then the email itself specifies the Discomfort Level forecasted.

The following graphics show as an example the trend registered in 2012 and the one registered in the very hot summer in 2003 (Figs. 6.12 and 6.13).

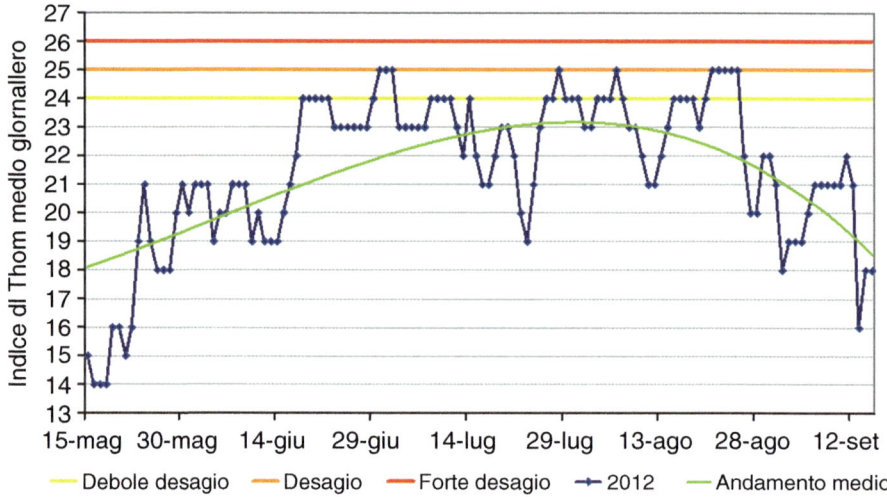

**Fig. 6.12** DI trend registered in 2012

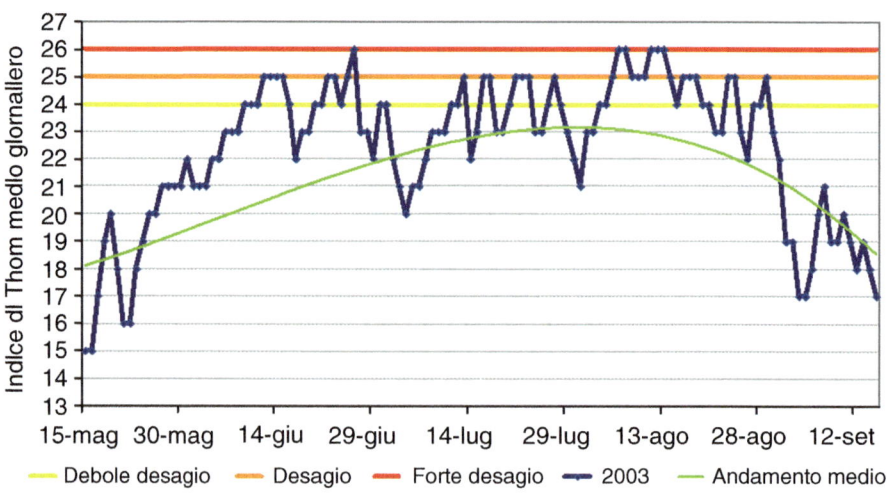

**Fig. 6.13** DI trend registered in 2003

### 6.2.5.2   Emilia-Romagna Regional Government Coordination Actions

Emilia-Romagna Regional Government coordinates actions to assist people groups who are most exposed to heat waves, through Health Care regional system, Civil Protection, non-profit sector. These actions are activated by the ARPA alert system and they include two main activities:

1. Home care assistance:

   – Set up and coordination of Local Networks of all the actors involved in home care assistance, i.e.: Local Health Authorities (AUSL), City Councils, Voluntary associations;
   – Map High Risk Groups to be monitored during Heat Waves Alert, i.e.:elder people [>75 Y/old] living alone at home (in particular Local Health Authorities are obliged to notify all cases of elder people discharged from hospital in summer period and living alone), disabled people;
   – To strengthen home care assistance services, using all possible collaborations with Civil Protection volunteers and non-profit associations active in welfare.

2. Information actions

   It has been shown that information on potential threats can be extremely useful in preparing the public for adverse events, as well as facilitating the response when such events occur. Starting form this assumption, a strong effort has been dedicated to produce an effective communication to citizens regarding heat waves effect on health and practical suggestions to prevent heatstroke.
   In particular specific information contents have been prepared on:

   – Heat effect on Health: direct effect (sunstroke, heatstroke, heat exhaustion, heat cramps), indirect effect (i.e. health condition worsening of people weakened by physical or psychological diseases), risk groups (elderly people, children, people affected by specific diseases – diabetes, cardiovascular disease...);
   – Practical suggestions to prevent Heatstroke: telling how to limit the discomfort (what to eat/drink and what not, most dangerous day time, how to dress, how to manage home air conditioning...); telling what to do in emergency case (symptoms, 1st aid, who to call...). All this information are spread out both through the specific web platform on Heatwave managed by ARPA ER (http://www.arpa.emr.it) and through specific awareness campaigns managed by Local Health Authorities, addressing the specific high risk groups. (i.e. leaflets have been created by local health governments and addressed to home care staff, informing them on what to do in case of heat waves, what to eat, how to dress, relation with medicines, how to behave at home and outside, early warning signs of a heat stroke).

#### 6.2.5.3    Proposed Pilot Action: Preliminary Assessment of a Possible Development of the Heat Risk Alert System Based on the Use of Emergency Ambulance Data

The suggested pilot action aims at verifying a possible improvement of the Heat Risk Alert System currently operational in the Emilia-Romagna region. The development would be based on the use of emergency ambulance data.

Today the discomfort index threshold levels have been identified via epidemiological analysis based on historical mortality data. A study developed by Arpa on all most important cities of the Emilia-Romagna region, except Rimini, evaluated the exposure–response curve of ambulance dispatches in relation to biometeorological conditions using time series techniques showing a strong correlation between ambulance dispatches increase and apparent temperature. The effects of high temperatures on health were evaluated for respiratory and cardiovascular diseases as well as for all non-traumatic conditions.

For apparent temperatures lower than 25 °C, ambulance dispatches were not affected by changes in biometeorological conditions. Above that threshold, an increase of ambulance dispatches associated with respiratory diseases has been found, while cardiovascular diseases remained unaffected by variations in apparent temperature. For apparent temperatures higher than 30 °C, the percent increases associated with each unit increment of apparent temperature became very high, with the main effect seen with cardiovascular diseases.

The findings of the study demonstrated the usefulness of ambulance dispatch data due to their strong link with heat and their real-time availability. As a matter of fact, these data are collected and stored with the same software and the same protocols across the region Emilia–Romagna, and are the only health data available in real-time.

Based on these reasons ARPA tried to test the possibility of the use of these data for a development of the heat risk alert system. The main goal of the analysis was to verify how strong was the relationship between bioclimatic discomfort conditions and increased ambulance dispatches across the Region on a daily basis.

ARPA defined as "alert day" a day when apparent temperature averaged among the main towns of the region is above 25 °C. The expected number of ambulance dispatches for each summer day was calculated averaging the daily ambulance dispatches in a 15-day period centered in the selected day over the years 2003-2006 (excluding all the alert days from the computation); a standard deviation was also computed as a variability indicator. Days with a number of ambulance dispatches exceeding the expected number for that day plus twice the standard deviation were considered as days with an elevated number of dispatches ("case day").

The strength of the correlation between case and alert days was tested using sensitivity, specificity and positive predictive value defined as follows.

| | | ON Alert Day (ONAD) | | |
|---|---|---|---|---|
| | | Yes | No | |
| Case Day | Yes | True Positive ( A ) | False Positive ( B ) | A + B |
| | No | False Negative ( C ) | True Negative ( D ) | C + D |
| | | A + C | B + D | |

Sensitivity (Se) = A/A+C

$$IC_{95\%(Se)} = Se \pm 1.96 * \sqrt{(Se * (1 - Se)/n}$$

Specificity (Sp) = D/B+D

$$IC_{95\%(Sp)} = Sp \pm 1.96 * \sqrt{(Sp * (1 - Sp)/n}$$

Predictive positive value (PPV) = A/A+B

$$IC_{95\%(PPV)} = PPV \pm 1.96 * \sqrt{(PPV * (1 - PPV)/n}$$

Sensitivity refers to the proportion of days showing elevated heat-related disease counts detected by the surveillance system during ON Alert Day - ONAD (reported cases correctly classified). Specificity refers to the proportion of days with normal numbers of heat-related diseases during Off Alert Day - OFAD. Positive predictive value (PPV) refers to the number of days with a significant count of ambulance dispatches during the ONAD among the total number of days with a significant count of heat-related ambulance dispatches. For example, a true positive is defined as the number of above-threshold days in terms of the number of ambulance dispatches during ONAD.

The fist results of our analysis were

| | | |
|---|---|---|
| **Sensitivity** | 0.19 | IC95 %(0.14, 0.26) |
| **Specificity** | 0.97 | IC95 % (0.92, 0.99) |
| **Positive predictive value** | 0.90 | IC95 % (0.76, 0.97) |

These findings shows a correlation in terms of specificity and positive predictive value: in fact, almost every day in which an elevated number of ambulance dispatches occurred was an alert day, i.e. a hot day. In addition, the high value of specificity shows that almost no false positive are produced by the model. On the other hand, low sensitivity shows that a relevant fraction of alert days (i.e. hot days) doesn't imply a large number of ambulance dispatches.

We also tested the calculation of the same indicators in single towns. The average value of sensitivity, specificity and positive predictive value obtained across the region were 0.13, 0.97, 0.83, respectively.

These preliminary results confirm the usefulness of a surveillance system based on ambulance data. An additional level of alert for the health prevention system would be suitable when alert days are associated with exceptionally elevated number of ambulance dispatches.

The most appropriate spatial scale for the alert system (e.g. single towns, "area vasta", whole region), a more sophisticated definition of hot days (with the inclusion of information on persistence), alternative definition of threshold values for apparent temperature are currently under investigation. The forthcoming availability of longer time series of ambulance dispatch data will also improve the robustness of the analysis.

The use of Ambulance dispatches to evaluate the risks associated with biometeorological discomfort has the following advantages respect the current use of mortality data:

- A higher number of data to correlate with Discomfort Index (10X), taking into account a wider range of the effect of heat on health.
- The ability to monitor the effect on health <u>during</u> heat episodes (real time monitoring)
- Additional information which can be gathered with ambulance dispatches (location of the calls) could allow spatial analysis to identify geographical areas at higher risk.
- A better categorization of diseases in ambulance dispatches could enable to better define effect of heat waves in human health (cardiovascular diseases, respiratory diseases...)

In conclusion, the encouraging results of this preliminary analysis point to the setting-up of a surveillance system, whose actual implementation should be arranged in cooperation with Health Authorities both at local and regional level.

### 6.2.6   Conclusions

The UHI project sought to boost transnational discussion among policy makers, local administrators and professionals in order to bring about developing policies and actions with the purpose of adapting and mitigating the natural and man-made risks arising from the UHI phenomenon. For this purpose the pilot action of "Villaggio Artigiano" was aimed at:

- providing a deeper knowledge on man-made risk of the UHI and its interactions with global climate change,
- setting up suitable strategies for the mitigation of - and adaptation to UHI,
- improving current land-use planning tools and civil management systems according to mitigation and adaptation strategies.

As explained above, the purpose of this case study was to find ways/rules that can mitigate and tackle the UHI phenomenon. Expected results are:

– mitigation of UHI phenomenon through the definition of appropriate construction requirements and standards (green roofs, reflective materials etc..) in accordance with urban quality,
– definition of guidelines to develop a specific project.

In this regard the Villaggio Artigiano is a suitable area to tackle UHI phenomenon, because it is part of a wider redevelopment context and regeneration process, strongly supported by public institutions and citizens.

The main objective of the regeneration plan is to allow a redevelopment of the housing stock in the Village, through a deep renewal of existing buildings, including demolition and reconstruction of a new building while maintaining the shape of the previous one, thus preserving the system of dimensional and volumetric relationships that characterizes the Village, and achieving a new building organism that continues and updates the typical evolutionary process of the "Villaggio Artigiano".

Given the new environmental context in which the Village is included, the plan entails interventions aimed at fostering high-performance at the overall urban but also environmental systems.

For this purpose, the Municipality of Modena has defined a set of environmental indexes focused on the assessment of 3 main affecting phenomena: the Urban Heat Island, the energy requirements and the hydraulic risk. Consequently, a new calculation methodology has been defined, to be tested in the redevelopment of the Village, capable to measure the environmental effects and the achievement of the planning targets and to estimate the benefit-cost ratio in the redevelopment of urban lots.

Identifying indexes to measure the multiple environmental effects of the urban transformations is the challenge for urban planners.

## 6.3  Next Steps

The City of Modena is collaborating with the EELab Department of the University of Modena, to refine the scientific approach adopted in the analysis of phenomena.

Furthermore, Modena is collaborating with engineers from ARPA, the Regional Agency for Environmental Protection, within the European project UHI, in order to analyze the correlation between types of intervention described in the index "rate" and the reduction of the Urban Heat Island phenomenon

The Municipality of Modena is awaiting the conclusions from the ongoing analysis, to finally validate the experimental indexes and put them into practice.

The indexes, evaluating the achievement of valuable solutions from the environmental point of view, could be useful to set the potential reward to be given to requalification projects. The municipality is considering to provide the use of the indexes within the redevelopment plan of the Village, so to regulate rewards, through discount on the contribute for requalification, or through the introduction of "bonus" related to the architectural and urban value of the adopted interventions.

The urban indexes represent an experimental approach with huge potential, a starting point to give the planner flexible and easy to use instruments.

# Appendices

## Appendix A

### Review of the Different Rules and Regulation Set Up by Emilia-Romagna Local Government

---

**Resolution of the regional council n. 344 of 14th March 2011** "Directive 2008/50/CE of the European Parliament on ambient air quality and cleaner air for Europe put into effect by D. Lgs (legislative decree) n. 155 of 13th August 2010 Request for extension of the deadline for accomplishment and dispensation from respecting specific limit values for nitrogen dioxide and PM10"

**Resolution of the regional Assembly n. 28 of 10th December 2010**

**Resolution of the regional Assembly n. 1570 of 26th July 2011**

**Resolution of the regional Assembly n. 50 of 26th July 2011**

**Resolution of the regional Assembly n. 156 of 4th March 2008**

**Regional Law n. 26 of 23rd December 2004 "Discipline of the territorial energy planning and other energy related provisions"**

**Regional Energy Plan "PER"** (Resolution of the regional Assembly n. 141 of 14/11/2007)

**Regional Territorial Plan "PTR"** (Regional assembly resolution n. 276 of 3/2/2010)

**Integrated Transports Regional Plan "PRIT" - 1998**

**Environmental Action Plan** – 3/12/2011

---

Resolution of the Regional Council n. 344 of 14th March 2011
"Directive 2008/50/CE of the European Parliament on ambient air quality and cleaner air for Europe put into effect by D. Lgs (legislative decree) n. 155 of 13th August 2010 Request for extension of the deadline for accomplishment and dispensation from respecting specific limit values for nitrogen dioxide and PM10".

This resolution passes the thematic cartographies regarding the areas in the municipalities where PM10 and NO2 value limits are exceeded. These cartographies are the knowledge basis used by the relevant authorities as far as the management of ambient air quality is concerned in order to detect and put in place the necessary actions to meet the limit values within the shortest period.

The resolution establishes that also the activities of regional planning must contribute to reaching the objective of air quality. Especially, as far as the definition of measures and interventions in the sectors of transportation, energy, industry, agriculture, construction and urban planning are concerned, Emilia-Romagna regional authority must take into consideration the necessity of meeting the limit values for nitrogen dioxide and PM10 as established by the Community laws.

Resolution of the Regional Assembly n. 28 of 10th December 2010
**Resolution of the regional Assembly n. 28 of 10th December 2010** for the implementation of the National Guidelines with regards to the identification of the areas available for installation of photovoltaic solar electrical plants. Regulation of the other kinds of renewable energy is intended to be addressed into a following resolution.

Resolution of the Regional Assembly n. 1570 of 26th July 2011
"Identification of the areas available for installation of wind power, biogas, biomasses and water electrical plants".

The national and Community legislator have stressed the importance of linking the principle of supporting the development of renewable energies to the other Community principle of land sustainable protection and enhancement in order to maintain the capacity of the territories of providing ecological, economic and social services and to maintain their functions, including the agricultural sector ones.

Resolution of the Regional Assembly n. 50 of 26th July 2011
"Second three-year implementation plan of the Regional Energy Plan (PER) 2011–2013". The first implementation of PER was carried out through the three-year plan 2008–2010, which was passed together with the PER itself (Resolution of the regional Assembly n. 141/2007).

The second implementation plan includes the objectives, commitments and programmes that have been agreed at the European and national level.

Energy efficiency and spare represent the first objective: Emilia-Romagna Region aims at building class A houses staring from 2014, at renewing the building stock, at reducing the emissions of motor vehicles, at spreading the cogeneration and the systems of distributed energy generation, at recovering heat from production activities and making them more efficient. The other main objective of the second implementation plan is the development of renewable energy.

Resolution of the Regional Assembly n. 156 of 4th March 2008
**Resolution of the regional Assembly n. 156 of 4th March 2008** which passes the "Orientation Act about the criteria of energy performance and about the procedures for the certification of energy performance of buildings".

Emilia-Romagna Region is one of the Italian regions which have fully and concretely implemented the Community directives on buildings energy performance.

This resolution regulates:

(a) The minimum qualifications for energy performance of buildings and of their energy plants
(b) The methodologies for the assessment of energy performance of buildings and of their energy plants
(c) The issue of the certificate of buildings energy performance

(d) The accreditation system of the building energy performance operators
(e) The maintenance of buildings and energy plants
(f) The regional information system for monitoring the energy performance of buildings and of their energy plants
(g) The measures supporting energy efficiency and the development of energy services for the regional population

The resolution was adopted according to: the Regional Law (L.R.) 20/00; art. 2 and 25 of the L.R. 26/04. It has implemented the provisions of directive 2002/91/CE and 2006/32/CE, also complying with the fundamental principles and the minimum performance standards set by the national legislator.

The provisions included in the resolution have come into force on 1st July 2008.

Regional Law n. 26 of 23rd December 2004 "Discipline of the Territorial Energy Planning and Other Energy Related Provisions"

This law promotes the sustainable development of the regional energy system and guarantees a matching among the energy which is produced, its rational use and the territory and environment carrying capacity.

Provinces are charged of:

(a) authorizing and implementing the energy spare and efficient energy use promotion plan, the promotion of renewable energies, the development of provincial energy plants and networks, also through the enhancement of existing buildings;
(b) authorizing the installation and the operation/practice of energy plants which are not covered by the State and regional scope.

Municipalities are charged of:

(a) authorizing programmes and implementing projects for energy qualification of the urban system, especially with regards to: intelligent energy use promotion, buildings energy spare, development of renewable energy plants, other actions and public services aimed at supply the demand of energy in urban areas, including district heating networks and public lighting also in the framework of urban regeneration programmes according to the current law;
(b) functions defined in art. 6 of law n. 10/1991, together with the other functions assigned by other specific laws.

Art. 5 "Tools for urban and territorial planning and adaptation of regulations on building issues" establishes that:

1. Local authorities operate through their tools of territorial and urban planning in order to guarantee the restraint of energy consumption in urban areas, promote renewable energies, promote the supply and usability of other energy related local services also in the framework of urban renovation interventions on existing buildings.

2. Territorial and urban planning:

   (a) set the energy local supply of public interest to be installed or renovated and the corresponding location
   (b) can decide to implement transformation interventions where there are infrastructures for renewable energy production, recovery, transportation and distribution or where their construction is planned

3. Municipalities, in the framework of their legislative powers on urban planning and construction activities, acknowledge the minimum qualifications for energy performance set by the regional council and can decide not to put them in place in case of the categories listed in art. 4 paragraph 3, directive 2002/91/CE.

4. Municipalities operate in order to:

   (a) in case of new urbanization interventions on a surface bigger than 1.000 mq, get the assessment of technical and economical feasibility for installing renewable energy plants, cogeneration, heat pumps, centralised heating and cooling systems;
   (b) in case of newly constructed buildings with centralised heating systems, requesting of putting in place systems for temperature control and heat accounting for each habitation unit;
   (c) in case of newly constructed public buildings or buildings used for public scopes, respect the obligation of using renewable energies and adopt electronic control systems;
   (d) in case of interventions according to art. 6 of the regional law n. 31 of 25th November 2002 (General construction discipline) on existing buildings with a surface bigger than 1.000 mq, improve their energy performance with the aim of meeting the minimum qualifications described in art. 25, paragraph 1, letter a) of this law and of creating the conditions to put in place systems for heat accounting for each habitation unit.

Regional Energy Plan (PER)

PER was approved by the Resolution of the regional Assembly n. 141 of 14th November 2007. The Plan complies with the general objectives of the energy policy of the Regional Law n. 26/04 and with the Community and State fundamental principles. It set the objectives, the tools and the guidelines for the actions to be carried out by Emilia-Romagna Region and the local authorities on its territory with regards to:

- energy spare
- renewable energies promotion
- improvement of the territorial energy performances, especially as far as the buildings, SMEs, mobility, distributed energy systems, agriculture and forest sectors are concerned
- improvement of security, continuity and cheapness of internal delivery

– usability, dissemination and quality of services for the public, especially as far as disadvantage areas and users
– improvement of environmental sustainability of energy supply
– reduction of greenhouse gas emissions

PER stresses the need of creating a context which can support the development of practices based on the principles of environmental and energy sustainability, aimed at the rational energy use and the use of renewable energy sources, together with the increase in the supply of high efficiency cogeneration services. This could provide the opportunity of fully using the produced thermic energy, also through the development of district heating networks for local communities. In this way the territorial planning would comply with the general objectives of energy sustainability set by LR n 26/04, art. 1, paragraph 3.

During the last years electric power plants based on renewable sources (water, wind, solar and biomasses) have grown in the regional territory. Nevertheless, they do not play a key role yet within the regional electrical balance. Among these plants, also thanks to the promotion politics that have been implemented both at national and regional level, the solar (in terms of number of plants) and the biomass (in terms of power) plants have had the biggest success.

Emilia-Romagna Region is aware of the need further promote renewable energy given that its exploitation contributes to climate change mitigation through the reduction of greenhouse gas emissions, sustainable development, security of supply and the development of a knowledge based industry, economic growth. According to directive COM (2008) 19, Italy's target for share of energy from renewable sources in final consumption of energy to be reached by 2020 is 17%, with an indicative trajectory set out to meet that share. Italy is the third producer of GHG in EU-27. In order to meet the "Kyoto goal", in 2012 Italy must reduce its emissions of 6,5% of 1990 level.

Regional Territorial Plan (PTR)
PTR was approved by Regional assembly resolution n. 276 of 3/2/2010 and it replaces the previous one approved in 1990. The regional law n. 20/2000 conceives it as the instrument to be used for setting the goals and ensuring social development and cohesion, improving regional competitiveness, ensuring social and environmental sources replicability and enhancement. The Plan does not include a list of issues to be ruled, but it sets economic, social, environmental and territorial goals and objectives to be reached.

According to the Plan, reduction of energy consumption can bring new chances not only for energy, plant and building related firms, but also for innovation of products and processes. This can help housing sector improve its quality and security. Emilia Romagna Region is already providing support in this sense through regional norms amendment.

PTR identifies bio-building and energy spare as the sectors where currently innovation mechanisms can be included in order to reach excellence, also at international level. In the framework of new energy politics, firms must play a leading role, both with regards to energy spare and renewable and clean energy production (green economy).

Issues such as environment and climate change are conceived by PTR as extremely relevant as far as connection between global and local dimension of environmental crisis are concerned. Problems exist of environmental quality at local level, such as constant air and noise pollution and the growth of cases of urban heat islands. It is of key importance that PTR sets not only the necessary politics for GHG emissions reduction, but also actions for adapting to climate change aimed at limiting the damages that can incur and to take any related opportunity. As far as urban issues are concerned, the Plan identify urban planning as a priority in order to put in place pilot actions of territorial management aimed at reducing the cases of heat islands. The current trend of setting oneself in low density new urban areas is causing an increase in non-renewable sources consumption and a progressive loss in environmental quality, which have relevant social consequences.

The Integrated Transports Regional Plan (PRIT) Approved in 1998
The plan has undergone a revision work and has been assigned by PTR the role of specifying the infrastructural and mobility arrangements providing coherence within the transports sector.

PRIT cannot directly rule urban mobility, but it can boost and promote good practices which can be integrated with regional politicise on suburban territory. PRIT'98 was not committed to reduce nor localize mobility infrastructures, but to maximize transports effectiveness while reducing its costs and environmental impact.

As far as environment is concerned, during the last years relevant results have been obtained with regards to air-quality. Monitoring figures show that regional and local authorities politics have had a positive impact on the levels of air pollutants. Emilia Romagna Region in committed to continue working on politics against atmospheric pollution, especially in urban centres. Focus will be both on private traffic regulation and enhancement –already ongoing- of incentives to sustainable mobility and improvement of public rail and road transport services.

As far as fundamental measures against pollution are concerned, among the main actions there are the renewal of buses and trains (on railways within regional scope), the improvement of cycle lanes and of sustainable people mobility, the reduction of energy consumption in the production and civil sector, cars' fuel switching into lpg and methane.

Environmental Action Plan
The Plan was approved by the Regional Assembly on 3rd December 2011 and it is aimed to putting in place environmental projects. 150 million Euros have been budgeted for actions on biodiversity, separate collection of rubbish and rubbish traceability, water and air quality and sustainable mobility.

**Incentives, Financing and Regulatory Actions in Support of Environmental Restoration, Energy Conservation and Reduction of the Phenomena Related to Climate Change Put in Act from Your Local Authorities to Facilitate a Sustainable Land Use (Local Authorities Means the Communal Level)**

| |
|---|
| **Regional Law (LR) 20/2000** "General discipline on territory protection and use" |
| **Municipal Operational Plans " *POC* "** |
| **Urban and Buildings Regulation " *RUE* "** |
| **Municipal Structural Plan " *PSC* "** |

**Regional Law (LR) 20/2000** "General discipline on territory protection and use" puts municipalities in charge of a series of tasks in order to facilitate the sustainable use of territory and keep its changes under control as far as urban transformation issues, social, economical and environmental topics are concerned. These tasks are included in two main instruments. The first one is the **Municipal Structural Plan (PSC)** which acknowledges all prescriptions and orientations set at the national, regional and provincial level and elaborates the politics and objectives aimed ad promoting and improving environmental quality in the framework of territorial management and urban planning. The second one consists of operational instructions for short term transformation and preservation actions, for which the Urban and Buildings Regulation (RUE) provides instructions on the methodology for conducting sustainability and feasibility assessments.

In specific terms, operational contents for urban planning are contained in:

- The **Municipal Operational Plan (POC)** (art. 30 – LR 20/2000): it is the urban instrument that follows the PSC and which sets and regulates protection and renewal actions, territorial management and transformation actions to be put in place in a 5 year period. POC is conceived as a multiannual operational plan, it relies on the municipal multiannual budget and it is an orientation and coordination instrument for the 3 year public works programme and for the other municipal instruments set by national and regional laws.
- **Urban and Buildings Regulation (RUE)** (art. 29 – LR 20/2000): it is the urban instrument that regulates the typologies and methodology of transformation actions, together with the designation of the area for any specific function. The regulation also focuses on: constructions, physical and function transformations, buildings preservation including sanitary and building norms related issues, architecture and urban issues, green areas and other elements which characterize urban areas.
- Implementation Urban Plans (PUA) (art. 31 – LR 20/2000): they are detailed urban instruments that regulate new-urbanization and renewal works scheduled in POC.

PSC stresses the importance of actions aimed improving the quality of urban and periurban areas both with green areas and landscape re-design, through trees planting along the borders between city and countryside in order to make landscape more heterogeneous and to protect biodiversity.

PSC orientations also focus on the development and renewal of existing green areas, together with ancient public gardens and those of historical villas. These areas are conceived as urban centres of excellence and can represent the staring point for "green thematic routes" that are further developed by RUE and detailed planning.

Big parks can also play a key role as far as environmental quality and social development of urban areas and outskirts are concerned. Parks are conceived not as equipped free areas, but as locations which can boost social and cultural activities. As a matter of fact, they have become areas providing services and places for spare time and cultural activities. In addition, parks located along the rivers have also the function of an ecological network which connects the different areas within a city.

Finally, green areas acquire a fundamental task in the framework of the transformation processes of urban landscape planned by municipal policies and can have a positive impact on territory and environment.

In the framework of urban planning, wide natural matrices are enhanced through a better definition of the borders between artificial and natural areas and through the creation of wide connection infrastructures (e.g.: main roads and railways).

The work plan that comes out, according to PSC forecasts, presents the following strengths:

– a strategy for managing accesses to urban areas, with a particular focus on tourism, based on intermodality and public transport;
– the improvement of mobility safety conditions;
– the protection and development of pedestrian and cycle mobility;
– public transports promotion and development.

In conclusion, PSCs, following PTR orientations, contribute to reduction of phenomena related to climate change, supporting compact settlements and maintaining the usual dimension of the cities and villages, avoiding duplication of services, planning an ecological network at municipal level aimed at enhancing existing wide environmental matrices and which serves as connection among areas of key environmental interest.

As far as UHI phenomena are concerned, norms set by municipal authorities and relating to constructions and urban areas transformation are included in RUE and focus on: air and water quality, air and water pollution prevention, water cycle management, reduction of noise and electromagnetic pollution; safeguard of land permeability and ecological rebalancing of urban environment, separate rubbish collection.

It is important to note that the Department for Urban Quality of Emilia Romagna Region has recently set a list of technical requirements on sustainable construction. They are updated according to the most recent guidelines on environmental protection and energy spare and they are currently tested on a voluntary basis by some municipalities, as the regional law does not compel to adopt such norms. Among these requirements, standard I 2 (Urban scale 2) focuses on monitoring sun exposure, because if this is not duly taken into consideration it can concur in causing UHI phenomena.

## Implementation of the Directive 2001/42/EC on the Assessment of the Environmental Effects of Plans and Programmes

| |
|---|
| **Regional Law n. 6 of the year 2009** "Government and renewal of territory" |
| **Regional Law n. 20 of the year 2000** "General discipline on territory protection and use" |

Sustainable development is among the priorities of the EU.

At the beginning of the XXI century, the European environmental policy makers must face hard challenges and must especially put in place the decisions of the Amsterdam Treaty concerning the integration of environmental economical and social politics.

In its conclusions of the Helsinki Summit in 1999, the European Council asks to the Commission to prepare a proposal for the Sixth Environmental Action Programme including a long term strategy to integrate sustainable development policies from an economical, social and ecological point of view. The programme "Environment 2010: our future, our choice" was approved in January 2004.

The sixth Programme consists of a short strategic document including the priority actions for environmental politics at European level for the next ten years, with a special focus on the environmental problems.

Following this Programme, the Commission issued the Communication "A Sustainable Europe for a Better World: A European Union Strategy for Sustainable Development" (COM 2001, 264) which highlights the integration of environmental protection in all the actions and politics within the environmental field.

According to the EU orientations, the economical objective of competitiveness, the social objective of employment and the improvement of an effective use of resources, must be integrated in a unitary strategy which crosses all the levels of territorial planning, from Provinces to Regions and States. This should also involve firms, consumers, economical institutions, fiscal and monetary authorities and the whole society because the general consensus is necessary to support sustainable development.

As a matter of fact, sustainable development is the only success strategy to increase competitiveness together with employment and to develop eco-efficient technologies, dematerialisation strategies, environmentally-friendly policies within all economic sectors.

In Emilia-Romagna Region, the concept of sustainable development is of key importance to evaluate and select the policies and actions to be included in the instruments of territorial planning.

The Environmental Evaluation (VAS) was introduced by the Directive 2001/42/CE of 27th June 2001 concerning the assessment of the environmental effects of plans and programmes. This has been followed by the Decision 871/CE of 20th October 2008 on the VAS Protocol.

Italy has regulated VAS through the legislative decree (D. Lgs) n. 152 of 3rd April 2006 "Environmental Norms" which has been replaced by D. Lgs. N.4 of 16th January 2008 "Further corrective and integrative regulations on D. Lgs. n. 152 of 3rd April 2006". This law focuses on procedural aspects: in order to guarantee a high level protection of environment it does not set limits to be respected, but it establishes that impacts on environment must be taken into consideration during the elaboration of plans and before their approval.

Ahead of European law on VAS, Emilia-Romagna Region has approved in 2000 the law n. 20 "General discipline on land protection and use" which introduced, among other innovations, the "pre-emptive evaluation on environmental and territorial sustainability" (VALSAT) as a constitutive element of approved plans.

After that, Emilia-Romagna Region has acknowledged the legislative decree n 4/2008 with the Regional Law n. 6 of the year 2009 "Government and renewal of territory".

Regional Law n. 6/2009 has stressed the importance, already expressed in the previous law n. 20/2000, of the sustainable territorial and urban planning introducing the following issues:

- Starting from the initial elaboration phases and until their approval, plans must take into consideration the impacts that their implementation can have on the environment and territory.
- In the annexes of the approved plan, a specific document must be included where potential impacts on environment of the implementation of the plan are identified, described and evaluated. Besides, the necessary actions to avoid, reduce or compensate these effects must be mentioned, taking into consideration the characteristics of the territory and the area.
- In order to avoid any duplications in the evaluation, the results of the higher level plans and of those ones that are intended to be changed.
- The regional and provincial authorities, as competent bodies, express their opinions on the environmental evaluation of, respectively, the provincial and municipal plans.
- The regional, provincial and municipal authorities must also monitor the implementation of their plans and of their effects on the environment and territory, also with the aim of revising or updating them if needed.

In addition, minimum fulfilments are set for the implementation of the environmental evaluation of the plans. Especially, in order to guarantee the transparency of the decisional process, the completeness and reliability of the information used within the evaluation, some obligations are set:

- Both environment related experts and the general public (citizens) must be consulted in the evaluation.
- Detailed explanations on how environmental issues have been taken into consideration during the elaboration of the plan and a monitoring programme during the plan implementation phase must be provided.
- Environmental documents used during the evaluation, the expressed advices and the final decision must be communicated and must made be available.

Finally, it is important to mention art. 6 of the Regional Law n. 20/2000 which acknowledges the development of mobility infrastructures –especially railways- as conditions that the plans for the transformation of the territory must comply with. Consequently, the concept is stressed once again that territorial planning must link the works for the development of settlements with the relevant conditions that reduce their impacts and that make them compatible with the contexts where the works are implemented. This also includes interventions to mitigate negative impacts such as the construction of mobility infrastructures and especially the public transports on railways.

## *Appendix B*

### Technical Content of the Requirement on Control of Solar Energy Intake

As part of the Emilia Romagna Region's General Directorate Territorial and negotiated planning, agreement, the Urban Quality Department, has recently prepared a list of technical standards relating to building sustainability.

These standards, updated to the latest guidelines in terms of environmental and energy savings, are currently tested voluntarily by some municipalities in the region.

Regional Administrative Arrangements in fact, does not oblige the adoption of similar rules, but leaves it to each municipality, the autonomy of decision regarding the environmental and urban planning.

The Standard I 2 attached, addresses the issue of controlling exposure to the sun, that if not properly designed, can cause the urban heat islands (UHI).

**STANDARD I 2** *(for residential and commercial areas and city neighbourhoods)*

**The requirement is part of the Family**: Energy Efficiency

Need
Contribute to a rational use of climate resources and energy by controlling access of the sunbeams to the building structures, to the active solar and passive systems and outdoor living spaces through the use of an integrated design approach that controls the solar energy and direct and indirect effects that can generate to outward microclimate and buildings.

Scope
Uses: for all purposes (residential and commercial buildings, city neighbourhoods)

Performance Level
The performance levels are reported separately within the winter and summer sunshine and in compliance with the methodology of integral design; the solutions must meet both conditions.

Summer Sunshine

In order to contain the phenomenon of "Urban Heat Island", and the resulting over-heating in the summer, it is necessary to simultaneously control the shading and manage a strategic relationship between the paved and built areas and green spaces, their position with reference to constructed and which finishing materials the outer surfaces have been chosen, within those with high reflectance characteristics of solar radiation. The shell of the buildings must be protected from the effects of solar radiation with specific solutions, such as ventilation of the same or with a double ventilated outer covering, green roofs, etc.

The external parking spaces and pedestrian paths, should be properly shaded. An obvious shielding effect is given by trees and vegetation. It's important to choose the essences in terms of their form and content of their character but also of their cast shadow. The beneficial effect of shading is more significant if the trees provide shade in the heated season, especially for deciduous plants that do not interfere with the winter sunshine. The use of green roofs is an excellent solution to reduce the load on summer thermal cover and to limit the "heat island phenomenon" in a neighbour-hood next to the intervention. And through the appropriate placement of plants, the local microclimate can be optimized by choosing the type of paved surfaces. The surfaces with which the user can come into contact have to submit poor attitude to overheating, through a feature of high reflectance of solar radiation and emissions.

Winter Sunshine

The only access to the building structures and outdoor spaces (in particular the stop-ping places) must be carefully controlled in relation to any external obstructions.

If there are areas devoted to active or passive solar house systems is required to control the sun exposure of the same. It is required that is guaranteed exposure to the sun more than 80 % occurred at 12 pm on December 21.

Method of Verification at the Design Stage

*Technical report* which explains the design process carried out, with reference to the performance specifications as above and giving reasons for decisions taken.

This report will demonstrate the control design through the use of daylight control tools (for instance: solar axonometriy) to analyze and document the effect of strategies on the control of energy intake on the aggregation of buildings and on outdoor spaces.

To control the effects of shading is to analyze the shielding (artificial plants or mixed) that restrict access of direct solar radiation on the outdoor areas of the site and on the fronts of buildings and roofs of the various projects.

The distance between the buildings or the placement of other obstructions induced by the intervention should be calculated on the basis of the above effects.

The verification tools with daylight control is mandatory in the case are provided for passive and active solar systems, the "solar access" must be assessed in places (roofs, roofs, etc. …) in which it is expected they will be installed. This verification will be aimed to the definite project of every single building of the area.

Method of Verification at Work Completed

*Declaration of conformity* of the work carried out with respect to the project.

## *Appendix C*

**POC MOW URBAN UPGRADING OPERATING PLAN FOR THE WEST
MODENA DISTRICT ARTISAN VILLAGE GUIDELINES
FEASIBILITY STUDY OF ENVIRONMENTAL QUALITY INDEXES
TO BE APPLIED TO BUILDING LOTS**

**MUNICIPALITY OF MODENA
URBAN IMPLEMENTATION SERVICE**

**Managing Director**

Engineer Marcello Capucci

**Experts in Charge**

Engineer Filippo Bonazzi
Surveyor Catia Rizzo

**External Collaborators**

Engineer Emilio Lucchese
Ingegneri Riuniti S.p.A.

Engineer Alberto Muscio
University of Modena and Reggio Emilia
"Enzo Ferrari" Department of Engineering

Engineer Sara Toniolo
Municipality of Modena – Environmental Sector

Resources, Territory and Civil Protection Service

# Table of Contents

# 1
# Introduction

On the main goals of the Artisan Village Upgrade Plan is to renew the buildings in the Village, by means of a deep restructuring of the existing edifices, respecting the size and volume relations among them and producing a new estate body, which carries on and updates the typical evolution process of the Artisan Village.

Taking into consideration the modified environmental conditions of the area, the Plan aims to promote measures envisaging high-level performances, in order to ensure the environmental as well as the urban upgrade of the area.

The Plan may envisage the granting of awards in case of upgrade actions which can guarantee such quality levels as to enhance the environmental value of the structures and the recognizability of the urban context. In this respect, the devised plan envisages the implementation of a procedure ("BAF", see below) for assessing the improvements achieved during the planning phase.

Later, some in-depth analyses were carried out on this index and similar indexes (described below), in order to be sure about the best tool to include in the award attribution Plan.

At the same time, we started to consider what the main environmental issues to work on could be, by means of building procedures. In this phase, a role of the utmost importance was played by the introduction of the Plan among the "pilot cases" of the "UHI" European project, which aimed both to analyse the phenomenon of urban heat islands and the prepare decision-making support tools in order to overcome this phenomenon.

As these issues were studied more and more in depth, we realized that each urban index analysed had certain peculiarities, which could meet specific requirements of the Plan, but at the same time specific issues came up which would limit their applicability.

The more in depth the approaches implemented by the various administration were studied, the more we could outline the main environmental issues to take into consideration in the planning of a building lot and, therefore, what behaviours were to be promoted in order to reduce the building impact. This has also made it possible to define the main characteristics to include in the Plan guidelines, then it was decided to try to develop a new index, an experimental one within the framework of the Plan, to be used both as a synthesis of the guidelines and as an assessment tool. Later, the single environmental issues were analysed, in collaboration with experts from the administrative bodies (Hydraulic Works office, Environmental Sector of the Municipality) and well known professionals (EELab Department, University of

Modena). Also due to some further contributions given by the various experts involved in the UHI project (external professionals or bodies, such as the Province and ARPA), and a comparison with the other European subjects involved, a first validation of this approach was reached.

This report summarises the main features of the environmental method prepared by the administration. It is worth stressing that there are various analyses still under way for the final validation of various technical and procedural issues; as a consequence, although the structural characteristics have already been defined, the approach implemented may still be modified or emended.

# 2
# Existing Environmental Quality Urban Indexes

In order to devise a method for assessing the "environmental" performances of a building lot, several towns in Italy, in Europe and worldwide have included standardized procedures ("urban indexes") in their building regulations, in order to calculate the general impacts of town-planning activities.

In general, for calculating the indexes, it is necessary to know the types of materials/actions present in the building lot and the area they cover (m₂). By applying different "weights" to every material/action, a value is obtained, proportional to its "environmental quality", through the use of simply calculation algorithms. These values make it possible to obtain a number, which expresses the overall characteristics of a specific building lot. In order to ensure a suitable performance level, it is generally necessary to plan an activity in such a way as to get an index not lower than a certain threshold.

The common characteristics of the indexes is to provide a relatively simple and clear calculation methodology, aiming to plan building areas without any fixed schemes, since it is based on the achievement of minimum performance levels, which let you free to choose the various materials/actions to apply within a batch.

### Biotope Area Factor (BAF) Index – Berlin

As early as in 1990,[1] the city of Berlin adopted an index called Biotope Area Factor (BAF), in order to reduce the impacts already present in the town centre and to facilitate an ecological upgrade of the urban context. The BAF is the ratio between ecologically "effective" surfaces and the total area of the land.

From a qualitative point of view, the index aims to safeguard the microclimate, to control the use of the land and water, to improve the quality of plants and the habitat for animals, to improve the vital space for human beings by means of the creation of yards with green areas (or areas with a certain permeability, such as self-blocking ground or gravel), green roofs, green walls or by means of infiltration of rain water to surfaces with extended vegetation.

---

[1] "The Biotope Area Factor as an Ecological Parameter" – Berlin, 1990.

From a quantitative point of view, it provides a value between 0 and 1 and represents the part of the area for plants and other functions of the ecosystem. The higher the value, the better the result obtained in the building lot planning.

The index applies to residential, commercial and infrastructural areas (whether existing or newly made): depending on the various actions to implement, minimum values of the index are set, in a range from a minimum of 0.30 to a maximum of 0.60.

**Green Space Factor Index (GSF) – Malmö**

This index was designed in 2001, partially drawing inspiration from the BAF in Berlin.

From a qualitative point of view, the index aims to measure the ecological value of a settlement, on the basis of the presence of vegetation and permeable areas. The index takes into consideration the types of surfaces envisaged by the BAF (yards with green areas or in any case permeable areas, green roofs or green walls) and integrates them with water surfaces or rainwater collection systems, quantifying the contributions of trees and bushes and promoting urban agriculture. As regards the quantity, it provides a value between 0 and 1. The minimum value to reach is 0.6.

What is peculiar in Malmö's approach is the requirement, on the part of planners, to implement the verification of the GSF index by envisaging a certain number of measures, called Green Points, which are pre-determined for certain types of biotopes, animal habitats or urban agriculture.[2]

Malmö, along with some other European organizations, belongs to the Grabs project (Green and Blue Space Adaptation for Urban Areas and Eco Towns). aiming to integrate strategies to adjust to climate changes within the framework of regional planning tools.[3]

**Green Factor Index – Seattle**

On the basis of the experience achieved in Berlin and Malmö, Seattle also designed its own index in 2006. Initially, it was only applied to commercial areas, later it was extended to residential areas. From a qualitative point of view, this index aims to increase the aesthetics of buildings, to increase permeability, to improve energy efficiency of buildings and to reduce the urban heat island.

(continued)

---

[2] Grabs Expert Paper 6 - The Green Space Factor and the Green Points System – 2011.

[3] The project involves the following countries: Austria, Greece, Italy (Genoa, Catania), Lithuania, Netherlands, Slovakia, Sweden, United Kingdom.

It uses the same action categories envisaged by the BAF (yards with green areas or permeable areas, green roofs or green walls) and those added by the GSF (rainwater collection, trees and bushes, urban agriculture), enlarging them by means of in-depth technical details. From a quantitative point of view, it is absolutely similar to the BAF, although it has a much more specific approach (the administration provides a spreadsheet for processing the index).

The peculiarity of the Green Factor is to carry out a cost analysis relative to each single action, in order to simplify the cost/benefit estimate, and it attributes bonuses for certain actions, such as improving the landscape publicly visible.

Seattle was the first American city to apply an index of this type, later it has served as a model for other cities, such as Bellingham, Portland, Chicago, DC, Bewark (etc.).

**Building Impact Reduction Index (RIE) – Bolzano**

Since 2004, the Municipality of Bolzano has been applying a Building Impact Reduction index (RIE), which defines the ratio between green areas and non-green areas, within the framework of its Building Regulations. The index applies to all newly built elements and on actions involving existing buildings, as well as actions which involve external surfaces exposed to rainwater. From a qualitative point of view, the index aims to reorder rainwater and lower urban temperatures. The index is mostly based on permeability of materials, assessing the presence of green surfaces (agricultural, non-agricultural or for sports facilities), trees, water bodies, partially permeable grounds and green roofs. From a quantitative point of view, the calculation algorithm is slightly more complex than the one used for the previous indexes, and results in a value between 0 and 11 (approximately,[4] the administration provides a spreadsheet for calculating the index). The higher the value, the better the result obtained in the building lot planning. Depending on the various actions to implement, minimum values of the index are set between a minimum of 1.5 to a maximum of 4.0.

What is peculiar in the RIE is the technical analysis of various types of green hanging covers on roofs (synthetically: intensive and extensive).

---

[4]"Manuale d'uso del foglio di lavoro Excel per il calcolo del RIE" – Comune di Bolzano – Ufficio Tutela dei Beni Ambientali.

**Building Impact Reduction Index (RIE) – Bologna**
Since 2009, the Municipality of Bologna has also been applying the RIE index in its Building Regulations. The index applies to both existing buildings and newly constructed ones, excluding those with a ratio between building lot and land area higher than 0.5. The algorithm used is fundamentally identical to that of Bolzano (the administration provides a spreadsheet for calculating the index).

Although it implements a coefficient which takes into consideration the pith of non-green areas, the spreadsheet provided for the calculation is designed in such a way as to attribute a single conventional value to the pith of all surfaces.

The peculiarity of the approach of Bologna is the presence of improvement performance levels aiming to promote building actions such as to enable an improvement of sustainability of the buildings: in addition to basic performances, identical to those defined by the RIE of Bolzano (minimum values of the index ranging between 1.5 and 4.0), "improvement" performance level (index minimum value between 2 and 5) and "excellence" ones (index minimum value between 2.5 and 6) are also envisaged, associated with certain incentives.

**Other Italian Experiences**
Several towns in Italy have implemented various tools aiming to enhance the environmental value of buildings. In particular, the experiences of Brescia, Florence and Rimini are explained below.

What is different in the approaches of these administrations is the application of incentives which may result in lower taxes (economic incentives) or in the adjustment of town-planning parameters by means of correction coefficients (procedural incentives).

The **Municipality of Brescia**, in the building regulations of 2008 (guidelines, chapter called "Nature") dealt with the use of green walls, introducing minimum quantitative parameters depending on the possibility of receiving a series of benefits (both procedural and economic).

The **Municipality of Florence**, in the 2007 building regulations, included a series of indications on the use of green areas, promoting the implementation of measures aiming to "decrease the heat island effect" (by controlling the pith of the ground, the use of urban green areas, designing proper positions for summer shades and the planting of trees and bushes).

The incentives envisaged are both economic and town-planning in nature, they apply to actions on new structures, newly-built edifices, urban restructuring actions and building restructuring (percent reduction of concession expenses and application of adjustments which take into consideration the increased extent on the S.U.L.).

(continued)

The **Municipality of Rimini**, in 2006, introduced a series of incentives for the sustainability of new actions and restructuring works in its building regulations. This document calls them "Biobuilding actions", and they introduce the issue of the use of green on façades and in covers. In the chapter "Quality of life", under the entries "Urban quality" and "Architectural quality", green actions which benefit from incentives are listed (for instance, a reduction of up to 50 % of secondary urbanization charges).

These procedures are a concrete approach which has been taken into consideration in the planning of buildings for years, promoting their environmental sustainability. In particular, the indexes or procedures designed by these administrations highlight the issues which may affect the environmental quality most: the implementation of actions which enhance the quality of buildings and the application of procedures which promote their completion.

Briefly, the favourable features of the indexes or procedures analysed so far are the following.

**Actions**

- Building of green areas, namely yards with green areas, green roofs (RIE) or green walls (BAF, GSF, GF).
- Building of permeable grounds (BAF, GSF, GF, RIE).
- Planting of trees and bushes (GSF, GF, RIE).
- Reuse/collection of rainwater (BAF, GSF, GF, RIE).

**Procedures**

- Implementation of incentives such as to promote high-performance actions (RIE, Municipalities of Brescia, Florence, Rimini).
- Implementation of procedures such as to highlight the quality/price ratio of each type of action (GF).
- Implementation of bonuses relative to actions which increase the quality of the visible landscape (GF).
- Integration of the indexes with a list of actions which describes, from a qualitative point of view, how to manage specific issues, providing planners with a series of potential solutions (Green Points, GSF).

An analysis of the indexes, in addition to these positive features, also shows some limits. As already mentioned, the indexes analysed show various approaches, relative to the specific urban contexts they are applied to. These indexes are structured in such a way as to promote the implementation of actions with a particular environmental value, but in some cases no solutions which would be equally profitable are envisaged, as summarized in the table below.

| | Limits |
|---|---|
| **BAF – Berlin** | The indexes do not take into consideration actions not focusing on "green areas", for instance Cool Roofs (important in order to lower the heat island effect), capable of exerting a positive effect on the climate and comfort of spaces. |
| **GSF – Malmö** | They do not quantify a physical measure attributable to a tangible phenomenon. |
| **GF – Seattle** | |
| **RIE – Bolzano** | The indexes only take into consideration the horizontal surfaces of the building lot, excluding the possibility of assessing the contribution of a "green" wall, and not just to qualify the positive effects of the materials used for building Cool Roofs.[5] |
| **RIE – Bologna** | They do not quantify a physical quantity attributable to a tangible phenomenon. |

In order to overcome these limits, it would be more suitable to implement the above-mentioned features:

- Actions not on "green areas", for instance Cool Roofs, as "high-quality" actions to promote a re-designing of the building lot.
- Definition of specific measurement units attributable to tangible physical phenomena for the calculation of the index, which enable to quantify the environmental sustainability of actions.

---

[5] http://urp.comune.bologna.it/portaleterritorio/portaleterritorio.nsf/a3843d2869cb2055c1256e63 003d8c4e/200cfbd63f33a6aac1257671004e6018?OpenDocument

# 3
# Design of Experimental Environmental Quality Indexes

In order to implement the procedures analysed in the previous chapter, it was deemed useful to design calculation methods:

- capable of analysing all the surfaces in the building lot, yards, walls and roof,
- capable of analysing the various types of surfaces, whether green or not,
- capable of keeping the entry of data as easy as in the existing indexes,
- capable of applying the typical approximations of town-planning indexes,
- capable of highlighting environmental performances on the basis of indicators which assess tangible physical phenomena,
- capable of implementing UNI values, procedures already defined by laws or municipal rules.

In order to assess the environmental quality of a building lot, an attempt has been made to design an approach based on technical and scientific consistencies, identifying specific indicators capable of highlighting the physical properties of the building lot which significantly affect the environment. Lest the typical approach of town-planning indexes be changed excessively, it was decided to apply suitable approximations, such as to not complicate the data entry phase performed by users (appointed technicians or owners).

The main environmental features which are affected by the types of actions which can be carried out within a building lot are the effect of a heat island (outside the building), energy saving (inside the building) and hydraulic risk (outside the building lot). As a consequence, on the basis of a technical analysis of these features, the indicators chosen to characterise the building lot are the following:

- thermal energy absorbed by the surfaces (yard, walls, roof), with reference to both the heat island effect and energy saving, which shows the extent of solar radiation withheld by the building lot;
- the amount of water leaving the building lot (yards, roof or any tanks), with reference to the hydraulic risk, which shows how much water due to rainfalls is released into the sewers.

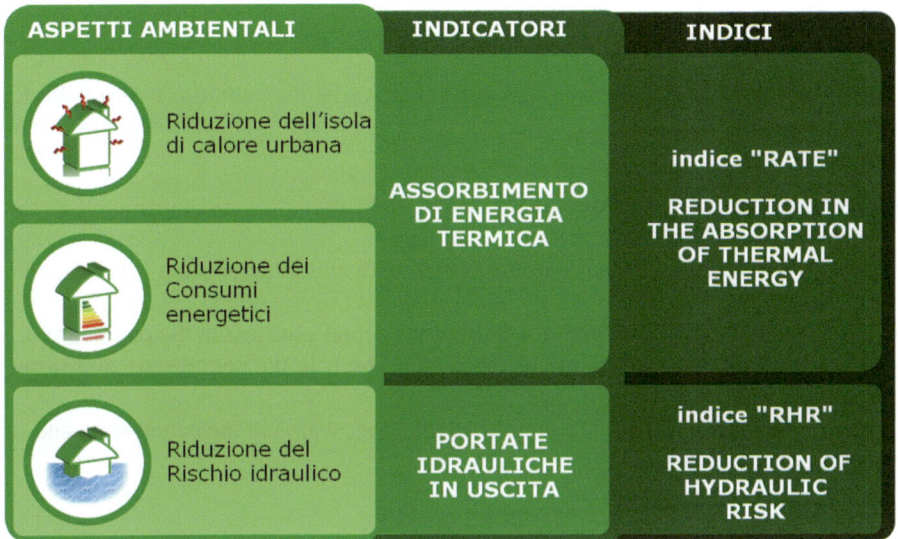

Conceptual framework applied in the designing of the indexes

The specific analysis of the two indicators resulted in two different calculation procedures and, therefore, two indexes. All the algorithms implemented by the indexes are based on validate procedures or procedures being validated by experienced technicians in this field.

In compliance with the preliminary requirement specified above, the complex calculations made in order to calculate the indexes do not make the user's data entry work more complex.

*Note: The indexes listed below are calibrated in such a way as to be applicable to the Artisan Village, with the approval of the POC MO.W plan envisaged by the municipal administration. It is believed that this experimental application may provide useful assessment elements to extend the application of the indexes to any urban context. In this respect, applications to parts of existing towns are particularly interesting.*

The index was designed in four phases.
1. In the first phase, a list of building typologies was made, including the typical materials present in the existing context and used for new actions.
2. In the second phase, a physical measurement to be used as an indicator and a calculation method were identified.
3. In the third phase, the technical and physical characteristics of each construction typology and the context necessary to assess the indicator were analysed.
4. In the fourth phase, the typical approximations of town-planning indexes were applied.

## 3.1   Product Characteristics

The data required for processing the indexes describe the conditions before works (actual state) and the conditions after works (planned state). The following information is required:

- the value of the Land Area in the actual state (which is also attributed to the planned state),
- the SC (covered area) value in the actual state and planned state (they may be different),
- the average height of the volume built in the actual state and in the planned state,
- the length of the longest side of the volume built in the actual state and in the planned state,
- the inclination southward (azimuth) of the longest side in the actual state and in the planned state,

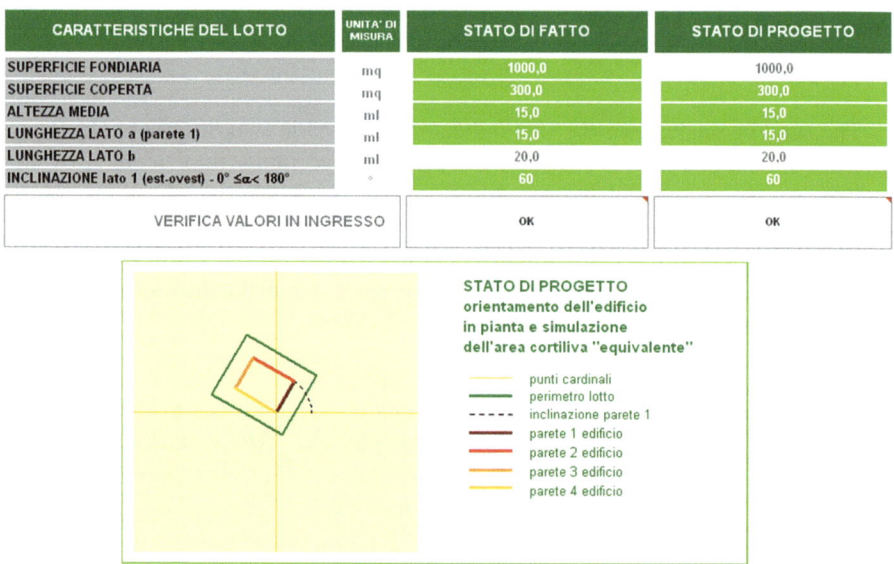

Software user's layout

In order to enable the spreadsheet to describe the hydraulic characteristics of the action area, the following information is also required:

– the hydraulic load class, as defined in the Knowledge Framework of the Municipal Structural Plan,
– the type of action (new building or restructuring).

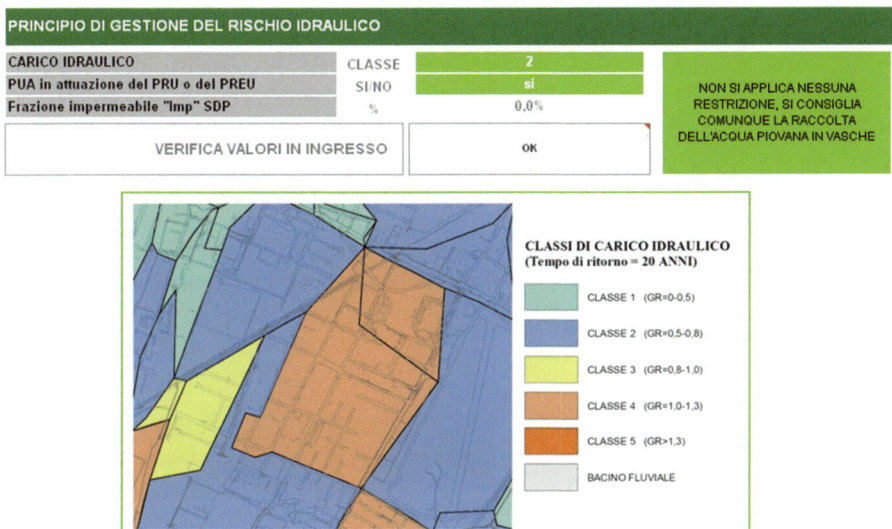

Software user's layout

With this information, the building lot is analysed as a geometrically simplified one: in the actual state, just like in the planned state, the building volume is considered as a regular parallelepiped placed in the centre of an "equivalent" yard, in which the distance between the walls of the building volume and the limit of the property are constant on all four sides.

The inclination of the building volume makes it possible to acquire more information about the amount of solar radiation touching each wall (for this feature, see the following in-depth analyses).

The last two pieces of information (hydraulic load class and type of action) make it possible to define what principle to apply for managing the extents generated by the building lot.

*Note: it is necessary to geometrically simplify the building lot lest the calculation of the index be overloaded with excessive input. By exaggerating in the definition of specific features, the index might be too articulated, thus losing its synthetic capacity. Indeed, it should be borne in mind that the aim of the index is not to create a simulation model, but to implement those already in existence.*

*However, subsequent implementations more specifically on the structure of the software than its logical setting cannot be excluded.*

## 3.2 Types of Actions

The characteristics of the building lot are assessed on the basis of the various typologies of surfaces which characterise the actual state and the planned state. Twenty possible actions have been identified, in order to summarise the possible technical solutions used in the territory.

*Note: If experts in this field tested the index, the envisaged typologies might be validated even further.*

| YARDS | | |
|---|---|---|
| 1 | gardens/meadows | Meadows and areas to be turned into meadows, playgrounds, etc. They trigger photosynthesis and evapotranspiration processes capable of reducing the effects of the heat in the environment. They make it possible to preserve permeability of the ground. The planting of trees and bushes promotes the establishment of air movements and inhibits the discomfort which might be generated in large empty areas due to evapotranspiration processes. |
| 2 | Trees/bushes | They increase the surface of plants capable of producing photosynthesis or evapotranspiration processes. They enhance shading, resulting in the formation of areas protected from direct radiation. They can trigger air movements with local recirculation. Their positions must be considered depending on the position of buildings and main winds, in order to avoid the formation of "barriers" and interfer-ences with the natural circulation of the air. |
| 3 | self-blocking | It makes it possible to make an area suitable for vehicles, while keeping part of the area permeable green. |
| 4 | meadow suitable for vehicles | It makes it possible make a yard suitable for vehicles without ruining the aesthetics of the garden and just partially reducing the permeability of the area. In these areas, however, planting trees or bush-es is not allowed. |
| 5 | with "cold" asphalt | It is a light-colour asphalt used for city street infrastructures. It limits overheating due to solar radiation, but inhibits permeability of the ground. |
| 6 | with normal asphalt | It makes large areas suitable for vehicles, but inhibits permeability of the ground and causes overheating due to solar radiation. |
| 7 | with gravel | Ground with inert material which makes an area suitable for vehicles, limiting overheating due to solar radiation, while keeping a good part of permeability. |

| VERTICAL SURFACES | | |
|---|---|---|
| 8 | green with a frame on the wall | It is made by applying supports to vertical walls, adhering to them or at a certain distance, in order to facilitate air circulation. It is easy to make and takes advantage of the properties of certain ram-bling plants to grow upwards up to 20 metres of height, thus simplifying maintenance and creating an easily manageable aesthetic and protective effect. It enhances shading of the walls it covers and triggers photosynthesis and evapotranspiration processes capable of lowering the effects of heat in the environment.

It is possible to write the year of construction on the wall, in order to keep track of the thermal resistance of the wall. |
| 9 | green integrated in the wall | It is made by applying supports to the vertical walls, at a certain distance in order to facilitate air circulation. Inside the frame, some bags of earth are inserted, containing suitable plants. A watering system makes it possible to irrigate yearly and provides for the necessary fertilisation/watering to enrich the ground 3-4 times a year.  It enhances shading of the walls it covers and triggers photosyn-thesis and evapotranspiration processes capable of lowering the effects of heat in the environment.

It is possible to write the year of construction on the wall, in order to keep track of the thermal resistance of the wall. |
| 10 | ventilated with a frame on the wall | It is a coating system laid dry on new buildings or existing constructions, which creates an air chamber between the wall and the coating. The energy benefits are mainly relative to the interior of the building rather than its outside; the ventilated wall lowers the energy load affecting the building in summer (thus lowering air conditioning expenses) and keeps the heat inside the building in winter (thus lowering heating expenses). The specific effect on the heat island can be compared, to a lesser extent, to that of wet surfaces in Japanese experimental buildings. It is possible to write the year of construction on the wall, in order to keep track of the thermal resistance of the wall. |
| 11 | plastered painted with a light colour | This is a standard plastered surface, with a colour which partially reduces absorption of thermal energy due to solar radiation.

It is possible to write the year of construction on the wall, in order to keep track of the thermal resistance of the wall. |
| 12 | plastered painted with a dark colour | This is a standard plastered surface, with a colour which does not particularly reduce absorption of thermal energy dueto solar radiation.

It is possible to write the year of construction on the wall, in order to keep track of the thermal resistance of the wall. |
| 13 | visible bricks | This is a standard surface, with a colour which does not particularly reduce absorption of thermal energy due to solar radiation.

It is possible to write the year of construction on the wall, in order to keep track of the thermal resistance of the wall. |

| HORIZONTAL SURFACES | | |
|---|---|---|
| 14 | "extensive" green | It is a garden, generally not accessible and particularly suitable for covering large areas, whose plants have the function of holding the ground. Maintenance is limited and the watering system is simple. Its thickness is approximately 5-12 cm, with a corresponding overload of about 60-250 kg/m$_2$. It triggers photosynthesis and evapotranspiration processes and can keep part of the rainfalls. In case of accessible green areas endowed with trees and bushes, the right term would be "intensive" green, but it is not taken into consideration by this study, due to its high costs. It is possible to write the year of construction on the wall, in order to keep track of the thermal resistance of the roof. |
| 15 | "cold" | It inhibits the amount of thermal energy going into buildings and its accumulation, with subsequent release into the environment. It is made by applying specific coatings (plasters, resins, paints, ceram-ics, etc.) and does not result in significant overloads. It is mainly applied to flat surfaces in buildings for productive activities. It is possible to write the year of construction on the wall, in order to keep track of the thermal resistance of the roof. |
| 16 | "cold" tiles | It inhibits the amount of thermal energy going into buildings and its accumulation, with subsequent release into the environment. It is made by using specific tiles and does not result in significant over-loads. It makes it possible to also apply the "cold" roof technology to buildings covered with tiles. It is possible to write the year of construction on the wall, in order to keep track of the thermal re-sistance of the roof. |
| 17 | photovoltaic (roof tile or else) | It does not particularly inhibit the amount of thermal energy going into buildings and its accumulation, with subsequent release into the environment. Solar, photovoltaic or transparent roof tiles are an alternative to solar or photovoltaic panels, both in the construction of the roofs of new buildings and in their restructuring, since they make it possible to keep the house cover aesthetically pleasant, without giving up the possibility of using the solar energy for producing electric power or hot water. As an alternative, the use of photovoltaic panels ap-plied to the cover makes it possible to optimise results, since they are placed according to the function of the specific latitude. The effect can be compared to that of a cold roof. It is possible to write the year of construction on the wall, in order to keep track of the thermal resistance of the roof. |
| 18 | tile roof | This is a standard surface typical of houses, made of materials which do not particularly reduce absorption of thermal energy due to solar radiation. It is possible to write the year of construction on the wall, in order to keep track of the thermal resistance of the roof. |
| 19 | light colour flat roof | This is a standard surface typical of production buildings, with a colour which partially reduces absorption of thermal energy due to solar radiation. It is possible to write the year of construction on the wall, in order to keep track of the thermal resistance of the roof. |
| 20 | dark colour flat roof | This is a standard surface typical of production buildings, with a colour which does not particularly reduce absorption of thermal energy due to solar radiation. It is possible to write the year of construction on the wall, in order to keep track of the thermal resistance of the roof. |

## 3.3   Data Entry (Actual State – Planned State)

Having identified the various types of actions characterising the Actual State (SDF) and the Planned State (SDP), the number of square metres relative to each surface typology is to be entered.

Software user's layout

For trees, it is necessary to enter the total number of those already present ("tree" means an element of approximately 3 m of height, with a crown of approximately 8 m², which is deemed to be the equivalent of 4 bushes of about 1.5 m of height and covering a surface of approximately 4 m²).

For the specific calculation of thermal resistance of the walls and roof, it is necessary to enter the year when the walls and roof were built. For walls or a newly-built roof, by selecting the "ex lege" option, the minimum parameters required by law are set, whereas by selecting "per detrazioni", the legal parameters which may benefit from fiscal deductions are set.

In the data entry phase, it is also possible to specify if a rainwater collection tank is present, as defined by the voluntary requirement of the Building Town-Planning Regulations (RUE, No. XXVIII.3.2) of the Municipality of Modena. It is possible to specify the cubic metres of the tank or to calculate the cubic metres necessary in a specific context (by filling in a simple table, based on the provisions in the RUE).

Module implemented in the software for calculations relative to the rainwater collection tank, as defined in the RUE

# 4
# "Rate" Index (Reduction in the Absorption of Thermal Energy)

In days characterised by anticyclonic conditions in the air and strong stability on the ground, a vertical temperature profile is established, with Thermal Inversion (thermal inversion on the ground, countryside). The heat produced by the buildings tends to contrast the vertical thermal inversion, without totally breaking it (thermal inversion in the air, town), creating an air dome whose maximum height corresponds to the zone with the highest concentration of buildings.

The inversion layer creates a barrier which prevents a redistribution of vertical air all over the atmosphere layer available. This creates a heat island, which therefore depends on the season, geographic position of the town and its characteristics.

This "blanket" withholds the heat, thus raising the temperature (for instance, minimum temperatures at night).

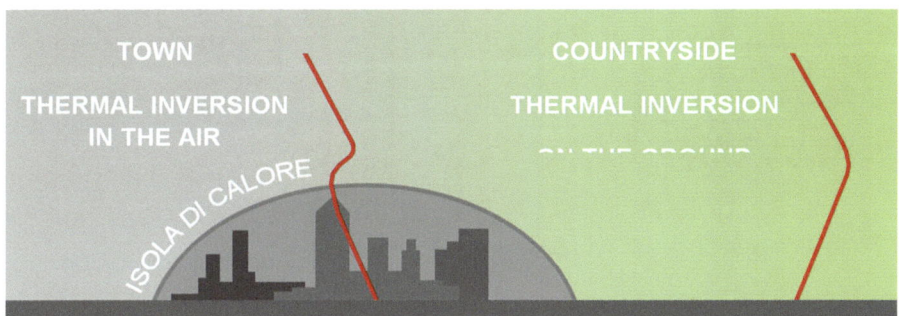

Graph showing the Urban Heat Island phenomenon

The presence of materials endowed with little capacity to reflect solar radiation (solar reflectivity), in conditions of strong solar irradiation, results in high thermal loads on the surface in general. If this surface has a scarce thermal isolation capacity (thermal resistance), it tends to warm up and, in its turn, if it does not have a good capacity to release this heat by irradiation (emissivity), it will take a certain period of time to return to ambient temperature. Continuing to release heat, even during the night, the material will affect the rising of the temperature within the heat island.

The capacity of materials to reflect solar radiation (solar reflectivity), to release the heat absorbed (emissivity) and to isolate thermally (thermal resistance) make it possible, at the same time, to reduce the heat entering buildings (in summer) and reduce thermal dispersion from the inside to the outside (both in summer and in winter), thus making it possible to reduce both summertime consumption due to air conditioning systems and winter consumption due to heating systems, improving energy saving.

## 4.1 Phenomena Analysis Methodology

The physical properties of the materials which make up the building lot (solar reflectivity, thermal resistance and emissivity) can be combined along with incident solar radiation in order to obtain a single value expressing the thermal energy absorbed [kWh/m$^2$y].

Summary of the methodological approach

For each type of floor, wall and roof (reported in the table above), the respective identification parameters were obtained (solar reflectivity, thermal resistance and emissivity).

In order to take into consideration the various irradiation conditions of the roofs and walls, energy values per unit of surface were used, obtained by applying norm UNI 10349 (for walls this value was attributed as a function of the azimuth angle). For assessing irradiation of yards, starting from the energy associated to the horizontal surface, the amount of energy lost due to the shading caused by the "equivalent" building lot was taken into consideration. For this calculation, a calculation file already in existence is used,[6] based on the UNI norms in force.

---

[6] "SOLE – Stima Ombreggiamento Locale Edifici" – Dott. Ing. Giulio de Simone – Dipartimento di Ingegneria Meccanica – Università degli studi di Roma "Tor Vergata" – Excel file.

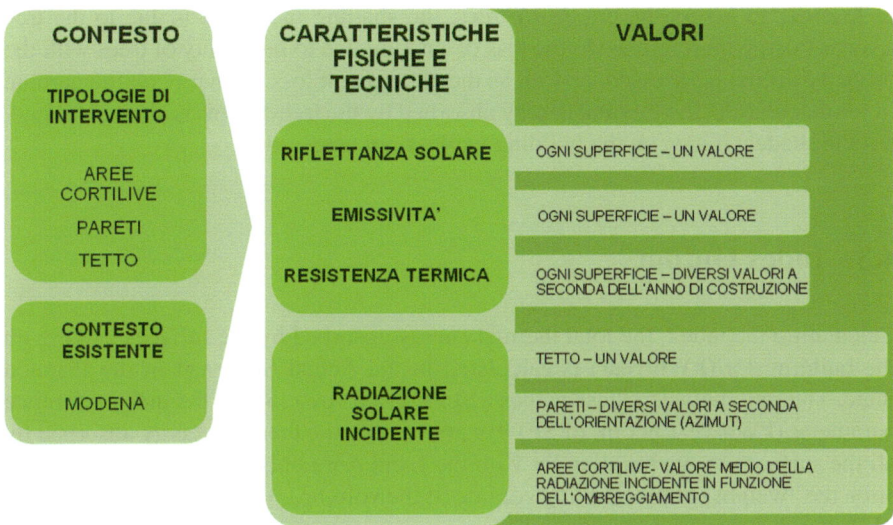

Summary of the methodological approach

> *Note: deterioration of the technical characteristics of materials and degradation of plant elements are not taken into consideration. For instance, in case of building defects, ageing of surfaces, lack of watering of green areas or increased volume of trees, the typical parameters of each single surface and, as a consequence, their relative indexes might vary.*
>
> *Note: the yards simulated by the "equivalent" lot do not take into consideration the real position of the various grounds; the average value of radiation on the yard is associated with them.*

## 4.2   Simulation of Plants

For calculating the thermal energy absorbed by trees or bushes, by using parametric coefficients, the part of solar radiation used for the photosynthesis (about 5 %) and which is released as latent heat in evapotranspiration processes (about 45 %) were taken into consideration.

In order to assess the effects of trees, the shading in the yards due to the tree crowns (also with reference to the loss of shading due to proximity of trees with the property limits) is assessed, as well as the consequent loss of energy on the garden. In the calculation, the thermal energy absorbed by the tree crowns is also calculated, but the shades, if any, on the building walls is not assessed.

## 4.3 Index Output

Calculating the index, the total thermal energy absorbed by the surfaces making up the building lot (kWh/year) is calculated in the Actual State and in the Planned State. The ratio between these values and the Land Area (SF) is the index reference parameter (kWh/year per $m_2$ of SF). By associating the thermal energy absorbed by all the surfaces in the building lot with the Land Area only, we take into consideration the "weight" of the building volume (if the volume grows, the index grows, the SF being the same).

The output value is the measure of specific "performances" of the building lot and can be highlighted depending on various features.

- the class of the planned scenario (SDP) identifies the quality of the action;
- by processing the different classes of energy performance which output values are associated with, the number of reduced classes between the initial scenario (SDF) and the planned scenario (SDP) highlights the improvements achieved;
- the total reduction of thermal energy absorbed (kWh/year) in the planned scenario (SDP), in comparison with the initial scenario (SDF) highlights the extent of the improvements with reference to the building lot size.

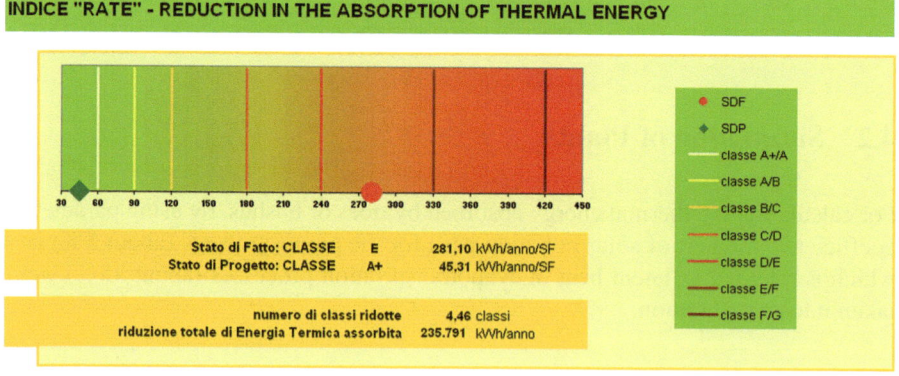

Software output

*Note: air circulation is not taken into consideration, therefore the natural movements of the air and the "barriers" which might be caused by the positions of buildings and trees are not assessed. For these issues, see the relevant guidelines.*

*Note: the canyon effect due to the presence of adjacent buildings is not taken into consideration, therefore it has to be dealt with, from a qualitative point of view, by referring to relevant guidelines.*

*Note: the energy performance classes used so far are approximate, they will have to be adjusted by analysing various study cases. From a merely qualitative point of view, the class typology used draws inspiration from the one used to classify energy performances of buildings (there is no relevance from a quantitative point of view).*

*Note: the assessments were made with reference to the yearly solar radiation, but the software is also set to calculate the summer solar radiation. In order to enhance improvements even more, in terms of energy consumption reduction (relative to the yearly solar radiation) or reduction of the heat island effect (more relative to the summer solar radiation), some further software implementations are being designed.*

# 5
# "Hyper" Index (Hydraulic's Permeability and Elements of Retention)

A building lot capable of ensuring proper disposal of surface water and rainwater, avoiding their stagnation and promoting a controlled outflow, contributes to mitigation of floods in urban area and non-urban areas, thus helping to reduce the hydraulic risk. The technical characteristics of building lots, necessary to meet these conditions, are defined in the municipal Building Town-Planning Regulations (RUE, binding requirement No. XXVIII.3.14).

The RUE define various principles to manage the hydraulic risk in the territory. For defining these principles, several data are necessary: the size of the surface of the building lot analysed, the fraction of the total area deemed to be impermeable, the hydraulic load class (which makes it possible to differentiate the various basins as "critical" and "non-critical") and the type of action (new building or restoration of areas already urbanised).

The applicable principles require a maximum value to be applied for the outgoing extent from the building lot in the planned scenario; they may be the following:

- principle of controlled hydraulic increase: it allows a possible specific extent increase of up to 100 %, 50 % or 30 % of the specific value of the outflow of the area in question, as it was before the start of the action;
- hydraulic invariance principle: it keeps the specific value of the outflow of the area in question unaltered, as it was before the start of the action;
- principle of hydraulic attenuation: it requires a specific outflow extent reduction of at least 30 %, 40 % or 50 % of the specific value of the outflow of the area in question, as it was before the start of the action.

All the principles are based on the specific value of the outflow of the area in question unaltered, as it was before the start of the action. This value is usually calculated when the hydraulic sheet of the building lot in question is drafted and requires the number of square metres of the various yards and roofs to be specified. These surfaces, associated with specific outflow coefficients and the hydrological parameters defined in the RUE, make it possible to determine the size of the rainwater drainage network (with a return time of 20 years).

For calculating the index, a calculation algorithm was implemented which, on the basis of the principle of hydraulic risk management to be applied in the building lot, defines the maximum value allowed for the outflow extent due to the SDP (Qsdp,max). If the outflow extent generated by the input action typologies for characterising the SDP (Qsdp) exceeds this value, the index requires a check of this extent and imposes the Qsdp=Qsdp,max equation, envisaging the presence of a lamination tank in the line where exceeding extents flow, then it calculates the volume of the reservoir. The tank is calculated on the basis of a return time of 50 or 100 years (depending on what is envisaged by the RUE).

If the user has chosen to prepare a rainwater collection tank, and the index has calculated that it is necessary to also envisage a lamination tank, then the square metres of the rainwater tank are deducted from those of the tank.

## 5.1  Index Output

In the calculation of the index, the extent (l/s) in the Actual State and in the Planned State is defined. The ratio between these values and the Land Area (SF) is the index reference parameter (l/s per $m_2$ of SF).

Since the hydraulic risk management principles "calibrate" the plan extent (Qsdp) on the basis of the extent in the actual state and the maximum plan extent acceptable (Qsdp,max), the classes defined for assessing the output have been associated to the Qsdp,max values required by the various principles. In this case, unlike what was done for the "RATE" index, the output produced by the Planned State was not compared with that of the Actual State, but with the maximum one in the Planned State (determined by the principle applied).

The output value is the measure of specific "performances" of the building lot and can be highlighted depending on various features.

• the planned scenario (SDP) class identifies the quality of the action with reference to what is already required by the principle applied (SDPmax);
• by processing different classes of energy performance output values are associated with, the number of reduced classes between the planned scenario (SDP) and the maximum planned scenario (SDPmax) highlights the improvements achieved;
• the total reduction of extent (l/s) in the planned scenario (SDP), in comparison with the maximum planned scenario (SDPmax) highlights the extent of the improvements with reference to the building lot size.

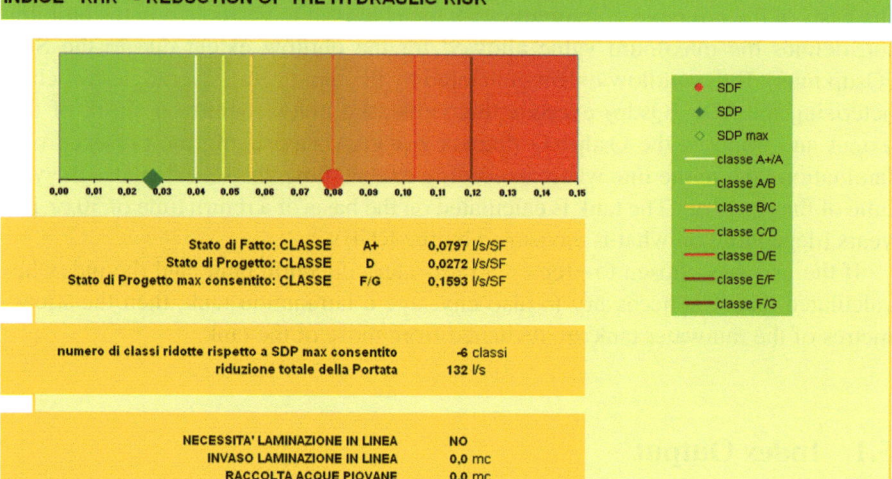

Software output

*Note: energy performance classes defined in this way are variable, since they are associated with values depending on the extent of the Actual State in the specific building lot analysed. A different approach from the one applied for the "RATE" index is therefore used. For the validation of this approach, see further analyses.*

# 6
# Indexes as Guidelines for the Planning Phase

The analysis of the indexes already used by other administrations, the analysis of the possible solutions to implement in yards and to cover buildings, and the analysis of the respective physical and technical characteristics have made it possible to obtain a complete and rational list of types of action, to be included in the experimental indexes. This list is the starting point for the various technical criteria explained in the previous chapters for calculating the output of the indexes, and it has been designed in such a way as to guide the planning in the implementation of new buildings and in the upgrading of existing buildings. By defining the characteristics of the existing context and the possible planning solutions, planners receive output which enables them to check the overall improvements achieved (in terms of reduction of the RATE or HYPER index).

If planners intend to modify the overall improvements, they can change the combination of types of planned actions and obtain new outputs, to be compared with the previous ones, in order to refine the planning and increase its environmental value.

The types of actions included in the indexes represent, therefore, a reasonable and functional synthesis of planning guidelines, which can provide both support to the architectural planning and an assessment of the impacts generated immediately and with suitable precision.

## 6.1   Indexes and Economic Assessment of Planning Solutions

As an integration to the planning support tools mentioned above, the indexes also implement an assessment of the parametric costs of the various types of action envisaged. By applying the same approach used for the assessment of impacts produced by the planning solution envisaged by the planner, a dynamic table has been designed, relative to the parametric costs of planning actions.

In this way, planners check the types of action chosen which might be applied in order to find the right balance between costs and resources. Moreover the planner is able to verify the quality/cost ratio of each type of action.

In order to promote such analyses, the software shows the synthesis of all the information relative to the two indicators and the parametric costs. By reading the information relative to each type of action, it is possible to easily check and compare the impacts caused and the relative costs.

# 7
# Further Guidelines Supporting
# the Planning Phase

In order to design the indexes, the need to implement a greater number of potentials, for calculating the environmental value of the building lot while trying to keep its use suitably simple, was discussed for a long time. While designing the software, it was decided to make some simplifications and not to implement certain physical aspects. The indexes and calculation methods implemented could be improved even more, in order to remove the simplification applied and refine the processing, also including all the necessary physical issues; in this way, though, the immediate applicability of a town-planning index would be lost and you would have a calculation software programme with an enormous potential, but only to be used by suitably trained experienced users. Although at the moment it is not possible to exclude such a future development, for the time being the structure of the indexes has willingly been processed in such a way as to make it as simple and self-explanatory as possible. Although the indexes require much more complex processes than those developed for similar indexes already in existence, in the data entry phase the approach has not been altered. Like in the other indexes, observations are put forward about a building lot, the materials present in it and their specific sizes. This approach identifies the building lot as an independent item, without putting it into the existing urban context.

This approach is a non-banal simplification of the context analysed, which has two limits:

- since the context in which the lot lies is not known, it is hypothesized that the solar irradiation conditions in the lot are not affected by elements outside the lot itself, like buildings, rows of trees, etc. This simplification makes it possible to apply the same theoretical maximum irradiation to each lot and, as a consequence, provides output values comparable among the various lots. The negative side of this approach is not seeing a lot within its specific context; for instance, although the indexes can assess the positive effect due to photovoltaic panels, installing them on a building would make no sense if that building stood in the shade of a skyscraper in the hours of the day when sun exposure is maximum;
- although the context in which the building lot lies is not known, the indexes will merely be used for analysing the impacts caused by the lot "within the lot" and do not allow to check the relationship between impacts caused by the lot and the surrounding lots. Therefore, the indexes cannot simulate how proximity to other

194

buildings, roads or other structural elements of the town may affect the indicators taken into consideration on the whole.

These issues show that the use of the indexes cannot ignore knowledge of the territory in which a general action takes place. The data produced by the indexes must be supported by further assessment elements, to be processed on the basis of a careful analysis of the urban context.

As for the Artisan Village, these elements are represented by the guidelines envisaged *ad hoc* within the framework of the EU "UHI Project", in which the Artisan Village Upgrade Plan is a pilot study case. These guidelines have revealed the urban and building characteristics, the main environmental issues, morphological, climatic and architectural aspects of the town structure already in existence and the possible planning features within the framework of the Artisan Village.

By processing this content even further, the experimental indexes design phase will be completed, and a list of guidelines will be drafted to support the planning. By taking into consideration the various territorial contexts within the Village (for instance: lot on the side of a street, lot inside other lots with buildings of the same height, etc.), these guidelines will be useful for the planning phase.

For example, in this way such phenomena as the Urban Canyon and the air circulation dynamics will be taken into consideration and, before deciding what planning actions to enter into the software programme to calculate the indexes, it will be possible to have a further tool for analyses, which can help planners to choose the best types of actions.

Later on, the main features which can affect the planning phase (presence of roads, presence of buildings, presence of green barriers, etc.) will be extracted from these guidelines, and a list of best practices will be drafted, to be applied in the planning of these features. In this way, by following and implementing the experience of the Green Points of Malmö, a list of actions will be drafted, which planners will have to respect in the planning of a building lot, having checked the surrounding context.

# 8
# Incentives, Bonuses and Awards

The analysis of the urban indexes in existence, which has served as a basis for this study, initially aimed to define a measurement tool such as to quantify the improvements made by high-quality planning solutions, such as green walls or green roofs, and such as to provide a calculation method for assessing the awards to attribute to these types of actions, taking into consideration the environmental benefits brought about and the relative costs.

The improvement of the indexes, achieved by defining the two experimental indexes, also thanks to an assessment of the parametric costs of single actions, has resulted in a tool capable of describing the quality/price ratio of a planning solution.

Within the framework of the Artisan Village Upgrade Plan, the incentive procedure envisages the granting of discounts for upgrade expenses. These discounts will be

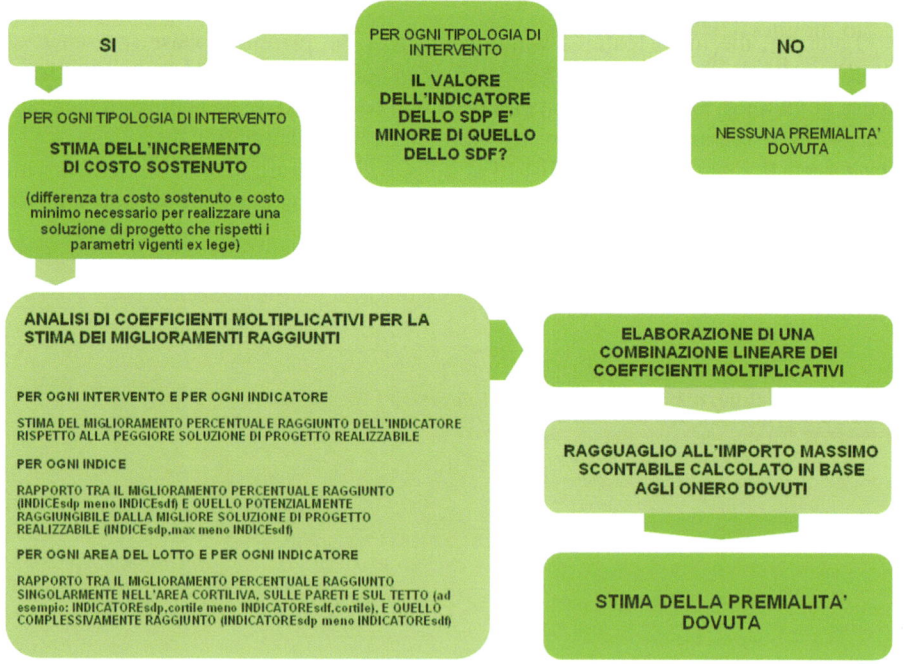

Calculation of awards – methodological setting

calculated on the basis of the quality achieved by planning actions which, in its turn, will be assessed on the basis of bonuses and awards defined for each type of action.

For calculating awards, the methodological setting shown in the table above is used. Also for this calculation, the software programme has been designed in order to automatically calculate all the various passages (except for the calculation of charges, which might in any case be implemented).

For calculating any bonuses which might result in further proportional discounts or one-off discounts, issues connected with the architectural and town-planning value of the actions implemented will be taken into consideration. Through these bonuses, the administrations will be able to manage the planning phase, within the framework of the Plan, towards solutions which do not just enhance the value of a single lot, but also result in a benefit for the whole district. These bonuses, which are still being designed, will focus on issues relative to the improvement of the landscape publicly visible, as it is envisaged by the Green Factor in Seattle. Along certain streets and in some other specific contexts some bonuses will be determined, such as to allow the administration to promote a coordinated and rational evolution of the urban landscape.

## 8.1   Limits to the Granting of Bonuses and Awards

For attributing the awards and bonuses, the subject promoting an upgrading action will have to concretely prove the will to implement the planning actions used for calculating the indexes.

General setting of procedures to attribute awards and bonuses

The check of the indexes and further guidelines enable planners to assess the actions which may ensure the best environmental value, the least costs and to assess the relative awards and bonuses. This phase precedes the planning phase and must be verified when the plan is approved. The municipal administration must require some suitable guarantees, in order to avoid the risk that certain actions envisaged by the project not be implemented or cease due to a lack of suitable ordinary maintenance works. In this respect, for each type of action, some necessary and sufficient binding characteristics will be defined in order to grant awards or bonuses. For instance, for the construction of yards, it will be necessary to make an automatic watering system, whereas for green walls or roofs, it will be necessary to prove that a maintenance contract of suitable duration has been signed, etc.

By means of a check system based on sample verifications, the administration will verify the upgraded areas, in order to ensure the preservation of the promoted features in the long run.

# References

**List of References**

"The Biotope Area Facto ras an Ecological Parameter" Principles for its Determination and identification of the target - Berlin, December 1990.

Grabs Expert Paper 6 - The Green Space Factor and the Green Points Sistem - 2011

"Manuale d'uso del foglio di lavoro Excell per il calcolo del RIE" Comune di Bolzano - Ufficio Tutela dei Beni Ambientali

"Recupero edilizio e bioclimatica" – Marco Sala – Sistemi Editoriali

"Spazi verdi urbani" – Sistemi Editoriali

"Gli strumenti normativi inerenti l'uso del verde in copertura e in facciata" Università Iuav di Venezia – Valentina Santi, Valeria Tatano – 2007

"Gestione della qualità dell'aria", McGraw-Hill, 2001

"Superfici vegetali applicate all'involucro edilizio per il controllo microclimatico dell'ambiente costruttivo", Dottorato in "Tecnologia dell'Architettura", Università degli studi di Ferrara, Dott. Olivieri Michele, 2009

"La pianificazione delle infrastrutture verdi nelle strategie di adattamento ai cambiamenti climatici in ambito urbano", Dottorato in "Analisi, pianificazione e gestione integrata del territorio", Università degli studi di Catania, Dott. Anna Maria Caruso


**Web Sites**

http://www.stadtentwicklung.berlin.de/umwelt/landschaftsplanung/bff/index_en. shtml

http://www.malmo.se/English/Sustainable-City-Development/Green-and-Blue. html

https://www.seattle.gov/dpd/cityplanning/completeprojectslist/greenfactor/ whatwhy/

http://www.comune.bolzano.it/urb_context02.jsp?ID_LINK=512&id_context= 4663&page=11

http://urp.comune.bologna.it/portaleterritorio/portaleterritorio.nsf/a3843d2869cb2
    055c1256e63003d8c4e/200cfbd63f33a6aac1257671004e6018?OpenDocument
http://www.eelab.unimore.it/site/home.html
http://www.solarity.enea.it

## Software

"SOLE – Stima Ombreggiamento Locale Edifici" – Dott. Ing. Giulio de Simone –
    Dipartimento di Ingegneria Meccanica – Università degli studi di Roma "Tor
    Vergata" – Excel file.

# Chapter 7
# The Urban Corridor of Venice and The Case of Padua

**Marco Noro, Renato Lazzarin, and Filippo Busato**

**Abstract** Urban Heat Island effect was widely studied in large cities around the world, more rarely in medium size ones. The chapter reports on the study of the UHI phenomenon in Padua, a medium size city of the North-East of Italy, one of the most industrialized and developed parts of the country. Experimental measurements were carried out during 2012 summer, recording the main thermo-hygrometric variables (dry-bulb temperature, relative humidity, global solar radiation) by a mobile survey along an exact path crossing different zones of the city area: urban, sub-urban and rural. The analysis of the data highlights the presence of UHI effect with different magnitudes in function of the zone of the city. In the city centre, an historical zone, the effect was up to 7 °C. In the meantime, some measurements in situ were carried out in order to evaluate other thermal comfort indexes rather than air temperature and humidity only: wind velocity and mean radiant temperature (besides the other meteorological variables) in some characteristic sites of the city area like historic centre, high and low density populated residential zones, industrial zone, rural zone, were recorded. In particular, a very famous square of the city (Prato della Valle) was analysed: it can be considered representative of the phenomenon because of the size and the very different characteristics from the UHI effect point of view. RayMan simulation model was used to calculate some outdoor comfort indexes and Envimet model was further used to evaluate the effect of some mitigation strategies in characteristic sites of the city.

**Keywords** Urban heat island • Padua • RayMan • ENVImet • Outdoor thermal comfort

M. Noro (✉) • R. Lazzarin • F. Busato
Department of Management and Engineering (DTG), University of Padova,
Stradella San Nicola 3 – 36100, Vicenza (VI), Italy
e-mail: marco.noro@unipd.it

© The Author(s) 2016                                                        201
F. Musco (ed.), *Counteracting Urban Heat Island Effects in a Global Climate Change Scenario*, DOI 10.1007/978-3-319-10425-6_7

## 7.1    Urban and Environmental Framework

Veneto Region climate is characterized by different issues: the region is between 44.9° and 46.7° North latitude and it is between central Europe (where western and Atlantic currents are predominant) and southern Europe (where subtropical anticyclones and Mediterranean Sea influences are predominant). The presence of the Po Valley, the Adriatic Sea, the Alpi Mountains and the Garda Lake are further factors that affect the Veneto Region climate. In a simple way, the climate is characterized by two main regions: the alpine zone (central Europe mountain climate) and the Po Valley zone (continental climate), with two sub-regions with milder climate (Garda Lake and Adriatic Sea zones).

Padua stands on the Bacchiglione River, 40 km West of Venice and 29 km South-East of Vicenza, inside the eastern part of the Po Valley; it has a humid subtropical climate (Köppen climate classification Cfa) with cold winters and hot summers, frequently associated with air stagnation (respectively fog and sultriness). The city has an area of 93 $km^2$, a mean altitude of 12 m a.s.l. and a population of 214,000 (as of 2011); the population density is quite high for Italy, 2300 inhabitants/$km^2$, the higher in Veneto Region and not far from that of Rome. The city is picturesque, with a dense network of arcaded streets opening into large municipal squares ("piazze"), and many bridges crossing the various branches of the Bacchiglione River, which once surrounded the ancient walls like a moat. The industrial area of Padua was created in 1946, in the eastern part of the city; now it is one of the biggest industrial zones in Europe, having an area of 11 million $m^2$. Here the main offices of 1300 industries employing 50,000 people are located. It can be of some interest to give some feature (data of 2010 (ARPAV 2011)) concerning the situation of the energy consumption of the territory. In terms of subdivision in electricity, thermal and transport consumption, no data at Province level are available. On the regional level, the respective values are: 2724, 6042 and 3088 ktep. The Province of Padua consumes 5500 GWh of electricity (19 % on the Veneto Region basis, 1.8 % on Italy basis), a half is consumed by industry sector and 28 % by tertiary sector. Concerning the residential sector only, Padua city per-capita consumption of electricity is 1300 kWh $year^{-1}$ while the consumption of natural gas is around 700 $m^3$ $year^{-1}$ (the averages of all Italian capitals of province are respectively 1200 kWh $year^{-1}$ and 390 $m^3$ $year^{-1}$, data 2011 (ISTAT 2012)).

City of Padua is quite sensitive to initiatives concerning the protection of the environment, human health and energy saving. During the last decade, the Municipality has been involved in different European Projects (Life Siam – n. LIFE04 ENV/IT/000524, Life "South-EU Urban ENVIPLANS", Belief – Building in Europe local intelligent energy forums, LIFE-PARFUM); initiatives like the arrangement of the Energy Plan and the Climate Plan have been adopted in order to give practical tools to reduce energy consumption and to introduce adaptation and mitigation strategies to climate change.

All these climatic and environmental characteristics of the territory can put forward the interest in the study of the heat island effect of the city of Padua.

## 7.2 Pilot Areas Identification Methodology

The first phase of the study was carried out by choosing five areas of interest within the city of Padua, in which to make the analysis of urban planning and surveys on urban heat island. Such areas were selected on the basis of the location with respect to a transect north-west to south-east (Fig. 7.1) and compared to urban characteristics: historic centre, urban mixed, high density residential, low-density residential, industrial. The selection of the pilot area, between the five hypothesis, were done considering useful to study the typical settlement of the central Veneto Region, with the aim to replicate in other areas the results and the mitigation techniques studied.

## 7.3 UHI Phenomena in the Pilot Area

### 7.3.1 Mobile Surveys

The goal of the field survey was a first characterization the UHI phenomenon in a medium size city like Padua (Noro et al. 2015). Within the framework of the "UHI" Project the Authors used mobile surveys with the measurement instrumentation installed on a vehicle running through the territory from the rural to the urban zone, in order to log data continuously (Fig. 7.1).

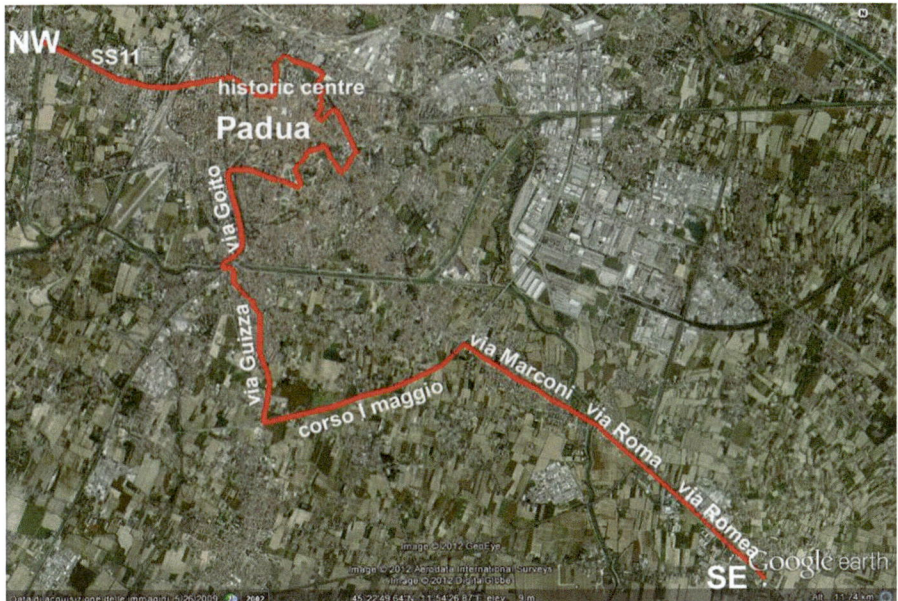

**Fig. 7.1** The path along which the mobile surveys were conducted, going from the NW to the SE of Padua and return (Google Earth)

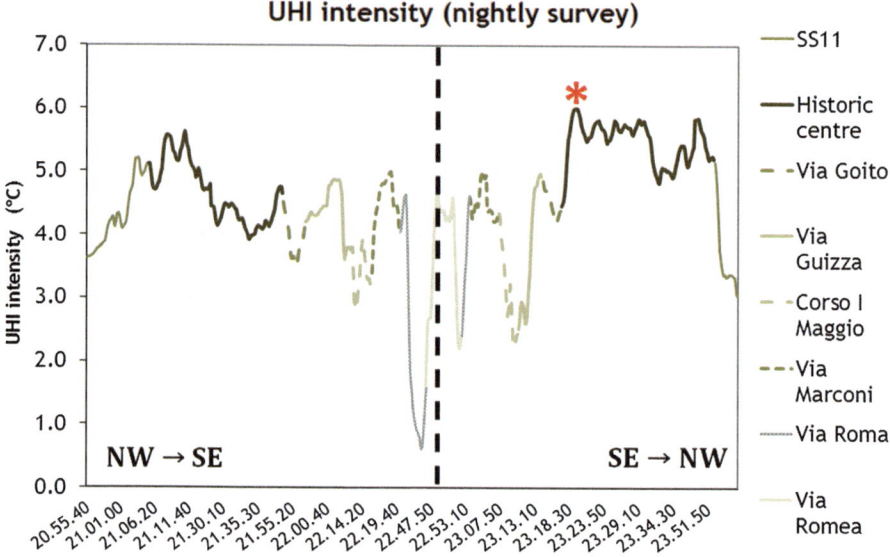

**Fig. 7.2** UHI intensity during night-time mobile survey of the 30th July, 2012. The vertical broken line indicates the U-turn during the survey (the names of the streets refer to Fig. 7.1). The *red star* represents Prato della Valle

Experimental data were logged by mobile surveys from 26 July 2012 to 9 August 2012, some of them in double sessions: day-time (during late afternoon) and night-time session (between 1 and 4 h after the sunset in order to investigate the phenomenon during its potentially maximum intensity). Dry-bulb air temperature, relative humidity and solar global radiation on the horizontal, with a time step of 5 s, were the main variables measured by the mobile station equipped on a vehicle. UHI intensity was determined by the difference between mobile measured air temperature and the value recorded at the same time by the reference ARPAV (Regional Agency for Environment Protection in Veneto) rural fixed meteorological station of Legnaro (rural zone, 8.5 km far from the city centre).

Details about the path, the instrumentation used and the measurement procedure are available in references (Noro et al. 2015). Here some very brief results are reported. Concerning air temperature, urban heat island intensity would be mostly present during night-time in the range of 3–6 °C, while the day-time would show a much less significant effect (1.2–2 °C, Fig. 7.2). Only the "return" (SE to NW) path was considered for the UHI intensity measurement because this was the most precautionary path, both in spatial terms –moving from the countryside towards the city centre – and in temporal terms – as the air was cooling. Considering the night-time surveys, the minimum UHI intensity was recorded in a lateral street (an unpaved dirt patch road in the countryside) of via Roma ("via" means street in Italian), crossed only during the NW to SE path. The maximum UHI intensity was

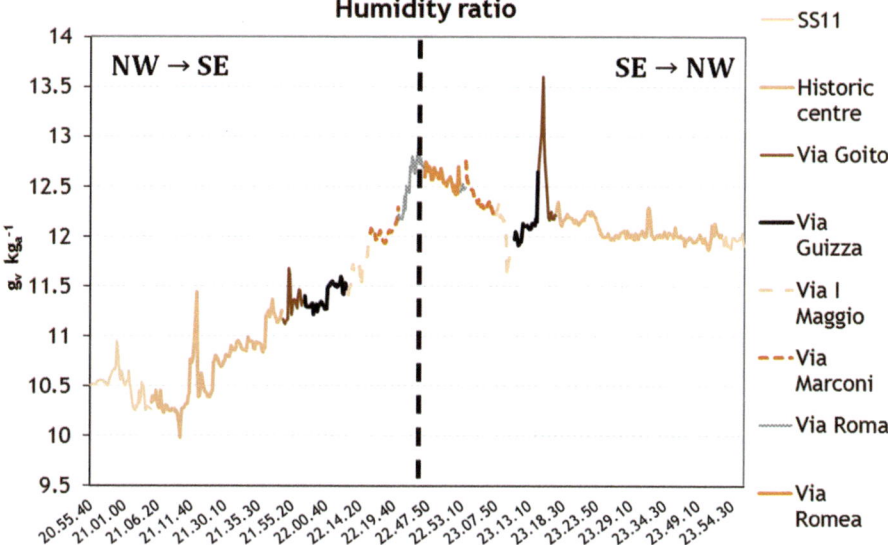

**Fig. 7.3** Humidity ratio during the night-time mobile survey of the 30th July, 2012. The vertical broken line indicates the U-turn during the survey (the names of the streets refer to Fig. 7.1)

recorded across the narrow streets (high H/W ratio) of the historic centre. As depicted in Fig. 7.2, going along via Guizza (a long road that traverse the sub-urban to inside of the city) UHI intensity would always increase, mainly due to the increase of the H/W ratio and the increase of the vertical and horizontal impervious surfaces facing the road. The subsequent stretch of path (via Goito) was along San Gregorio Canal, so the evaporative cooling effect of water would lead to a decrease of the temperature.

The graphs concerning humidity ratio (more significant with respect to relative humidity because it is independent on air temperature) show some particular situation along the path, with a presence of water and/or green areas, but the presence of a gradient directed from the countryside to the city centre (from 12.8 to 12 $g_v$ $kg_a^{-1}$ in Fig. 7.3) is apparent. Differences in humidity ratio between rural and urban zones are more noticeable during the daily runs because of the higher evapotranspiration effect due to the presence of the solar radiation.

Prato della Valle, in the centre of the city, is a very famous and picturesque circle and the second largest (about 90,000 m$^2$) urban square in Europe (the first being the Red Square of Moscow), with a lot of water flows and green zones, so a decrease in nightly temperature would be expected, while a maximum shown as the red star in the historic centre in Fig. 7.2 was measured. As a matter of fact the path passes along the perimeter and not inside the circle: the spatial influence of the green would be limited and a deeper investigation within this zone will be presented in the next section.

**Table 7.1** Comparison between the three outdoor comfort indexes. Values between brackets in PMV column refer to the correspondent values of the other indexes

| PMV | PET (°C) | SET* (°C) | Sensation |
|---|---|---|---|
| >3 (3.5) | 41 | >37.5 | Very hot, great discomfort |
| 2 – 3 (2.5) | 35 | 34.5 – 37.5 | Hot, very unaccepTab. |
| 1 – 2 (1.5) | 29 | 30 – 34.5 | Warm, uncomforTab., unaccepTab. |
| 0.5 – 1 (0.5) | 23 | 25.6 – 30 | Slightly warm, slightly unaccepTab. |
| −0.5 – 0.5 | | 22.2 – 25.6 | ComforTab., accepTab. |
| −1 – −0.5 (−0.5) | 18 | 17.5 – 22.2 | Slightly cool, slightly unaccepTab. |
| −2 – −1 (−1.5) | 13 | 14.5 – 17.5 | Cool, unaccepTab. |
| −3 – −2 (−2.5) | 8 | 10 – 14.5 | Cold, very unaccepTab. |
| < −3 (−3.5) | 4 | <10 | Very cold, great discomfort |

## 7.3.2 In Situ Measurements

Besides the mobile surveys, in situ measurements were performed in some characteristic sites of the city area along the path, in order to measure air temperature and humidity, wind velocity and mean radiant temperature. Consequently these data were processed using the RayMan model (Matzarakis et al. 2007, 2010) in order to calculate some outdoor thermal comfort indexes: the Predicted Mean Vote (PMV), the Physiological Equivalent Temperature (PET) and the new Standard Effective Temperature (SET*). RayMan model is a simulation tool for the estimation of radiation fluxes and mean radiant temperature $(T_{mr})$ and other variables, compatible with Windows® that can analyse complex urban structures and other environments. The model requires only basic meteorological data (air temperature, air humidity and wind speed) for the simulation of radiation flux densities and common thermal indices for the thermal human-bioclimate.

Table 7.1 reports a comparison of the three thermal comfort index values. All these indexes were calculated by the mean radiant temperature $T_{mr}$, a physical quantity that accounts for the human-biometeorological influence of short – and long – wave radiation flux densities (Jendritzky and Nubler 1981). While PET and SET* have been specifically defined to assess outdoor thermal comfort (Matzarakis et al. 2007, 2010); (Mayer 1993); (Gagge et al. 1986); (Höppe 1999); (Mayer and Höppe 1987), the use of PMV is not universally recognized. Also the ISO 7730 standard focused on the use of PMV only as indoor thermal comfort index. Nevertheless some authors applied the use of PMV to outdoor environment (Matzarakis et al. 2007, 2010); (Berkovic et al. 2012); (Honjo 2009); (Jendritzky and Nubler 1981); (Thorsson et al. 2004). In this study the mean radiant temperature was measured using the globothermometer during night-time measurements only, when UHI intensity was higher. For this reason the use of PMV for outdoor thermal comfort assessment was judged to be suiTab. for this study.

Table 7.2, 7.3, 7.4, 7.5, and 7.6 provide the description of the five sites selected for the in situ measurements, representative of very different zones of the fabric of

**Table 7.2**  Data obtained in measurement sessions in via Rinaldi (historic centre) (see also Fig. 7.1). The "X" refer to the point of measurement. Thermal comfort indexes refer to a person with summer clothing (0.5 clo) and slight activity level (80 W above the basal metabolism) (Google Earth-RayMan)

Via Rinaldi (H/W = 1.8 - SVF = 0.18)

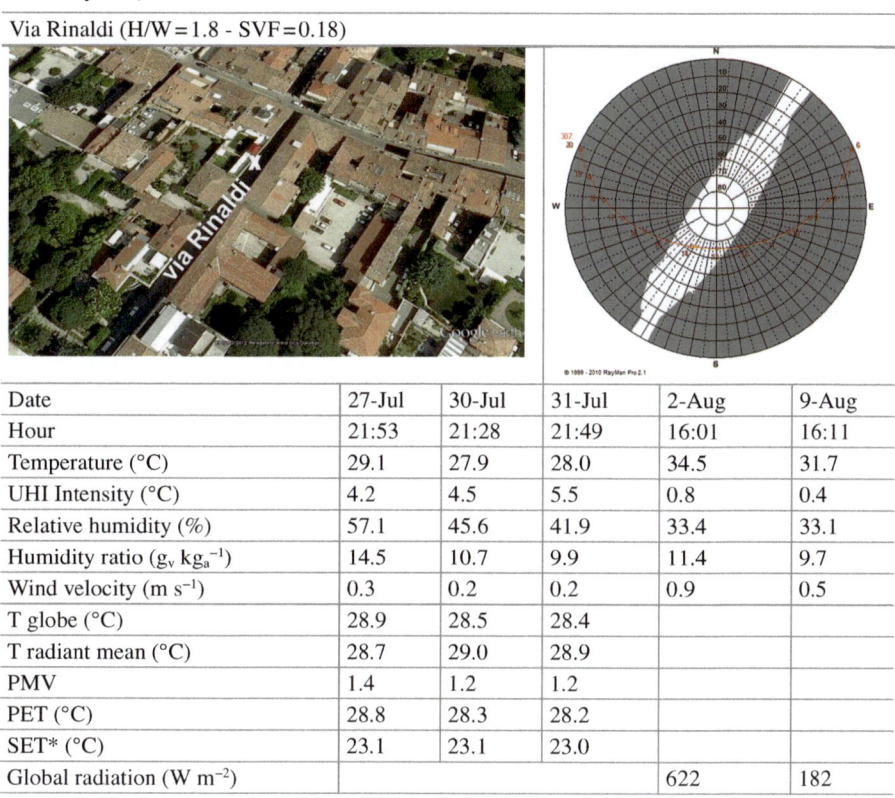

| Date | 27-Jul | 30-Jul | 31-Jul | 2-Aug | 9-Aug |
|---|---|---|---|---|---|
| Hour | 21:53 | 21:28 | 21:49 | 16:01 | 16:11 |
| Temperature (°C) | 29.1 | 27.9 | 28.0 | 34.5 | 31.7 |
| UHI Intensity (°C) | 4.2 | 4.5 | 5.5 | 0.8 | 0.4 |
| Relative humidity (%) | 57.1 | 45.6 | 41.9 | 33.4 | 33.1 |
| Humidity ratio ($g_v$ $kg_a^{-1}$) | 14.5 | 10.7 | 9.9 | 11.4 | 9.7 |
| Wind velocity (m s$^{-1}$) | 0.3 | 0.2 | 0.2 | 0.9 | 0.5 |
| T globe (°C) | 28.9 | 28.5 | 28.4 | | |
| T radiant mean (°C) | 28.7 | 29.0 | 28.9 | | |
| PMV | 1.4 | 1.2 | 1.2 | | |
| PET (°C) | 28.8 | 28.3 | 28.2 | | |
| SET* (°C) | 23.1 | 23.1 | 23.0 | | |
| Global radiation (W m$^{-2}$) | | | | 622 | 182 |

the city. The pictures show the main uses of territory, all the data measured during experimental sessions were reported in the Tab.s; the sites differ for the decreasing H/W ratio (height of buildings to width of street) and for the increasing SVF (sky view factor).

Via Rinaldi is an urban canyon in the historic centre of the city (Table 7.2). Street pavement was constructed by porphyry, buildings by brick walls and tiled roofs; no trees are present. It is characterized by a relative high H/W ratio (street width 5.5 m, buildings height between 8 and 12 m) and a small SVF (calculated by the RayMan model by position and height of buildings). The latter hampers the nightly cooling of surfaces causing the mean radiant temperatures to be higher or similar to air temperatures (respectively around 29 °C and between 28 and 29 °C). UHI intensity was always more than 4 °C during the measurement sessions. Such environmental

**Table 7.3** Data obtained in measurement sessions in via Pindemonte (see also Fig. 7.1). The "X" refer to the point of measurement. Thermal comfort indexes refer to a person with summer clothing (0.5 clo) and slight activity level (80 W above the basal metabolism) (Google Earth-RayMan)

Via Pindemonte (H/W = 1.2 – SVF = 0.29)

 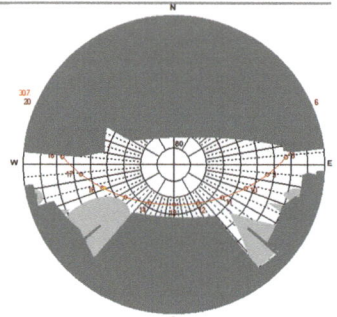

| Date | 27-Jul | 30-Jul | 2-Aug |
|---|---|---|---|
| Hour | 22:37 | 22:09 | 16:28 |
| Temperature (°C) | 27.7 | 27.2 | 34.2 |
| UHI Intensity (°C) | 4.2 | 4.8 | 0.8 |
| Relative humidity (%) | 56.6 | 50.6 | 33.0 |
| Humidity ratio ($g_v\ kg_a^{-1}$) | 13.2 | 11.4 | 11.2 |
| Wind velocity (m s$^{-1}$) | 0.6 | 0.3 | 2.1 |
| T globe (°C) | 27.9 | 27.5 | |
| T radiant mean (°C) | 28.2 | 27.9 | |
| PMV | 0.9 | 0.9 | |
| PET (°C) | 26.8 | 27.0 | |
| SET* (°C) | 21.1 | 21.7 | |
| Global radiation (W m$^{-2}$) | | | 548 |

conditions determines, for a person with summer clothing (0.5 clo) and slight level of activity (metabolic rate 80 W), slight thermal stress (PMV and PET) or thermal comfort (SET*). During day-time UHI intensity was small, about 0.4–0.8 °C. Lower values of UHI could be measured during the morning because direct solar radiation would reach the point of measurement only during central hours of the day (Table 7.2).

Two residential zones were studied: via Pindemonte (higher population density) (Table 7.3) and via San Basilio (lower population density) (Table 7.4). The first is a lateral street of via Guizza (Fig. 7.1) with high apartment buildings (18 m) and wide street (15 m) so H/W is lower (1.2) than that of via Rinaldi. The pavement is covered with asphalt and there is the presence of trees (Table 7.3). The SVF measured was slightly higher than via Rinaldi but mean radiant temperature values were near to dry-bulb air temperatures, respectively 28 and 27.2–27.7 °C. The lower values of

**Table 7.4** Data obtained in measurement sessions in via San Basilio (see also Fig. 7.1). The "X" refer to the point of measurement. Thermal comfort indexes refer to a person with summer clothing (0.5 clo) and slight activity level (80 W above the basal metabolism) (Google Earth-RayMan)

Via San Basilio (H/W = 0.4 – SVF = 0.75)

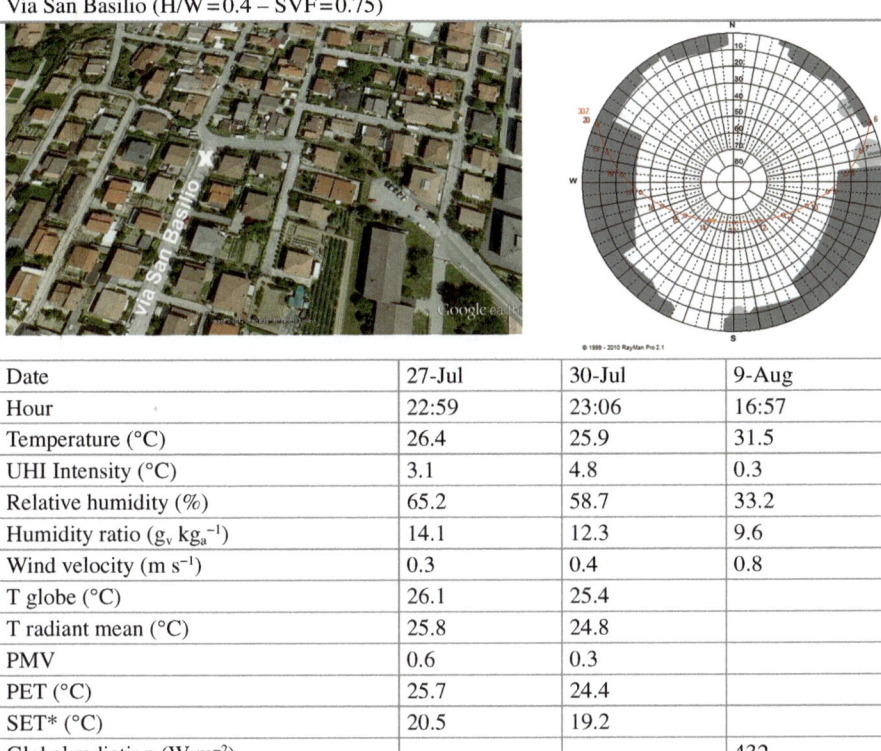

| Date | 27-Jul | 30-Jul | 9-Aug |
|---|---|---|---|
| Hour | 22:59 | 23:06 | 16:57 |
| Temperature (°C) | 26.4 | 25.9 | 31.5 |
| UHI Intensity (°C) | 3.1 | 4.8 | 0.3 |
| Relative humidity (%) | 65.2 | 58.7 | 33.2 |
| Humidity ratio ($g_v$ $kg_a^{-1}$) | 14.1 | 12.3 | 9.6 |
| Wind velocity (m s$^{-1}$) | 0.3 | 0.4 | 0.8 |
| T globe (°C) | 26.1 | 25.4 | |
| T radiant mean (°C) | 25.8 | 24.8 | |
| PMV | 0.6 | 0.3 | |
| PET (°C) | 25.7 | 24.4 | |
| SET* (°C) | 20.5 | 19.2 | |
| Global radiation (W m$^{-2}$) | | | 432 |

$T_{mr}$ and the higher values of wind velocity in via Pindemonte with respect to via Rinaldi resulted in lower values of the thermal comfort indexes.

Via San Basilio (Table 7.4) is mainly characterized by detached houses with height of 6–8 m and streets width of 18 m, so the H/W ratio is quite small (0.4). The lack of trees is also the reason of high SVF value (0.75). This is why the mean radiant temperature measured was always 0.5–1 °C lower than air temperature, so PMV and PET indicate full thermal comfort situation; nevertheless UHI intensity measured during evening sessions was quite significant, about 3–5 °C. During the day no significant UHI effect measured.

In order to compare environmental conditions of urban and sub-urban sites with a rural site, measurements in a lateral unpaved road of via Roma were carried out (Table 7.5, Fig. 7.1). The road winds through agricultural fields with no obstacles nearby (SVF = 1); mean radiant temperature was 2–3 °C lower than air temperature, due to the very high SVF and the lower temperature of the agricultural surface (characterized by higher emissivity, lower thermal inertia and higher water storage capacity with respect to urban surfaces).

**Table 7.5** Data obtained in measurement sessions in a lateral dirt road of via Roma (see also Fig. 7.1). The "X" refer to the point of measurement. Thermal comfort indexes refer to a person with summer clothing (0.5 clo) and slight activity level (80 W above the basal metabolism) (Google Earth)

Via Roma (SVF ≈ 1)

| Date | 30-Jul | 31-Jul | 2-Aug |
|---|---|---|---|
| Hour | 22:46 | 22:52 | 16:53 |
| Temperature (°C) | 21.4 | 22.9 | 33.6 |
| UHI Intensity (°C) | 0.5 | 0.4 | 0.8 |
| Relative humidity (%) | 80.8 | 61.2 | 35.8 |
| Humidity ratio ($g_v$ $kg_a^{-1}$) | 13.0 | 10.7 | 11.6 |
| Wind velocity (m $s^{-1}$) | 0.0 | 0.2 | 2.4 |
| T globe (°C) | 19.0 | 21.6 | |
| T radiant mean (°C) | 18.4 | 20.6 | |
| PMV | −0.5 | −0.4 | |
| PET (°C) | 21.6 | 21.4 | |
| SET* (°C) | 15.8 | 17 | |
| Global radiation (W $m^{-2}$) | | | 502 |

Finally, a deeper analysis of the thermal comfort in Prato della Valle was performed. This very popular circle is characterized by a central green island delineated by a channel decorated by statues and encircled by a wide asphalt circle road (Table 7.6). Because of the extension of the square, six different positions were fixed for the measurements.

The results indicate a difference of 0.5–1 °C in the air temperature between position 4 (on the green) and external positions near urban streets (1, 2, 3, 6); a quite larger difference in mean radiant temperature was measured, with a maximum of 7 °C, because of the different surface characteristics (emissivity, thermal inertia and water storage capacity) and SVF (higher) at the centre of the square with respect to outer positions.

**Table 7.6** Data obtained in measurement sessions in Prato della Valle (dimension of ellipse are 180×230 m) (see also Fig. 7.1). Thermal comfort indexes refer to a person with summer clothing (0.5 clo) and slight activity level (80 W above the basal metabolism) (Google Earth)

Prato della Valle

|  | Pos.1 | Pos.2 | Pos.3 | Pos.4 | Pos.5 | Pos.6 |
|---|---|---|---|---|---|---|
| Date | 2-Aug | 2-Aug | 2-Aug | 2-Aug | 2-Aug | 2-Aug |
| Hour | 21:53 | 22:03 | 22:14 | 22:25 | 22:35 | 22:44 |
| Temperature (°C) | 28.4 | 28.4 | 28.1 | 27.0 | 26.9 | 27.4 |
| UHI Intensity (°C) | 3.2 | 3.9 | 4.3 | 3.8 | 3.7 | 4.7 |
| Relative humidity (%) | 48.0 | 48.2 | 49.1 | 52.9 | 53.5 | 51.5 |
| Humidity ratio ($g_v$ $kg_a^{-1}$) | 11.6 | 11.7 | 11.7 | 11.8 | 11.8 | 11.7 |
| Wind velocity (m $s^{-1}$) | 0.8 | 0.8 | 0.8 | 0.8 | 0.7 | 0.7 |
| T globe (°C) | 27.8 | 28.1 | 27.2 | 24.8 | 25.3 | 26.5 |
| T radiant mean (°C) | 26.5 | 27.4 | 25.2 | 20.1 | 22.2 | 24.8 |
| PMV | 0.8 | 0.8 | 0.6 | 0.0 | 0.1 | 0.5 |
| PET (°C) | 26.1 | 26.5 | 25.3 | 22.5 | 23.3 | 24.7 |
| SET* (°C) | 20.1 | 20.5 | 19.3 | 16.5 | 17.6 | 19.0 |

The thermal comfort indexes calculation had reflected these considerations: PMV and PET had indicated a slightly warm environment at outer positions while on the green there was a neutral situation.

It is interesting to compare Pos. 4 and 5: the first was on the grass, the second was on a dirt patch near a fountain. Even so, same air temperature (27 °C) was measured at these locations. This was mainly due to the lower $T_{mr}$ for Pos. 4 and to the evaporative cooling effect for Pos. 5.

## 7.4  Feasibility Study: UHI Mitigation Strategies by Simulations

Many mitigation measures can be adopted and have been proposed by various researchers, which could be classified as measures that could only be implemented during the design and planning stage (eg sky view factor and building material etc.) and those that could also be implemented after the design and planning stages (eg green areas and roof spray cooling) (Rizwan et al. 2008). RayMan model was used in order to quantify possible increasing in thermal comfort as a consequence of some possible mitigating measures of both types. Main inputs of the model relate to the outdoor environment conditions: dry-bulb air temperature and RH, wind velocity, Bowen ratio (ratio of sensible over latent heat flux in evapotranspiration, fixed at 1.5) and cloud cover (fixed at 1 okta). Other inputs are the albedo and emissivity of surfaces, fixed respectively at 0.30 and 0.95, typical values of urban environment.

The following limitations in RayMan analysis must be highlighted:

– emissivity is considered the same for all the different kinds of surfaces;
– consequences of higher albedos cannot be correctly evaluated: the lower surfaces temperature would not be estimated as it is given by the air temperature (input of the software).

For these reasons next simulations concern topology modifications only (height and distance of buildings, presence of green); obviously these are mitigation strategies that can be implemented during the design and planning stage only.

As described by Noro et al. (2015) and Busato et al. (2014), a slightly warm PMV was obtained for via Rinaldi (Padua old town) and in via Pindemonte, the pilot area (high density population residential zone): for the latter the modification in thermal comfort with some characteristics of the site was evaluated. Table 7.7 reports the results considering different layout of buildings:

– considering the actual situation;
– increasing the street width from 15 to 25 m;
– limiting the maximum height of buildings to 12 and 6 m;
– having a garden in front of the point of measurements instead of an apartment building.

Every simulation was repeated using the same values of environmental variables (air temperature and RH, wind velocity) as measured during the experimental sessions. Results in Table 7.7 show that an increase in Sky View Factor (SVF) thus allowing a more effective nightly cooling of surfaces and a decrease of $T_{mr}$. In particular, limiting the maximum height of buildings to 6 m would be the action with the most relevant effect. The mean radiant temperature was shown to decrease by 2.6 °C and PMV by 0.2. Anyway, the night effects of an increased SVF were probably underestimated by RayMan, because the mean radiant temperature and so PMV and PET were calculated by knowledge of air temperature (input) that is

**Table 7.7** Calculated $T_{mr}$ and thermal comfort indexes in via Pindemonte (Bowen ratio = 1.5, cloud cover = 1 okta, clothing = 0.5 clo, activity level = 80 W above the basal metabolism) for different disposition of buildings

| | | SVF | Date | $T_{mr}$ | PMV | PET |
|---|---|---|---|---|---|---|
| Actual situation | | 0.29 | 27-Jul | 24.9 | 0.6 | 25.3 |
| | | | 30-Jul | 24.3 | 0.5 | 25.3 |
| Street width enlarged by 10 m | | 0.38 | 27-Jul | 24.2 | 0.5 | 25 |
| | | | 30-Jul | 23.6 | 0.5 | 24.9 |
| 12 m buildings height max | | 0.45 | 27-Jul | 23.6 | 0.5 | 24.7 |
| | | | 30-Jul | 22.9 | 0.4 | 24.6 |
| 6 m buildings height max | | 0.6 | 27-Jul | 22.3 | 0.4 | 24.2 |
| | | | 30-Jul | 21.7 | 0.3 | 24.1 |
| Green in front | | 0.56 | 27-Jul | 22.7 | 0.4 | 24.4 |
| | | | 30-Jul | 22.1 | 0.3 | 24.2 |

actually expected to decrease when SVF increases. Also the effect of having the green (last solution of Table 7.7) was probably underestimated because the model does not consider the cooling effect due to evapotranspiration.

The measures just described can be suggested both in new and built areas, in case of a redevelopment of an existing urban area. Within the frame of the European Project "UHI", the Authors conducted simulations using the ENVImet model (Bruse and Fleer 1998) in order to quantify the effect of selected mitigation actions (usable in already built areas) in the pilot are of via Pindemonte.

ENVI-met is a three-dimensional microclimate model designed to simulate the surface-plant-air interactions in urban environment with a typical resolution of 0.5–10 m in space and 10 s in time. The model area is described in Fig. 7.4 and Fig. 7.5. The main area is a $99 \times 70 \times 30$ grid (in a x,y,z tridimensional reference system), with a $5 \times 5 \times 3$ m grid dimension. An appropriate number of nesting grids (five) has been set in order to minimize boundary effects. Four specific points of interest have been identified in the zone to characterize the air temperature (at 2 m above ground) during 24 h, from 6 am to 6 pm. Simulations lasted 72 h, but only the last 24 h have been considered for the results. The daily mean air temperature of the days before the start simulation ones (Table 7.8) have been used as initial air temperature at 6 am of the first day. As requested by the Project, four characteristic days representative of the four seasons were considered. Simulations used the default values of ENVImet except for the ones reported in Table 7.8.

To measure the UHI intensity, simulations have been extended in the rural zone just outside Padua (Via Roma), in the same point where the previously described experimental measurements have been conducted during 2012 summer. Four scenarios were supposed besides the actual one ("AsIs" scenario):

– "Green ground": increasing the pervious surfaces of the area from 18 % to 23 % by planting trees, 10 m height, within the urban canyon and the main road of the area, and converting an impervious zone – eg asphalt car park surface – to a pervious zone by planting grass. The main effects were: Sky View Factor decreases for the presence of trees along the streets; impervious surface fraction decreases (and pervious surface fraction increases) because green area increases; albedo slightly increases; other thermo-physical properties of the surfaces/materials remain quite the same;
– "Cool pavements": substituting all the traditional asphalt (albedo 0.2) and concrete (albedo 0.4) (roads and pavements) with "cool materials", that is materials with an higher albedo (0.5). The main effects were that albedo significantly increases while other properties remain the same;
– "Cool roofs": using of "cool materials" for the horizontal impervious surfaces of roof. In particular albedo has been increased from 0.3 to 0.6 for roofs;
– "Green ground + cool pavements": scenario with both the mitigation actions just described simultaneously.

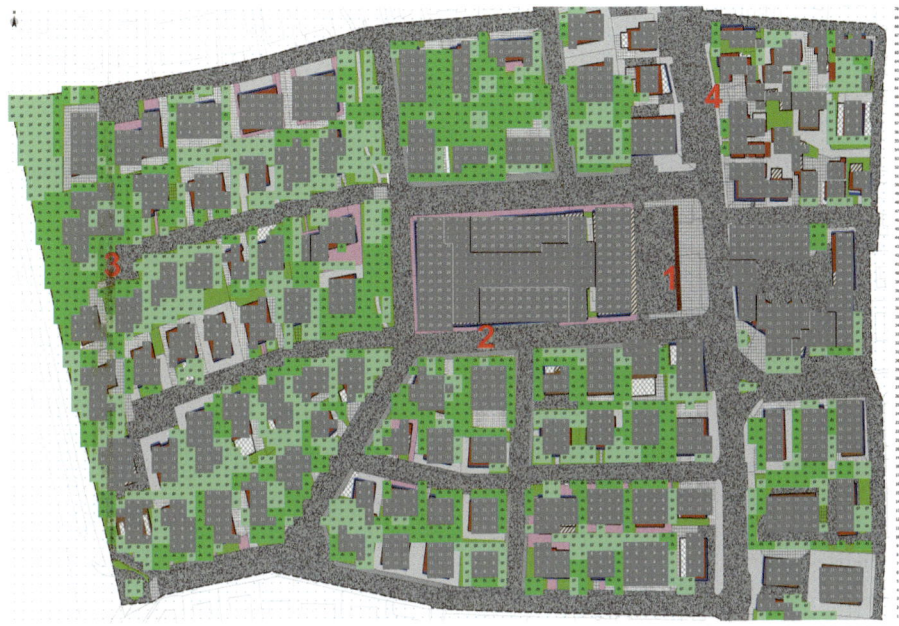

**Fig. 7.4** The model area in ENVImet used for the simulations of the "AsIs", "Cool pavements" and "Cool roofs" scenarios. The red numbers identify four characteristic points for which air temperature at 2 m above ground has been considered in the study

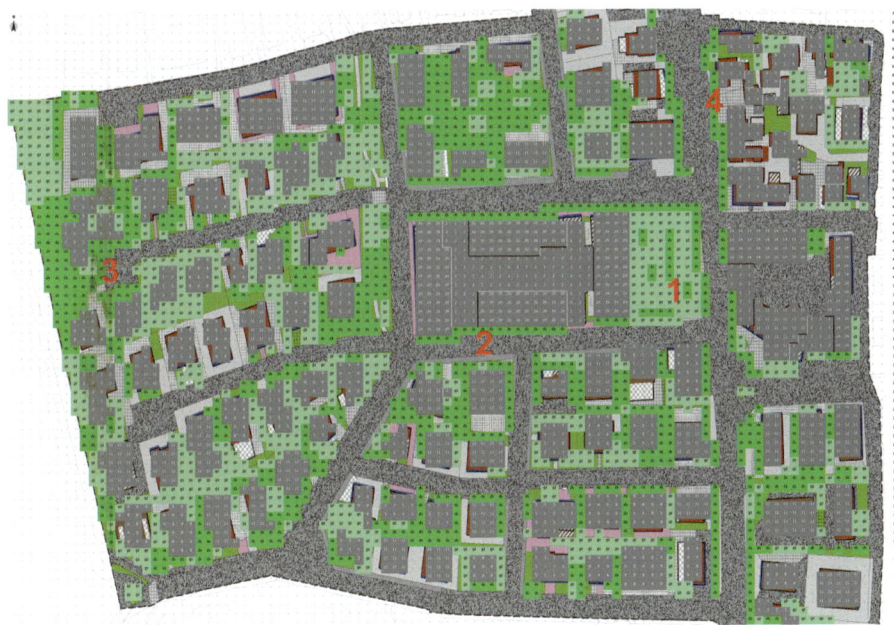

**Fig. 7.5** The model area in ENVImet used for the simulations of the "Green ground" and "Green ground + Cool pavements" scenarios. The red numbers identify four characteristic points for which air temperature at 2 m above ground has been considered in the study

**Table 7.8** Configuration values in ENVImet

| |
|---|
| Simulation tool: ENVI-met 3.1 |
| Start Simulation at Day (DD.MM.YYYY): |
| 02.02.2012 Winter; 02.05.2012 Spring; 27.07.2012 Summer; 03.11.2012 Autumn |
| Start Simulation at Time (HH:MM:SS)=06:00:00 |
| Total Simulation Time in Hours=72.00 |
| Save Model State each ? min=60 |
| Wind Speed at 10 m ab. ground [m s$^{-1}$]=3 |
| Wind Direction (0:N.. 90:E.. 180:S.. 270:W..)=90 |
| Roughness Length z0 at Reference Point=0.1 |
| Initial Temperature Atmosphere [K]=279 K Winter; 290.9 K Spring; 300 K Summer; 282.2 K Autumn |
| Specific Humidity at 2500 m [g$_{water}$/kg$_{air}$]=7 |
| Relative Humidity at 2 m [%]=50 |
| Output: air temperature at 2 m above ground |
| Building properties |
| Inside Temperature [K]=298 |
| Heat Transmission Walls [W m$^{-2}$ K$^{-1}$]=1 |
| Heat Transmission Roofs [W m$^{-2}$ K$^{-1}$]=2 |
| Albedo Walls=0.2 |
| Albedo Roofs=0.3 (all scenarios except "Cool roofs"); 0.6 ("Cool roofs") |
| Albedo pavements=0.4 (all scenarios except "Cool pavements"); 0.5 ("Cool pavements") |
| Albedo roads=0.2 (all scenarios except "Cool pavements"); 0.5 ("Cool pavements") |

Results, in terms of 24 h of air temperature at 2 m above ground and UHI intensity with respect to Via Roma (rural zone) are reported in Fig. 7.6 for the point 1 of Fig. 7.4 (as representative of the area) and for the four characteristic days of the year. Summarizing, the main results were:

- The higher UHI intensity (difference between the "AsIs" and "Via Roma" curves per each hour of the day) is obviously obtained in summer. During the other seasons the UHI phenomenon is much lower and negligible in winter during the central hours of the day;
- Talking about the summer, the higher UHI intensity (9–10 °C) is noticed after the sunset (8 pm) and till the first sunrise (4 am);
- The same value of 9–10 °C for the UHI intensity during the day is noticed only for point 2 (1 pm), that is a street canyon (characterized by low SVF and impervious surfaces); for the other points the maximum daily UHI intensity is always lower than 9 °C;
- The best UHI mitigation strategy is the "Green ground+cool pavements" (Scenario 4) that allows 2 °C decrease in UHI maximum intensity (but till nearly 3 °C decrease in Point 2);
- The "Cool pavements" mitigation strategy allows between 1 and 2 °C decrease in UHI maximum intensity in all the points;

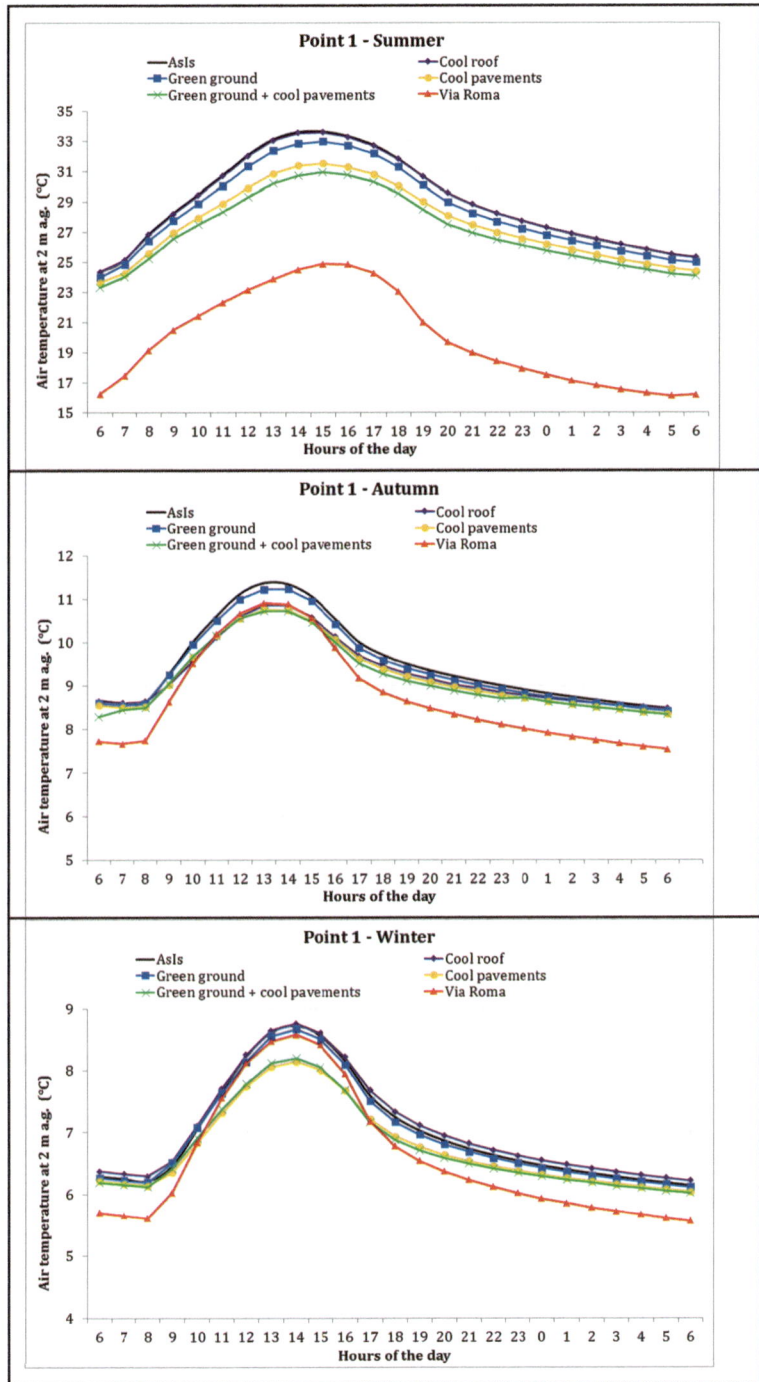

**Fig. 7.6** Air temperature at 2 m above the ground during one typical day (from 6 am to 6 am of the day after) for each season at point 1 of the pilot area modelled by ENVImet (asphalt car park surface in front of a flat block). The differences between the upper curve (the actual scenario "AsIs") and the other ones depict the UHI intensity mitigation for the strategies described in the text

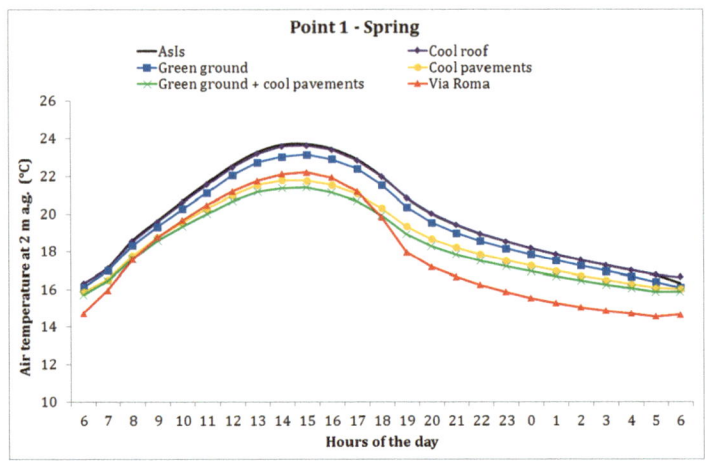

**Fig. 7.6** (continued)

- The "Green ground" mitigation strategy allows an appreciable UHI maximum intensity decrease only in point 2 (around 1 °C): in all other points the positive effect of the action is quite negligible;
- During the day (afternoon) the most effective mitigation actions are "Green ground" and "Green ground + cool pavements": they allow till 3 °C decrease in UHI intensity;
- During all the other seasons, but especially in winter, the most effective UHI mitigation strategies ("Cool pavements" and "Green ground + cool pavements") cause a negative effect during the central hours of the day: UHI intensity becomes negative, that is air temperature in urban zone id colder (about 0.5 °C) than in rural zone.

**Acknowledgement** This chapter was based on the work undertaken as part of the UHI Project implemented through the CENTRAL EUROPE Programme 2007–2013 co-financed by the ERDF (grant number 3CE292P3). The authors are grateful to Mr. Pierpaolo Campostrini and Mr. Matteo Morgantin (CORILA) for the support. Thanks to ARPAV and to Mr. Matteo Sottana for the use of meteorological data. Thanks to Prof. A. Matzarakis (Meteorological Institute – University of Freiburg) for the use of RayMan simulation model and to Prof. M. Bruse (Inst. of Geography – University of Mainz) for the use of ENVI-met simulation model.

# References

ARPAV Regional Agency for Environment Protection in Veneto (2011). Bilancio Energetico della Regione Veneto con dettagli provinciali sulla base dei dati disponibili fino all'anno 2010 (Veneto Region Energy balance with province details on the basis of data available till 2010). Technical Report.

Berkovic, S., Yezioro, A., & Bitan, A. (2012). Study of thermal comfort in courtyards in a hot arid climate. *Solar Energy, 86*, 1173–1186.

Bruse, M., & Fleer, H. (1998). Simulating surface-plant air interactions inside urban environments with a three dimensional numerical model. *Environmental Modelling and Software, 13*, 373–384.

Busato, F., Lazzarin, R., & Noro, M. (2014). Three years of study of the Urban Heat Island in Padua: Experimental results. *Sustainable Cities and Society, 10*, 251–258, http://dx.doi.org/10.1016/j.scs.2013.05.001.

Gagge, A. P., Fobelets, A. P., & Berglund, L. G. (1986). A standard predictive index of human response to the thermal environment. *ASHRAE Transactions, 92*, 709–731.

Honjo, T. (2009). Thermal comfort in outdoor environment. *Global Environmental Research, 13*, 43–47.

Höppe, P. (1999). The physiological equivalent temperature-a universal index for the biometeorological assessment of the thermal environment. *International Journal of Biometeorology, 43*, 71–75.

ISTAT Italian Statistical Institute (2012, July 30). Indicatori Ambientali Urbani (Urban Environmental Indexes). Technical Report. Available online, http://www.istat.it/it/archivio/67990. Accessed 1 Mar 2013.

Jendritzky, G., & Nubler, W. (1981). A model analysing the urban thermal environment in physiologically significant terms. *Meteorology and Atmospheric Physics, 29*, 313–326.

Matzarakis, A., Rutz, F., & Mayer, H. (2007). Modelling radiation fluxes in simple and complex environments – application of the RayMan model. *International Journal of Biometeorology, 51*, 323–334.

Matzarakis, A., Rutz, F., & Mayer, H. (2010). Modelling radiation fluxes in simple and complex environments – basics of the RayMan model. *International Journal of Biometeorology, 54*, 131–139.

Mayer, H. (1993). Urban bioclimatology. *Experientia, 49*, 957–963.

Mayer, H., & Höppe, P. (1987). Thermal comfort of man in different urban environments. *Theoretical and Applied Climatology, 38*, 43–49.

Noro, M., Lazzarin, R., & Busato, F. (2015). Urban heat island in Padua, Italy: Experimental and theoretical analysis. *Indoor and Built Environment, 24*(4), 514–533. doi:10.1177/1420326X13517404.

Rizwan, A. M., Dennis, V. C. L., & Liu, C. (2008). A review on the generation, determination and mitigation of urban heat island. *Journal of Environmental Sciences, 20*, 120–128.

Thorsson, S., Lindqvist, M., & Lindqvist, S. (2004). Thermal bioclimatic conditions and patterns of behaviour in an urban park in Göteborg, Sweden. *International Journal of Biometeorology, 48*, 149–156.

# Chapter 8
# Mitigation of and Adaptation to UHI Phenomena: The Padua Case Study

Francesco Musco, Laura Fregolent, Davide Ferro, Filippo Magni, Denis Maragno, Davide Martinucci, and Giuliana Fornaciari

**Abstract** Elaborating solutions to counteract UHI effects can represents a relevant challenge for spatial planning and urban design. A specific experimentation has been developed on the city of Padua, analysing different scenarios of urban warming and using specific monitoring tools (Lidar/aerial survey) to define a DIM (Digital Surface Models) providing local situation in terms of green quality and extension, solar incidence/radiation, sky view factors, building materials. This chapter reconstruct the methodology followed during the survey and the elaboration of specific solutions to counteract UHI accordingly different scenarios.

**Keyword** Urban planning • Local plans • Digital surface models • Climate adaptation

## 8.1 Introduction

The test developed by the Veneto Region and by the working group of the Iuav University of Venice as part of the European project "UHI – *Development and application of mitigation and adaptation strategies and measures for counteracting the global Urban Heat Islands phenomenon*" is based on the territorial peculiarities of the Veneto region's lowlands, mostly characterized by small sized historical centers and the widespread settlements that have developed around them over the last 40 years. This urbanization occurred without strategies or rules as a summation of individual initiatives that amalgamated residential forms and functions with large thoroughfares, as well as production and commercial areas (Selicato and Rotondo 2003).

F. Musco (✉) • L. Fregolent • D. Ferro • F. Magni • D. Maragno
D. Martinucci • G. Fornaciari
Department of Design and Planning in Complex Environments, IUAV University of Venice, S.Croce, 1957, 30135 Venice, Italy
e-mail: francesco.musco@iuav.it; climatechange@iuav.it;
laura.fregolent@iuav.it; fmagni@iuav.it; dmaragno@iuav.it

© The Author(s) 2016
F. Musco (ed.), *Counteracting Urban Heat Island Effects in a Global Climate Change Scenario*, DOI 10.1007/978-3-319-10425-6_8

In some ways, this process broke the environmental balance of the medieval towns (which were designed keeping in mind local microclimate regulation) often creating an artificial barrier around them, suffocating them, and contributing greatly to raise the amount of impermeable surfaces to the detriment of permeable ones. In recent years, the relationship between urban planning and architecture paid a price for the rigidity dictated by Local Strategic Plans conforming to homogeneous and repetitive rules rather than adapting to the peculiarities of the various land areas (Samonà 1980). As a result, we are dealing with areas that are already rigid and intensely anthropized, with a paucity of characterizing settlements, whose development in the near future may be expected to involve mostly the transformation of the existing tissues. Our test focused on the connection between local climate, urban structure and the emergence of the urban heat island effect, with the purpose of providing land management guidelines for the near future (Musco et al. 2014). Within this framework, we singled out a section of Padua's metropolitan area for analysis and planning, with the intent of applying the results of our tests to the rest of Veneto's central area. Often, the cause for urban heat islands are specific factors (such as large paved areas), which are directly connected to widespread systemic factors (such as the nocturnal dispersion of the heat absorbed by peripheral urban tissues, or pollution from production areas, again located in the suburbs). Such a plurality of causes leads to studying heat islands from different points of view, which are both horizontal and vertical.

Oke's model (2006) approaches this phenomenon by analyzing different urban climate scales, where diverse climate events occur that influence each other:

- Horizontal scale: Microscale, Local scale and Mesoscale;
- Vertical scale (according to different UHI types): *Air* UHI (*Urban Canopy Layer* UCL, and *Urban Boundary Layer* UBL), *Surface* UHI, and *Subsurface* UHI.

The *Urban Boundary Layer* (UBL) encompasses the urban cover layer above the average height of buildings, whereas the *Urban Canopy Layer* (UCL), encompasses the urban cover layer below the average height of buildings. After considering the goals of our projects, namely to analyze the causes of this phenomenon at the microscale level with the intent of coming up with accurate mitigation measures, we proceeded by considering the heat island on the vertical scale encompassed between ground level and the average height of buildings, that is, in the *Urban Canopy Layer*.

This microscale level can help verify the relationship between urban form, roofing materials and UHI, with particular reference to the vegetative cover, soil permeability and albedo of materials. Within this context, the following factors influence microclimate at different urban scales in a significant way: orientation of buildings, surface covering, Sky View Factor (SVF), solar incidence, materials used, and shape of buildings. For example, where building facades are too close to each other, temperatures are affected by the SVF, i.e., they heat up more than other facades located on more open and ventilated roads (which are perhaps no more than thirty or forty yards away).

A recent study shows how urban microclimate affects the functions of buildings in terms of thermal performance, proving that urban form has an effect on the UHI phenomenon (Wong and Chen 2009). Speaking specifically of Italian cities, heat islands are not caused by the anthropogenic heat produced by human activities, but rather by the heat stored by urban surfaces (buildings, roads, parking lots) during the day and then released gradually at night. This effect generates a nocturnal heat island, insofar as the heat released does not allow the city to cool down as much as the rural environments external to it.

The complexity of the UHI phenomenon is directly related to the relationship between city and atmosphere; urban climate and atmospheric climate affect and influence each other (Oke 2006b).

Usually, the aspects that influence climate, generating an urban microclimate that is different from the climate of the atmosphere, (Shahmohamadi 2012) are the following:

– amount of grass, permeable soil, trees, asphalt, and concrete;
– artificial heat released from buildings, air-conditioning systems, cars, and production areas;
– surface water storage and lamination in favor of underground canals and drains;
– air pollution;
– urban ventilation.

Urban heat islands arise from extensive anthropization, or rather, it could be said that the fewer the ecological properties of a city, the greater its heat island will be. It is no coincidence that this effect had already been observed by meteorologist Luke Howard for the first time in 1818, in London, at the height of that city's expansion. At the time, it was not identified as a heat island[1]; the term "island" was coined when isotherms were used to map the city. When air temperatures are mapped through isotherms, the city appears like an island compared to the surrounding rural areas, which are differentiated by lower temperatures. On these bases, we started our project by studying the different behaviors of the urban heat island of pilot experiment city Padua vis-à-vis its different urban contexts. We chose this area also taking into account our purpose of drafting an urban planning manual for the Veneto Region to be delivered to municipal administrations in order to support their future strategic choices in terms of mitigating the UHI phenomenon and adapting vulnerable urban areas to climate changes. Therefore, we picked this area for our study also because it conforms to the urban and spatial characteristics of other cities and areas of the Veneto region.

To conduct urban heat island effect analyses and surveys, we then went on to selecting five pilot areas in the municipality of Padua, based on their location with respect to a survey transect crossing the city along the north-west/south-east axis

---

[1]The term "urban heat island" was coined in 1958 by Gordon Manley in an essay found in the *Quarterly Journal of the Royal Meteorology Society.*

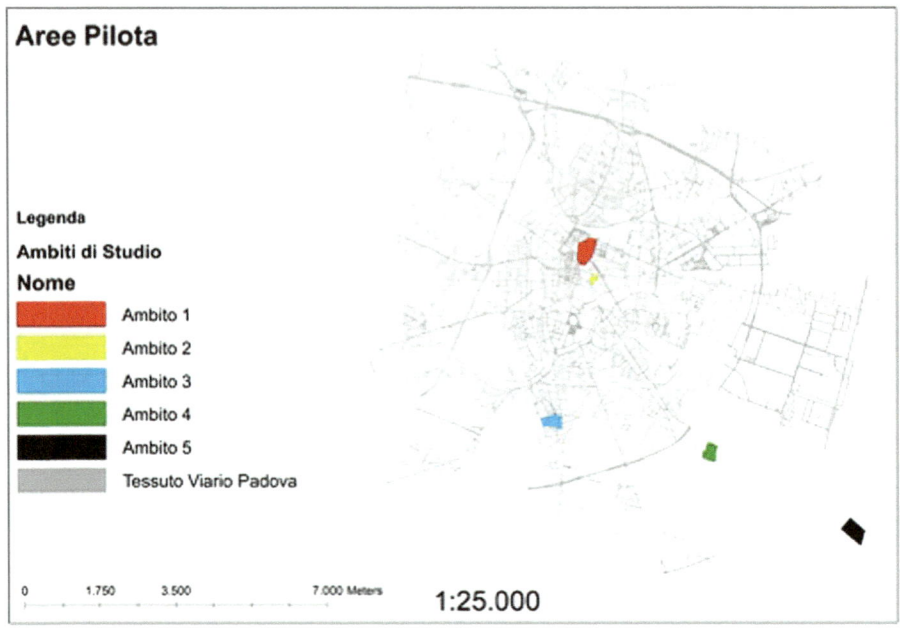

**Fig. 8.1** Pilot Areas in Padua

and the intrinsic features of their settlement structure. These features are the following (Fig. 8.1):

**Area 1, a dense urban area located inside the medieval historical center;**
**Area 2, a mixed-use area, ranging from a major river to a large parking lot;**
**Area 3, a "high density" residential area built in the 60/70s;**
**Area 4, a "low density" residential area, also built in the 60/70s, located in the first outer ring of the city and consisting of free-standing 1–2 storey buildings;**
**Area 5, a production area located outside the municipality of Padua.**

## 8.2 Analysis Methods: Traditional Surveys and Remote Sensing

The first part of our project concerned the implementation of an efficient urban heat island study method. We focused our attention right away on producing an urban planning manual that could be used by other municipal administrations. With this in mind, we sought to adopt simple but effective methods. An ideal process of analysis would require atmospheric temperature readings throughout the urban environment.

Temperature represents an important descriptor of UHI behavior; unfortunately however, detectors often are not spread evenly through the urban environment. In Italian cities, the location of temperature and humidity detection devices is often

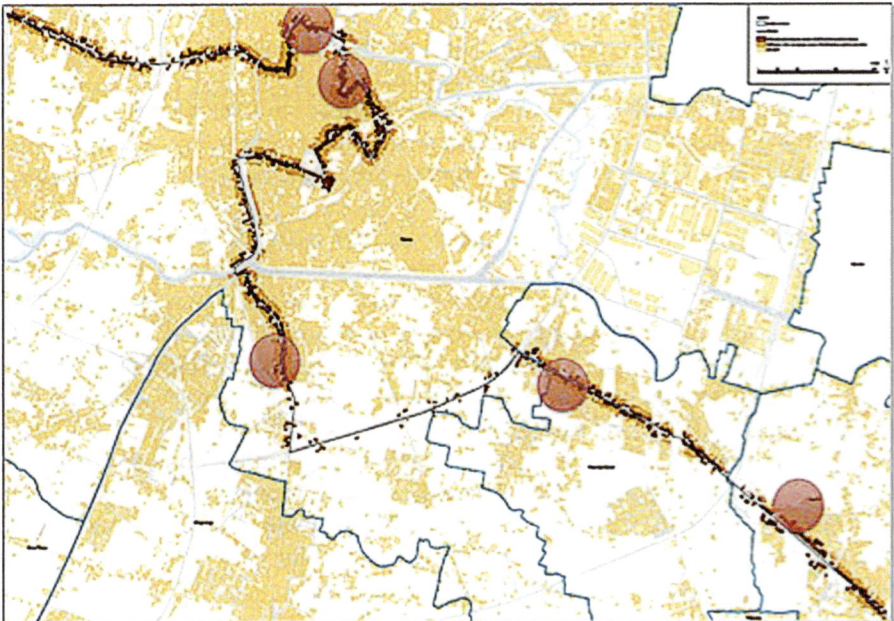

**Fig. 8.2** Urban transect used for UHI survey

organized around the monitoring of pollution rather than the microclimate. Due to this lack of information, it was not possible to build a homogenous framework capable of bringing out the causes of urban overheating for the various scales. In addition, some land management data usually available to public administrations do not consider the variables used to identify this phenomenon. As part of our project, the University of Padua research unit measured urban environment atmospheric heat, working to determine the heat island within the city by paying specific attention to the five selected areas described above (Fig. 8.2).

The picture that emerged from our analysis revealed a significant difference in nocturnal temperature between the urban area and the rural area peripheral to the city. A nocturnal heat island that becomes more intense as dawn approaches is already a strong indicator of the causes of urban overheating, mainly due to its morphology and surface types.

Scientific literature (Oke 1982; Santamouris 2005) states that heat islands are caused by anthropogenic factors when developing gradually from late afternoon to night time (namely, resulting from human activities), whereas if they are detected during the night, their formation factors are dependent on the ratio between permeable and impermeable surfaces, materials used and urban ventilation (Papadopoulos 2001).

It was therefore clear from the outset that our analyses and possible strategies should address built-up areas rather than the human activities in and around them.

Our analysis of the existing urban fabric followed the guidelines of the Technical University of Vienna, which has proposed a number of indicators for weighing and quantifying the various UHI production factors (Table 8.1).

We quantified these indexes and compared them to our temperature readings to obtain cogent results that would approximate the real state of things. Thanks to this approach we were able to evaluate the various urban microclimates of the selected areas and single out incidence factors.

For example, the first area that we analyzed showed that heat island production factors are mostly ascribable to the low ratio between permeable – impermeable surfaces and the sky view factor, whereas the fourth area (which we later picked as final pilot area) showed that heat production is mostly ascribable to the type of materials used for buildings. Therefore, it seems obvious that mitigation strategies (and the urban planning tools for implementing project interventions) must be different for the two areas in question.

In order to ensure and maintain the high effectiveness of the proposed interventions and solutions, it is essential that the different overheating causes and issues of each area be carefully identified so as to come up with site-specific strategies. The necessary information for evaluating (and then monitoring) the resilience of an urban area to heat waves were the following:

**Paved surface areas;**
**Permeable surface areas;**
**Built up surface;**
**Sky View Factor (SVF);**
**Urban compactness;**
**Solar incidence;**
**Reflectance/albedo of materials;**
**Thermal conductivity of materials;**

Due to the great number of details provided by all this information, we had to come up with an appropriate data collection method for our analysis. Two alternative methods were used; one was a traditional analysis on the field that classified ground covering and building types as well as the height of buildings, the other used remote sensing and three-dimensional data processing from LiDAR[2] and very high resolution orthophotos.

The traditional method allowed us to map the urban tissue and determine the types of materials of all the surfaces, as well as their thermal properties. This activity required a lot of time, spent mainly on the field, but yielded a complete and current set of facts for the area.

---

[2] Il LiDAR (Laser Imaging Detection and Ranging) performs remote sensing to determine the distance of an object or surface through the emission of high frequency laser pulses by a flying sensor (plane or drone). The distance of an object is given by the length of time elapsed between emission and reception. Very high frequency pulses bouncing from objects or the ground are converted into geo-referenced and dimensioned points, thus giving rise to a "point cloud" from which the exact reconstruction of an area can be created in the form of three-dimensional digital models.

**Table 8.1** Main physical parameters and indexs to analyse UHI phenomena

| Geometric properties | Symbol | Unit | Rage | Definiation |
|---|---|---|---|---|
| Sky View Factor | $\Psi_{sky}$ | – | 0–1 | Mean value of the fraction of sky hemisphere visible from ground level |
| Aspect ratio | H/W | – | 0–3$^+$ | Mean height-to-width ratio of street canyons, consider length of streets as a weighting factor |
| Built area fraction | $A_b/A_{tot}$<br>$A_b$: building plan area [m$^2$]<br>$A_{tot}$: total ground area [m$^2$] | – | 0–1 | Ration of building plan area to total ground area; fraction of ground surface with building cover |
| Unbuilt area fraction | $1-A_b/A_{tot}$ | – | 0–1 | Ratio of unbuilt plan area to total ground area; fraction ground surface without building cover |
| Impervious surface fraction | $A_i$ | – | 0–1 | Ration of unbuilt impervious plan area (paved, sealed) to total ground area |
| Pervious surface fraction | $A_p=(A_e+A_g+A_{H2O})$ | – | 0–1 | Ration unbuilt impervious plan area (bare soil, green, water) to total ground area |
|  | $A_{e: earth}$ | – | 0–1 | Bare soil area |
|  | $A_{g: green}$ | – | 0–1 | Green area |
|  | $A_{H2O:water}$ | – | 0–1 | Water bodies area |
| Mean building | $I_c$<br>$I_c=V_b/A_b[m^3/m^2]$<br>$V_b$: built volume [m$^3$] | M | – | Ration of built volume (above terrain) to total building plan area |
| Built surface fraction | $A_s/A_b$<br>$A_s$: total built surface area [m$^2$] | – | >1 | Ration of total built surface area (above terrain) of building (walls and roofs) to total built area |
|  | $A_w/A_b$<br>$A_w$: total wall area [m$^2$] | – | >1 | Walls |
|  | $A_R/A_b$<br>$A_R=(A_{R,i}+A_{R,p})$<br>$A_R$: total roof area [m$^2$] | – | ~1 | Roofs |
|  | $A_{R,I}/A_b$<br>$A_{R,I}$: total impervious roof area [m$^2$] | – | ~1 | Impervious roofs |
|  | $A_{R,p}/A_b$<br>$A_{R,p}$: total pervious roof area [m$^2$] | – | ~1 | Pervious roof |
| Mean sea level | $H_{sl}$ | m | – | Average height sea level |

(continued)

**Table 8.1** (continued)

| Surface/material properties | Symbol | Unit | Range | Definition |
|---|---|---|---|---|
| Reflectance/albedo | $P_{sw}$ | – | 0–1 | Mean value of albedo (shortwave) |
| Thermal conductivity | $\lambda = (\lambda_i + \lambda_p)$ | W.m$^{-1}$.k$^{-1}$ | >0 | The property of a material's ability to conduct heat |
| | $\lambda_i$: impervious surface | W-m$^{-1}$.k$^{-1}$ | >0 | Thermal of a material's ability to conduct heat |
| | $\lambda_p$: pervious surface | W.m$^{-1}$.k$^{-1}$ | >0 | Thermal conductivity of pervious surface |
| Specific heat capacity | $c = (c_i + c_p)$ | J.kg$^{-1}$.k$^{-1}$ | >0 | The amount of heat required to change a unit mass of a material by one degree in temperature |
| | $C_i$:impervious surface | J.kg$^{-1}$.k$^{-1}$ | >0 | Specific heat capacity of impervious surface |
| | $C_p$:pervious surface | J.kg$^{-1}$.k$^{-1}$ | >0 | Specific heat capacity of pervious surfaces |
| Density | $\rho = (\rho_i + \rho_p)$ | Kg.m$^{-3}$ | >0 | The mass density of a material is its mass per unit volume |
| | $\rho_i$:impervious surface | kg.m$^{-3}$ | >0 | The mass density of impervious surfaces |
| | $\rho_p$: pervious surface | Kg.m$^{-3}$ | >0 | The mass density of pervious surfaces |
| Anthropogenic heat output | $Q_F$ | W.m$^{-2}$ | >0 | Mean annual heat flux density from fuel combustion and human activity (traffic, industry, heating and cooling of buildings etc.) |

Proposed variable for the specification of an urban unit of observation (U2O)
DOCUMENT WP5-UHI-01_112012
WP 5, authors: A, Mahdavi, K. Kiesel, M. Vuckovic (November 5th, 2012)
Main reference:

Mahdavi, A., Kiesel, K., Vuckovic M. (2013). *A framework for the evaluation of urban heat island mitigation measures.* SB13 Conference, Munich, Germany.
Stewart, I. D., & Oke T. R. (2012). Local climate zones for urban temperature studies. *Bulletin of the American Meteorological Society.*
Unger, J., Savic, S., & Gal, T. (2011). Modeling of the annual mean urban heat island pattern for planning of representative urban climate station network. *Advances in Meteorology, 2011*, 1–9.
Hens, H. (2007). *Building physics – Heat, air moisture.* Ernst & Sohn, Berlin.

The remote sensing method required less time to collect the data and yielded useful information to describe and map the phenomenon. Depending on the information, computers and technology available to individual local administrations, this method could be applied easily and quickly to the whole of the Veneto region.

Remote sensing analyses require LiDAR data and high resolution orthophotos (0.2–0.5 m per pixel), preferably including the infrared band, for the entire administrative area.

For each selected area, this methodology allowed us to find out the sqm of vegetation (divided by height), the ratio between permeable and impermeable surface, the incident solar irradiation, and the sky view factor (Berdahl and Bretz 1997). Technically, the analysis involved the creation of three-dimensional digital models of the terrain, DSM (Digital Surface Models) and DTM (Digital Terrain Models), which made it possible to identify and inventory the composition of urban surfaces. By adding the DEM (Digital Elevation Models) obtained by processing the LiDAR data with the multispectral orthophotos, we also got to an automatic breakdown of the horizontal surfaces of the city by type and height, resulting in an atlas of surfaces composed of green spaces, with their respective heights, and impermeable spaces (buildings, roads, parking lots).

Next, we used software like *LAStools*, *Saga Gis* and *eCognition* to produce sky view factor and solar irradiation maps, which most importantly provide essential information to determine the specific areas that require intervention, in addition to which interventions should be performed to mitigate the UHI phenomenon and adapt the urban environment to climate changes.

The key strength of these innovative analysis techniques is that they can be replicated over very extensive urban areas, whose level of detail would require months to obtain with traditional topographical detection methods. However, we must realize that not all areas are equipped with LiDAR or similar detection devices, which means this methodology is still used in limited areas despite the fact that it is innovative and efficient (Figs. 8.3, 8.4, 8.5, 8.6, 8.7, 8.8 and 8.9).

The collected information, converted into vector format, can be queried using height and covering type data. Breaking down the city in all its three dimensions, we can identify the areas that are most vulnerable to heat waves, and also adapt portions of the city to the extreme weather phenomena, suggesting some possible strategies to achieve that goal. So as to test and evaluate the efficacy of the interventions, we then proceeded to build four different transformation scenarios of the area under study.

The four scenarios, and their specific interventions, which resulted from the integration of accurate temperature readings and the indicators research, were then processed using the *ENVI-met* software, which simulates air temperature changes based on the physical changes proposed within a selected area. It can therefore verify and indicate mitigation strategies for the UHI phenomenon by showing the results of the proposed interventions. For example, this simulator can show what benefits would be derived from adding trees to an actual area or modifying the albedo of some of its surfaces. *ENVI-met* can not only verify the effectiveness of an intervention but also the optimal location for its application.

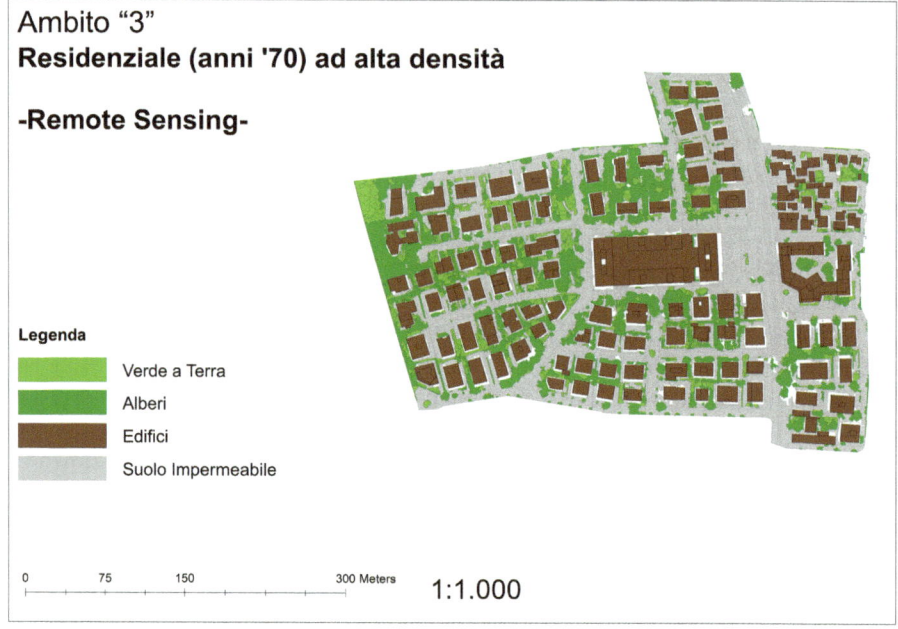

**Fig. 8.3** Hight desnity residential building neighbourhood (1970's) Remote sensing survey (green areas, trees, buildings, surfaces)

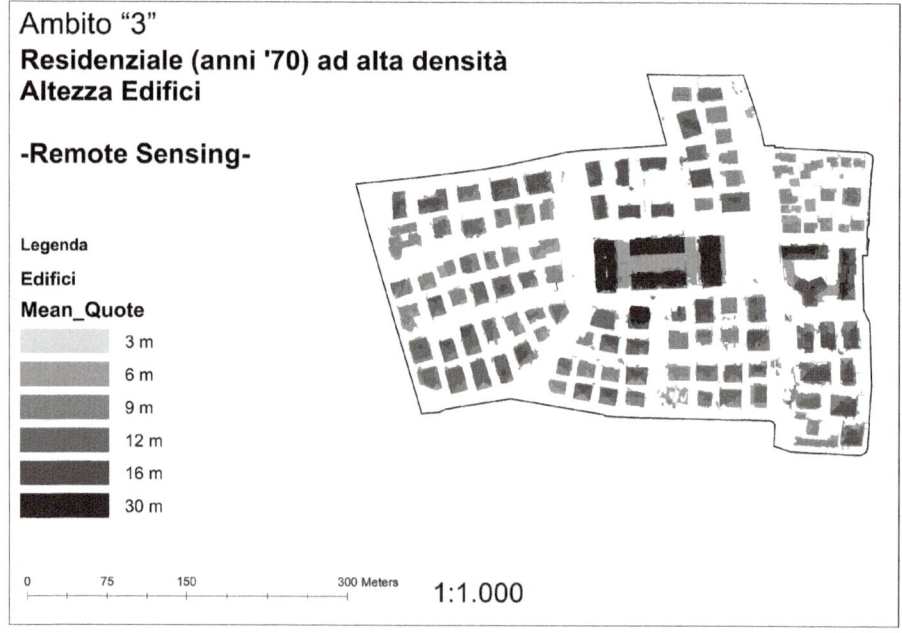

**Fig. 8.4** High density residential neighbourhood (1970's) Bulding Height

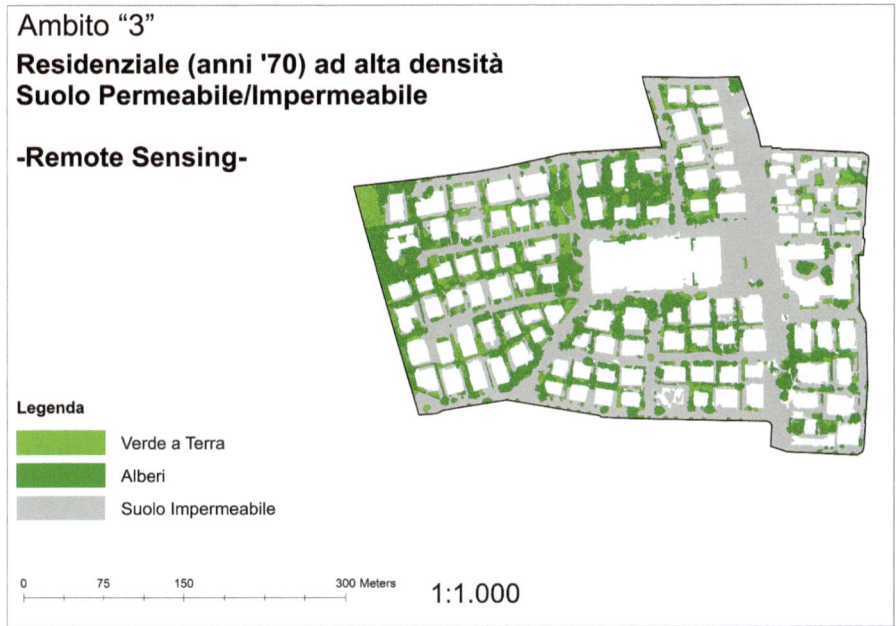

**Fig. 8.5** High density residential neighbourhood (1970's) Permeable/Water proof surfaces

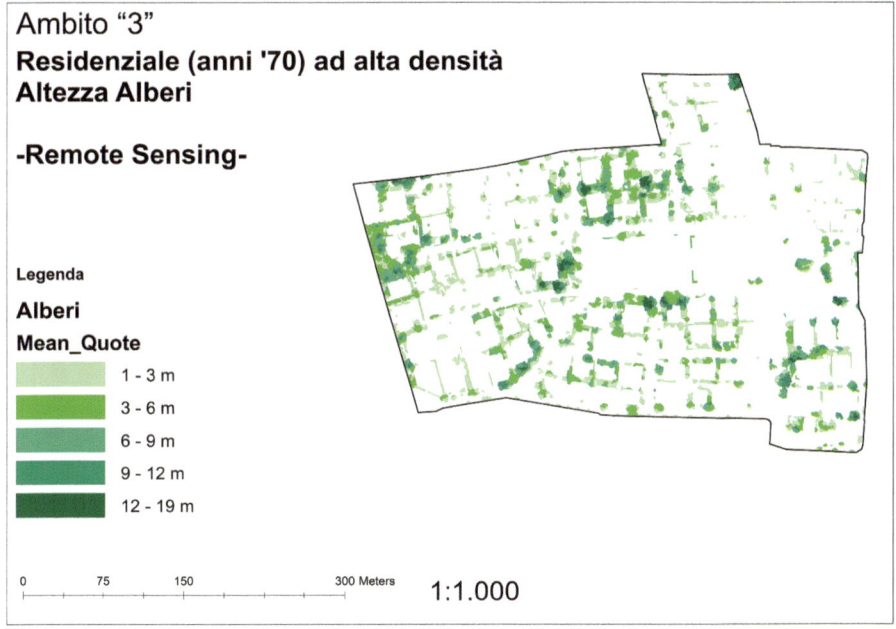

**Fig. 8.6** High density residential neighbourhood (1970's) Trees Hight

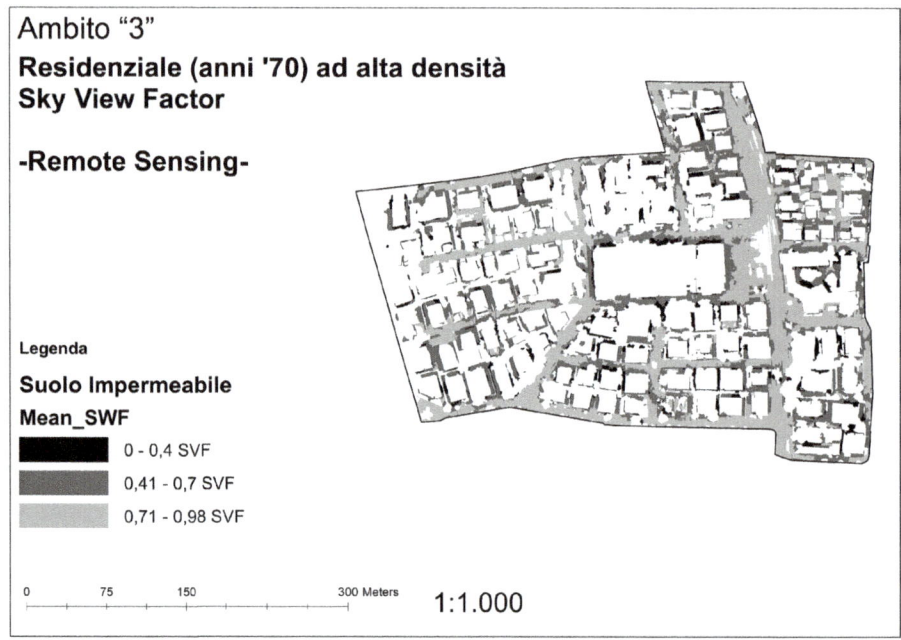

**Fig. 8.7** High density residential neighbourhood (1970's) Sky view factor

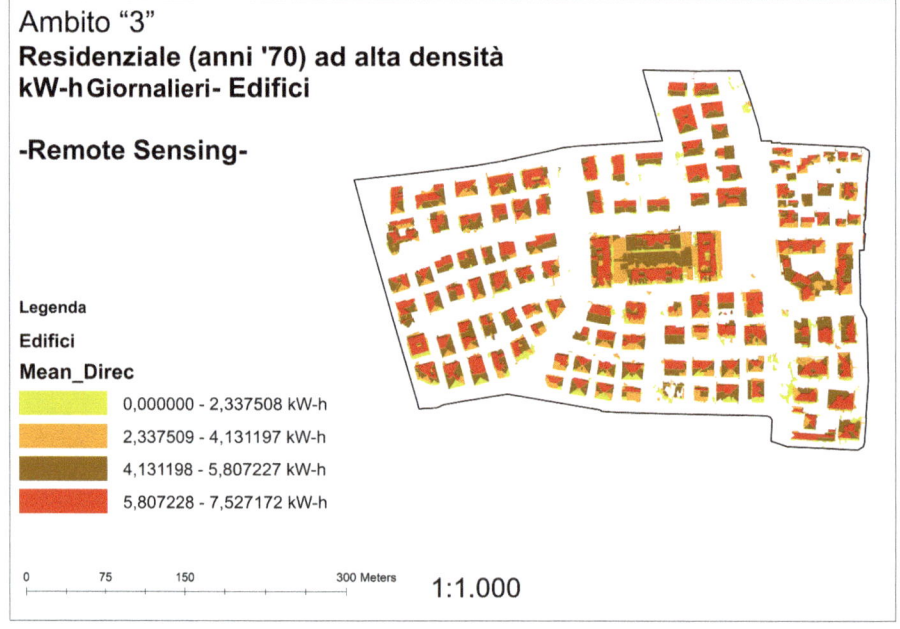

**Fig. 8.8** High density residential neighbourhood (1970's) Daily kW/h/buildings

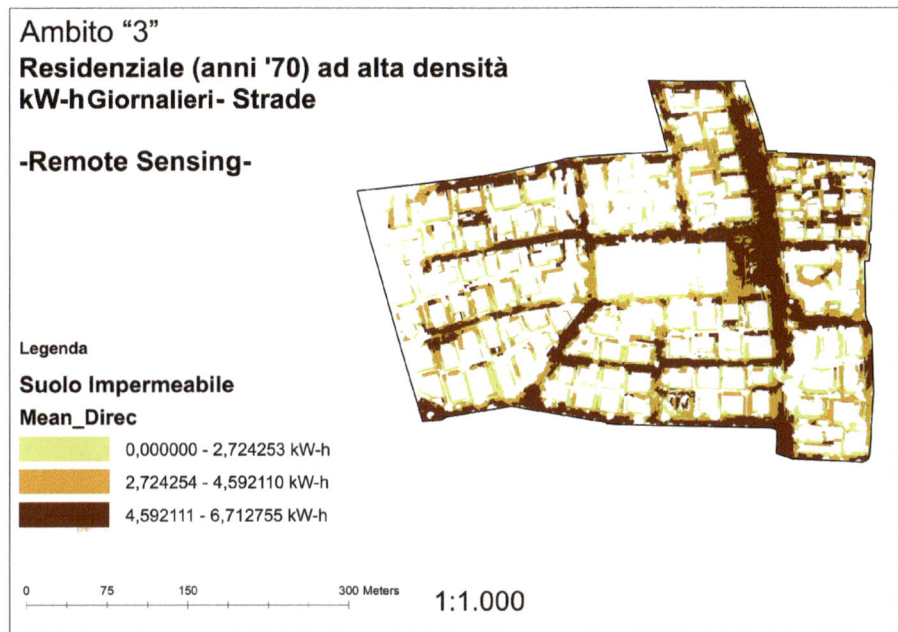

**Fig. 8.9** High density residential neighbourhood (1970's) Daily kW/h/roads

## 8.3   Feasibility Study

After processing the data pertaining to the areas inside the survey transept, it became necessary to understand what mitigation and adaptation interventions should be considered based on the morphological aspects of this specific neighborhood of the city of Padua, including buildings, outdoor public spaces, and private spaces.

The integration of these sets of interventions produced the "transformation scenarios" that were later tested for effectiveness with the *ENVI-met* software.

Through this process, we initiated an actual project for the pilot area, by adapting to it specific mitigation measures that until then had been generic proposals for other areas. As a result of the preliminary project test performed inside the pilot area, we came up with four different project scenarios as follows:

*"green ground"*: a scenario where the permeable surface of the area is increased (from 18 % to 23 %) by turning a paved parking lot into a grass surface and planting 10 m tall trees along the main roads of the area;

*"cool pavements"*: substituting the traditional paving material (0.2 albedo) and concrete (0.4 albedo) used on streets and sidewalks with high albedo (0.5) "cool" materials;

*"cool roofs"*: substituting the traditional tile or covered flat roofs with "cool" materials (0.3 to 0.6 albedo);

*"green ground+cool pavements"*: a scenario that adopts both these mitigation interventions simultaneously.

The precision of the digital terrain model, obtained through the use of *LiDAR* and orthophotos data, made it possible to increase the number of details with which to perform the effectiveness measuring simulations in terms of temperature reduction, using the various mitigation interventions considered for each of the four scenarios.

The subsequently performed simulations allowed us to close the working cycle through the virtual testing of the interventions under consideration, thereby assessing the best strategies for the pilot area.

The fourth scenario, *"green ground + cool pavements"*, tested with the *ENVI-met* software, provided the best results from the point of view of temperature reduction. Based on this, we used this scenario for our project.


## 8.4   The Pilot Area as Testing Ground for the Veneto Region

The scenario we selected for the pilot area is aimed at increasing resilience to negative externalities caused by climate variability. This urban green infrastructure plan becomes a driving force for the adaptation of urban and regional systems to climate changes.

A network of natural and semi-natural areas has a good ability to make the land more resilient; if well designed, green infrastructure can mitigate the effects of floods and the increasing droughts, improve water and air quality, effectively promote soil protection, and oppose hydrogeological instability. In addition, it ensures air filtration, erosion protection, water flows regulation, coastal protection, soil structure maintenance, and carbon storage.

The multiple benefits of green infrastructure are also set forth in the European strategy for green infrastructure published last year (EU 2013). For example, trees and green areas may prevent flooding in cities while reducing air pollution and noise levels. Furthermore, the use of natural systems can often be cheaper and more durable than a hard artificial structure.

However, we have yet to understand how to apply these changes to a real area. Veneto's central area, having been greatly transformed over the last 40 years, mostly through a series of small spread out interventions, requires a specific design approach. By the same talking, the pilot area transformation project produced by our feasibility study can be implemented through small interventions that will be made presumably over a period of about 20 years.

Our project and its graphical representations provide a potential scenario that can presumable be striven for. Its mitigation measures against the heat island effect can be effectively used in the first place through the adoption of appropriate land and urban management and planning tools that can implement the new adaptation priorities arising from climate changes.

Based on this, in order to make our mitigation measures as applicable as possible, our work group performed a survey of the existing land management and planning tools, linking each measure to a potentially modifiable planning tool (Table 8.2).

**Table 8.2** Ordinary urban planning and management tools: possible UHI moderation interventions

| GROUND SURFACES | Intervention | Main regulating body | Tool (for urban planning or managment) | Indication type | Notes |
|---|---|---|---|---|---|
| Management of the Reflectance and Emissivity of impermeable surfaces for public and private spaces | (1) Pigmentation type | Municipal Administrations | Municipal urban plan (the name will change according to the specific regional legislation) | Indications on the surfaces of each ordinary transformation area | The pigmentation of existing pavements should be modified gradually |
| | (2) Material type | | Ordinary and extraordinary maintenance plan | Reflectance parameters of existing surfaces | New surfaces should employ materials that combine greater reflectance and a low impermeabilization rate |
| | | | Infrastructure plan | Reflectance parameters of new infrastructure surfaces | |
| | | | Building regulation | Reflectance parameters of surfaces of new private and public buildings | |

Source: *IUAV data processing, 2014*

## 8.5   Possible Transformations of the Pilot Area

The possible transformations/interventions proposed below refer to the previously analyzed *"green ground + cool pavements"* scenario. We took the basic pattern used for *ENVI-met* modeling and came up with a number of potential transformations for the pilot area.

These possible interventions are not part of a single urban planning project, but they are structured rather as small interventions to be implemented through the use of the urban planning tools analyzed above (see Table 8.2).

It should be noted that the proposed interventions can be effective on their own in mitigating the heat island effect; however, more specifics are needed on the areas they are going to be applied to, so as to make effective and cost effective decisions. Maximum effectiveness can be reached when all of these interventions together become part of a general strategy of adaptation to climate changes combined with the more important urban and/or socio-economic concerns of a given area.

Intervention 0 actual conditions + summary of intervention 0 actual conditions.

### 8.5.1   Outdoor Public Spaces

The first intervention posits an increase in the reflectance of the road surface. This can be done by means of several types of materials. Two technical options may be considered: the more immediate one would be acting on the pavement's coloration/pigmentation, the other would involve a more structured approach of asphalt type modification. This type of intervention can be planned on a municipal scale over a set period of time, for example, the years it would take to pave everything over and remanufacture some types of streets signs. For the sake of economic sustainability, it would make sense to prioritize such interventions by area with the aid of specific maps. For larger cities, the mapping process can be integrated with urban heat studies, using direct readings or indirect photogrammetric data processing. Municipalities that do not have access to complex analyses of urban heating phenomena can prioritize their most densely occupied areas, and also apply the indexes suggested by the University of Vienna for this specific project (Figs. 8.10 and 8.11).

The second intervention also concerns the reflectance of impermeabilized urban surfaces; however, in this case, the spaces considered – sidewalks, parking lots, and city squares – do not involve car traffic. These surfaces, like the ones we just discussed, need to be approached according to a set of urban planning priorities. Modifications will necessitate the application of street furniture programs that will include reflectance limits to the repaving of city squares and sidewalks. For the application of this intervention inside the pilot area, we considered also parking lots and street side parking areas, which are normally paved, and which will have to be handled using a different approach, like more permeable materials for improved absorption of rainwater (Fig. 8.12).

**Fig. 8.10** Intervention 1 + summary of intervention 1
Modify the albedo of streets. Modify the reflectance of the road surface

**Fig. 8.11** Intervention 2 + summary of intervention 2
Modify the albedo of sidewalks and parking lots. Modify the reflectance of sidewalks and parking lots

**Fig. 8.12** Intervention 3 + summary of intervention 3
Add *green areas* on the ground of public spaces in addition to public trees

### 8.5.2   Public Green Spaces: Create New Traffic Islands and Plant New Trees

The third intervention focuses on increasing public green spaces. Here too, we are not talking about great new parks or large green areas; these are micro interventions that can be applied in a city with a consolidated infrastructure. Practically speaking, it is about creating new traffic islands and planting new trees. These interventions necessitate an innovative approach based on a new public space management vision. At present, for most Italian cities to add new trees and traffic islands in urban areas where everything has already been built could involve an increase in maintenance costs and a loss of needed urban space (for car, bicycle, and pedestrian traffic, parking, etc.). This is why this new approach would have to be adopted as part of a strategic paradigm shift in the general management of city spaces. Creating new green spaces inside the context of a built up city entails changing one's perception of street space as just for transit, parking, and car maneuvering. A new paradigm for the use of public green spaces requires a strategic rethinking of urban greenery, insofar as what it can offer in terms of the urban ecosystem adapting to climate changes. Going down this road means understanding and valuing the gamut of services that green spaces can offer to mitigate the heat island effect in addition to other negative externalities due to weather phenomena: lamination for the containment of water during extended rainfall, reducing air pollution, helping to reduce the speed of urban traffic, and even a general improvement in the environmental and aesthetic quality of public spaces. New traffic islands were added inside the pilot area along the streets whose width could reasonably be reduced or that could be switched from two-way to one-way traffic. We also added a green area on a space used as a square along Via Guizza.

## 8.5.3   *Private Outdoor Spaces*

The management of private outdoor spaces is also a major factor in determining the occurrence and intensity of urban heat islands. Depending on the type of settlement, a significant portion of Veneto's cities that is not covered by buildings is private property; in the case of Padua's pilot area, private property covers more than 1\3 of the total surface.

It is obvious that the management of these surfaces is greatly relevant to the mitigation of the heat island effect. However, in this case, even a simple technical solution (increasing the green surface and the reflectance of impermeable surfaces) must be justified at the management and legislative level; a cogent response is necessarily based on a general strategic vision that can harmonize individual management needs with the understanding of the importance of adapting to climate changes. Private outdoor spaces can easily be handled when it comes to new buildings, where building codes can set new extension and surface type parameters, but they require a much more thorough consideration for consolidated urban areas, as is the case of the pilot area, since it is more difficult to find legislative levers or incentives to modify already built property.

In these cases, implementation can be achieved through the following:

– come up with education programs for the city's inhabitants, so that the more sensitive section of them may be stimulated into performing ordinary maintenance of their private spaces;
– coordinate with the water department to establish incentive policies where bill payments reflect how much of a property is impermeabilized.

Within the pilot area, we proposed to modify only the reflectivity of the private surfaces currently impermeabilized by asphalt, concrete or the like with intervention 4, whereas with intervention 5 we proposed to replace them with greenery (Figs. 8.13 and 8.14).

## 8.5.4   *Post-scenario Intervention*

The last project intervention on the pilot area takes a step beyond the interventions encompassed by the feasibility study, proposing a more incisive transformation that involves buildings as well.

**Fig. 8.13** Intervention 4 + summary of intervention 4
Albedo on private paved surfaces. Modify the reflectance on private surfaces

**Fig. 8.14** Intervention 5 + summary of intervention 5. Put greenery on private surfaces
Add green areas on private surfaces

**Fig. 8.15** Intervention 6+summary of intervention 6. *Green roofs*: extra scenario intervention

This intervention pictures a gradual transformation of the roofs of private buildings, first by modifying the reflectance of flat surfaces and then by turning them into green roofs. A real expansion of green roofs can only occur if our cities will understand and assess their value in terms of the services that they can offer to the community: mitigation of the heat island effect, lamination of rainwater, improvement of air quality, and last but not least, recreational use.

This final scenario is not meant to offer a new utopian vision of what a city should be like, but it is rather an attempt to propose the application of solutions that would adapt the current state of affairs to a no longer so remote future of rapid climate changes.

To increase concretely the resilience of our urban system to climate changes, such as in the case of heat island mitigation demonstrated by this project, we must utilize several approaches: in-depth climate evolution studies on the regional and local scale, the use of climate modeling, the use of new technologies as a support for urban planning, the research of new building materials, the revision of the urban governance system, and above all the creation of a new strategy to review and harmonize all the aspects of the matter (Fig. 8.15).

# Bibliography

Betsill M., & Bulkeley H. (2006). Cities and multilevel governance of global climate change. *Global Governance: A Review of Multilateralism and international Organizations, 12*(2). Boulder: Lynne Rienner Publishers.

Bulkeley, H. (2006). A changing climate for spatial planning. In *Planning theory and practice.* London.

EU. (2013). *Green infrastructure (GI)*, COM(2013) 249 final, Brussels.

Foley, J. A., & DeFries. R., et al. (2005). Global consequences of land use. *Science, 309*(5734), 570–574.

Fung, W. Y., Lam, K. S., Hung, W. T., Pang, S. W., & Lee, Y. L. (2006). Impact of urban temperature on energy consumption of Hong Kong. *Energy, 31*, 2623–2637.

Nyuk Hien Wong, & Yu Chen. (2008). Tropical urban heat Islands: Climate, buildings and greenery. UK: Taylor and Francis.

Grimmond, C. S. B. (2007). Urbanization and global environmental change: Local effect of urban warning. *Cities and global environmental change,173*(1), 83–88.

Grimmond, C. S. B., Cleugh, H., & Oke, T. R. (1991). An objective urban heat storage model and its comparison with other schemes. *Atmospheric Environment, 25B*(3), 311–326.

Huang, L., Li, J., Zhao, D., & Zhu, J. (2008). A fieldwork study on the diurnal changes of urban microclimate in four types of ground cover and urban heat island of Nanjing, China. *Building and Environment, 43*, 7–17.

IPCC. (2001). *Summary for policymakers, climate change 2001: Synthesis report.* In S. Solomon et al. (Eds.). Cambridge: Cambridge University Press.

IPCC. (2007). *Summary for policymakers, climate change 2007: Synthesis report.* In S. Solomon et al. (Eds.). Cambridge: Cambridge University Press.

IPCC. (2012). Managing the risks of extreme events and disaster to advance climate change adaptation. Special report. In C. B. Field et al. (Eds.). Cambridge: Cambridge University Press.

Jusuf, S. K., Wong, N. H., Hagen, E., Anggoro, R., & Yan, H. (2007). The influence of land use on the urban heat island in Singapore. *Habitat International, 31*, 232–242.

Makar, P. A., Gravel, S., Chirkov, V., Strawbridge, K. B., Froude, F., Arnold, J., & Brooks, J. (2006). Heat flux, urban properties and regional weather. *Atmospheric Environment, 40*(15), 2750–2766.

Musco, F., Fregolent, L., Magni, F., Maragno, D., & Ferro, D. (2014). Calmierare gli impatti del fenomeno delle isole di calore urbano con la pianificazione urbanistica: esiti ed applicazioni del progetto UHI (Central Europe) in Veneto. [Moderating the impacts of the phenomenon of urban heat islands with urban planning: outcomes and applications of the UHI project (Central Europe) in Veneto]. In *Focus su Le città e la sfida dei cambiamenti climatici*, Ispra S/A52/14, Rome.

OECD. (2009). *Integrating climate change adaptation into development cooperation: Policy guidance.* Paris.

Oke, T. R. (1982). The energetic basis of the urban heat Island. *Quarterly Journal of the Royal Meteorological Society*, Vancouver.

Olgyay, V. (1998). *Arquitectura y Clima. Manual de diseño bioclimático para arquitectos y urbanistas.* [Architecture and Climate. Handbook of bioclimatic design for architects and urban planners]. Barcelona: Gustavo Gili.

ONU-Habitat. (2011a). *Planning for climate change. A strategic values based approach for urban planners.* Nairobi.

ONU-Habitat. (2011b). Global report on human settlements 2011: Cities and climate changes. Nairobi.

Vitousek, P., & Mooney, H., et al. (1997, July 25). Human domination of Earth's ecosystems. *Science, New Series, 277*(5325), 494–499.

Sánchez-Rodríguez, R. (2012). Understanding and improving urban response to climate change. Reflections for an operational approach to adaptation in low and middle-income countries. In D. Hoornewg et al. (Eds.), *Cities and climate change: Responding to an urgent Agenda* (Vol. 2). Washington, DC: World Bank.

Sánchez-Rodríguez, R. et al. (2005). *Science plan. Urbanization and global environmental change.* Bonn: International Human Dimensions Program on Global Environmental Change.

Seto, K., Sánchez Rodriguez, R., & Fragkias, M. (2010). The new geography of contemporary urbanization and the environment. *Annual Review of Environment and Resources, 35.* California: Palo Alto.

Svirejeva-Hopkins, A., Schellnhuber, H. J., & Pomaz, V. L. (2004). Urbanized territories as a specific component of the global carbon cycle. *Ecological Modeling,173*(2–3), 295–312.

UNDP. (2005). *Adaptation policy frameworks for climate change: Developing strategies, policies and measure.* New York: United Nations.

UNDP. (2010). Designing climate change adaptation initiatives. A UNDP toolkit for practitioners (Vol. 7, pp. 203–214). New York: United Nations

Wong, N. H., & Chen, Y. (2009). *Tropical urban heat islands: Climate, building and Greenery.* London: Taylor and Francis.

World Bank. (2011). *Guide to climate change adaptation in cities.* Washington, DC.

World Bank. (2012). *Building urban resilience: Principles, tools, and practice.* Washington, DC.

Berdahl P., & Bretz. (1997). *Preliminary survey of the solar reflectance of cool roofing materials.*

Bonafè, G. (2006). Microclima urbano: impatto dell'urbanizzazione sulle condizioni climatiche locali e fattori di mitigazione [Urban microclimate: Impact of urbanization on local weather conditions and mitigating factors]. Area Meteorologica Ambientale, Servizio IdroMeteorologico, ARPA, Emilia Romagna.

Mayer, H. (1992). PlannungsfaktorStadtKlima. *Muchner Forum, Berichte und Protokolle, 107*, 167–205.

Oke, T. R. (2006). *Initial guidance to obtain representative meteorological observations at urban cities.* Instruments and Observing Methods. Canada: WHO. No. 81.

Oke, T. R. (2006). Towards better scientific communication in urban climate. *Theoretical and applied Climatology, 84*, 179–190.

Papadopoulos, A. (2001). The influence of street canyons on the cooling loads of buildings and the performance of air conditioning systems. *Energy and Buildings, 33.*

Santamouris, M. (2007). Heat island research in Europe – State of the art. *Advances in Building Energy Research, 1*, 123–150.

Samonà, G. (1980). Come ricominciare. Il territorio della città in estensione secondo una nuova forma di pianificazione urbanistica [How to start over. The extended city area according to a new form of urban planning]. *Parametro, 90*, 15–16.

Shahmohamadi, P., Cubasch, U., Sodoudi, S., & Che-Ani, A. I. (2012). Mitigating the Urban Heat Island. Effects in Teheran Metropolitan Area, Chapter 11, pp. 282–283. In *Air pollution – A comprehensive prospective*. Intech.

Selicato, F., & Rotondo, F. (2010). *Progettazione Urbanistica Teorie e Tecniche* [Urban design theories and techniques]. Milan: McGraw-Hill.

Heat Island effect in the Teheran metropolitan area, Ch. 11 of Air pollution – A comprehensive perspective. Croatia, Edited by Budi Haryanto.

Wong, N. H., & Chen, Y. (2009). Tropical Urban Heat Island: Climate, buildings and greenery. London/New York: Taylor & Francis Press.

## Focus A: Energy and Urban Form

Filippo Magni
Department of Design and Planning in Complex Environments,
IUAV University of Venice,
S.Croce, 1957, 30135, Venice, Italy
fmagni@iuav.it

**Keywords** Urban form, Energy planning, Design and UHI, Climate change

Nowadays, the population which can be considered an urban population exceeds 50 % of the total with areas where the percentage reaches 80 %. This process of urbanization of the population, combined with the potential impacts of climate change induced by the anthropogenic component (IPCC 2007), provides a new impetus to efforts to understand how the forms, functions and resources interact within urban environments. In some cities, energy consumption per capita has grown at approximately the same rate as spatial growth (Baynes and Bai 2009). From a point of view of urban metabolism, energy therefore represents one of the most critical resource flows for the life of a city, since it is a primary factor in supporting physical and economic systems (Alberti 1994). Considering then that the growing global contribution to GHG emissions of cities (Bai 2007), addressing global climate change bringing it down to an urban level acquires strategic interest, as it provides greater effectiveness of intervention: energy consumption, urban form, density and morphology if correctly associated, may provide the opportunity to address the issue of climate locally. Much of the literature available, for example, Williams et al. (2000), Jenks and Burgess (2000) and Foley (2005), Oke (2006),

(continued)

focus on the issue of mitigation considering it a driver of urban sustainability. Better planning, better design of spaces and urban forms, should be able to both mitigate climate change, and ensure a gradual process of adaptation to reduce the direct and indirect impacts of climate change on cities.

The approach to mitigation carried out so far to reduce GHG emissions, has been focused primarily on the production of energy from renewable sources, energy savings of buildings, "green" technologies for industrial production, alternative fuels with a higher efficiency for vehicles, and an increase in public transport. It has focused less on the study of urban form and the role it plays in an energy strategy for the conservation and efficient use of this resource.

The globalization of the construction industry and the total delegation of the indoor systems and plants has in fact determined, in the last century, an increasingly pervasive realization of approved buildings and urban structures barely related to their climatic context, cultural material. "The same buildings can be found from Stockholm to Nairobi, from Shanghai to Sao Paulo, with age-old design principles simply eliminated" (Butera 2004): a challenge for nature set by man, to prove that he can live indifferently in any context and in any climate.

If we take a broader view, which embraces and considers the territory as a geometric area of energy consumption (Olgyay 1951), we must consider that urban planning aimed at energy saving and sustainability must be sensitive to local conditions and able to exploit the resources that the environment provides. The end result of this approach is expressed naturally in architectural forms and urban structures, contextualized by morphology, type, use of materials. This does not necessarily mean that they should be vernacular or traditional, given that typological, morphological and technical-constructive solutions evolve over time as new requirements emerge and new materials and new building systems are introduced. We must also take into account on the fact that the use (and waste) of energy does not only depend on the use of the individual buildings and their systems, but often on the way in which these have been designed and arranged in relation to each other. For example, the layout of a building on the land, its position in relation to the prevailing winds, the path of the sun and the reciprocal relationship with other surrounding buildings can prevent the sunlight needed from reaching it, creating barriers to hot winds and vice versa channelling the cold winds, leading to an inefficient use of energy. It is very rare that building regulations or urban-building standards for the implementation of planning regulations contain directives aimed at ensuring environmental conditions which are conducive to energy saving for temperature regulations.

Therefore, urban planning policies that are sensitive to reducing energy consumption and comfort, related to the use of the spaces within a city, must be based on a bioclimatic approach, which aims to simultaneously control three interconnected levels: environmental-climatic, typological and

(continued)

technical-constructive which, if studied in sufficient detail, can provide the following information:

- with regards to the control of aspects relating to the relationship between a building and the environment, planning and architecture (especially that related to temperate climates) have always had to deal with a climate characterized by significant seasonal changes (temperature, humidity, wind, solar radiation) which therefore encourages and imposes solutions capable of adapting to these seasonal variations. In addition to the climate, individual buildings must also respond to the microclimate of the area, i.e. the specific features of individual sites also in relation to the shape of the urban constructions or landscape (which influence, and sometimes change, the typical climatic conditions such as temperature, humidity, wind speed and solar radiation, distinguishing a single context with local conditions).
- with regards to the control of the typological aspects, buildings must be characterized by a search for balance between a compact form in winter (based on the more advantageous ratio between surface and volume in relation to heat loss) and a more open form in the summer (based on the possibility of favouring natural ventilation), with "open-closed" structured spaces for winter/summer use (porches, balconies, patios, filter spaces). For example, the typical Mediterranean building is a home with a patio, compact but "porous" (Olgyay, 1998). The in-line or terraced type house is equally effective, allowing compactness to be favoured (seen as support-type housing), but also to identify two preferable exposures, namely south facing (so that sunlight can be exploited in the winter months) and north facing (to have a "cool" side in the summer triggering the natural ventilation throughout the building).
- for the control of the technical-constructive aspects, an urban structure must be characterized by the passive use of energy thanks to the exploitation of sunlight both directly (windows) or indirectly (heat storage units) and by the presence of an adequate capacitive mass (and thermal inertia) to retain heat and mitigate temperature peaks (reducing and off-setting the introduction of the thermal wave) in the summer. Therefore, building orientation, building shape and cladding characteristics are the aspects on which designers should focus more carefully. A building which exploits the characteristics surrounding it is defined as "passive", which should be distinguished from those buildings which artificially (and therefore "actively") construct comfort within the rooms (not to be confused with the term "*passivhaus*", which refers to an energy standard). A passive urban structure combines the ability to use favourable climatic factors (capturing solar energy in winter, directing air flows in the summer) with the ability to maintain favourable conditions (storing heat in winter and night-time cold in summer) and hamper unfavourable conditions without resorting to costly and energy-intensive additions to the system.

(continued)

It is therefore the designer and/or planner who, on different levels, who is called upon to deal with the issues related to the regulations, the shape of the urban structures, the orientation of the buildings, the cladding and systems, and therefore work towards reducing energy consumption and ensuring suitable living comfort.

Therefore, designing buildings today, with a climate which is changing constantly, means understanding the reasons related to the microclimate, resources and local materials. Planning in these geographical areas does not require a strict adherence to architectural shapes of traditional buildings, but rather an innovative reinterpretation of the reasons which have "naturally" driven construction for centuries.

## Bibliography

Alberti, M., Solera, G., & Tsesti, V. (1994). La città sostenibile. Analisi, scenari e proposte per un'ecologia urbana in Europa, Franco Angeli, Milano.

Bai, X. M. (2007). Integrating global concerns into urban management: The scale and readiness arguments. *Journal of Industrial Ecology, 22*, 15–30.

Baynes, T., & Bai, X. M. (2009). Trajectories of change: Melbourne's population, urban development, energy supply and use 1960–2006 (GEA Working Paper).

Butera, F. M. (2004). Dalla caverna alla casa ecologica. Storia del comfort e dell'energia, Edizioni Ambiente, Milano.

Foley, J. A., & DeFries, R., et al. (2005). Global consequences of land use. *Science, 309*(5734), 570–574.

IPCC. (2007). *Summary for policymakers, climate change 2007: Synthesis report*. In S. Solomon, et al. (Eds.). Cambridge: Cambridge University Press.

IPCC. (2012). *Managing the risks of extreme events and disaster to advance climate change adaptation. Special Report*. In C. B. Field et al. (Eds.). Cambridge: Cambridge University Press.

Jenks, M., & Burgess, R. (Eds.). (2000). *Compact cities. Sustainable urban forms for developing countries*. London: Spon Press, Taylor and Francis Group.

Oke, T. R. (2006). Towards better scientific communication in urban climate. *Theorical and applied Climatology, 84*, 179–190.

Olgyay, V. (1998). Arquitectura y Clima. Manual de diseño bioclimático para arquitectos y urbanistas. Barcellona: Gustavo Gili.

London: Spon Press, Taylor and Francis Group.

Williams, K., Burton, E., & Jenks, M. (2000). Achieving sustainable urban form.

(continued)

# Focus B: Using Aerial Photogrammetry for Urban Sustainability Analysis

Denis Maragno
Department of Design and Planning in Complex Environments,
IUAV University of Venice,
S.Croce, 1957, 30135, Venice, Italy
dmaragno@iuav.it

**Keywords** Forecasting UHI, Monitoring climate change, Digital surface model

The ongoing climate change and energy issues are among the most important challenges that city planning, zoning, and architecture are now facing.

Limited resources and the economic crisis aggravate the difficulties of intervention, forcing one to identify multifunctional interventions for combined solutions to different problems and needs (Musco et al. 2013).

In recent years, there has been considerably growing interest in alternative and renewable forms of energy supply (Pearce 2002); at the same time, the difficulties experienced by cities in resisting the effects of climate change are steering Public Administrations toward formulating integrated policies in order to decrease $CO_2$ production and simultaneously increase the land's resilience to climate change (Musco 2008).

The ensuing difficulties lead one to analyze urban environments using the most advanced technology and the best tools available.

We introduce an example of applied methodology, where, using a point cloud generated by a photogrammetric technique, we will consider what buildings are most suitable for installing a PV system and which areas are most affected by solar incidence (Wilson et al. 2000).

Thanks to recent advances in photogrammetry Hardware and Software (Hirschmüller 2008), by using stereoscopic techniques it is possible to make 3D point clouds[3] and digital models of a terrain comparable in definition to those produced by active sensors (e.g., LiDAR). But unlike LiDAR data,[4] a

---

[3] These point clouds consist of geo-referenced and dimensioned points, with an average density of 12 points per $m^2$, which thus render the photographed area (or detected, in the case of LiDAR processing) in digital form and in three dimensions. Available algorithms, including through open source software, are able to analyze both geographically organized and dimensioned points, and render an exact 3D urban composition in numerical form.

[4] LiDAR (Laser Imaging Detection and Ranging) performs remote sensing to determine the distance of an object or surface through the emission of high frequency laser pulses by a flying sensor (plane or drone). The distance of an object is given by the length of time elapsed between emission and reception. Very high frequency pulses bouncing from objects or the ground are converted into geo-referenced and dimensioned points, thus giving rise to a "point cloud" from which the exact reconstruction of an area can be created in the form of three-dimensional digital models.

(continued)

point cloud created with the Dense Image Matching (DIM) method has a much lower cost, making it more accessible to Public Administrations.

To assess the solar incidence impacting the slopes of buildings and horizontal urban surfaces, we trace a methodology composed of three steps.

### Step One

In order to transform the point cloud generated by the stereoscopic technique, the first step is to transform it into a DSM[5] (Digital Surface Model), which is a raster file whose pixels contain the average dimensions of the points of the cloud.

The very high density of the cloud has allowed us to create a DSM with a 0.5 m² per pixel resolution. The conversion can take place with the aid of various software tools; for the prototype we used LAStolls. This application has a very simple command line interface, developed by Martin Isenburg for LiDAR data, but which could be used for this specific operation on photogrammetric data in the LAS format.

The command for the operation is the following:

las2dem -i name_file_input.las -o name_file_output.tif -step 0.5

By using this command on Command Prompt, the software automatically transforms the point cloud into the desired DSM.

### Step Two

The second step consists in producing information pertaining to solar incidence and the inclination angles of roof slopes. This operation was performed with the Open Source Software System for Automated Geoscientific Analyses (SAGA GIS). SAGA GIS has many algorithms, one of which can calculate the inclination angles of surfaces (by Olaya 2004), and another the potential solar incidence of surfaces (by Conrad 2010).

The information layer pertaining to the inclination angles of roof slopes will be crucial in the classification of roofs, as for the study in question it is important to distinguish flat roofs from sloping roofs. The inclination angles of roofs are also essential in selecting and suggesting an installation type.

---

[5] DSM is a digital elevation model in raster format. Each cell of the model, called pixel, holds a height number. Therefore, DSM represents the spatial distribution of the dimensions of an area or surface.

(continued)

The calculation of the solar incidence uses the three-dimensional knowledge acquired automatically by SAGA GIS from the input raster file: knowing the latitude and longitude supplied by the georeferenced raster file, the algorithm simulates the height of the sun and considers the dimensioned urban elements, projecting shadows, and thereby calculating the potential solar incidence for each pixel.

The output will therefore still be a raster file with a $0.5 \text{ m}^2$ per pixel resolution, where each pixel contains the information in kWh. The process has the option to choose the time of year to be analyzed (January rather than July), and the number of days (1–365).

## Step Three

The final step of the analysis compares the information obtained so as to identify the urban areas vulnerable to strong solar radiation, in order to come up with more accurate mitigating and/or adaptive solutions.

Making use of any GIS Software (for this example we used Quantum GIS) it is possible to analyze quantitatively the two dimensions under study.

As for identifying the roofs most suitable for a photovoltaic or solar thermal installation, one can simply evaluate information such as solar incidence, inclination angle, and slope size.

As for identifying hot urban areas (with greater solar incidence), one can simply ask Quantum GIS for the location of areas vulnerable to solar incidence (it will be better to convert the raster file into vector format for this purpose). The resulting zoning information will be of great help in choosing an intervention, where, depending on land area type (square, parking lot, sidewalk, intermodal station, etc.), this can be done with greater accuracy and solutions pertaining to shading in conjunction with renewable energy production can be easier to find.

## Bibliography

Hirschmuller, H. (2008). Stereo processing by semi-global matching and mutual information. *IEEE Transaction on Pattern Analysis an Machine Intelligence, 30*(2), 328–341.

Musco, F. (2008). Cambiamenti Climatici, politiche di adattamento e mitigazione: una prospettiva urbana, in Asur, Franco Angeli, Milan.

Musco, F., Maragno, D., Gariboldi, D., & Vedovo, E. (2013). Remote Sensign e Cambiamenti Climatici: rischi e opportunità nel riuso e riciclo delle città.

(continued)

In *Il Governo della Città nella Contemporaneità: La città come motore di sviluppo* (pp. 187–189). Rome: Edizioni INU, Salerno.

Pearce, J. M. (2002). Photovoltaics: A path to sustainable futures. *Futures, 34*(7), 663–674.

Wilson, J. P., & Gallant, J. C. (Eds.). (2000). *Terrain analysis – Principles and Applications*. New York: John Wiley & Sons, Inc.

## Focus C: Urban Sprawl and Measures for Environmental Sustainability

Laura Fregolent
Department of Design and Planning in Complex Environments,
IUAV University of Venice,
S.Croce, 1957, 30135, Venice, Italy
laura.fregolent@iuav.it

**Keywords** Urban sprawl, Urban growth, Carbon dioxide emissions, Welfare policies

## *Urban Sprawl: Characteristics and Impacts*

The contemporary city has increasingly been characterised by certain distinctive features, in particular by a progressive urban fragmentation and by a polarisation and specialisation of some functions and services outside the urban centres, which resulted in a growing increase of mobility.

The phenomenon, known as *urban sprawl*, is generalised and extended and has been studied extensively in its social, economic, cultural, political and institutional components[6] as well as in its causes, impacts,[7] formal expressions and local specificities, which have led to different manifestations of the phenomenon according to the different territorial and geographical contexts.

---

[6] About the effects of local scale regulations and sprawl and about the relationship between administrative fragmentation and sprawl increase, see: Pendall R. (1999), "Do Land-Use Controls Cause Sprawl?", *Environment and Planning B*, vol. 26/4, pp. 555–571.

[7] Between the various studies, see Ewing R., Pendall R., and Chen D. (2002), *Measuring Sprawl and Its Impact*, Smart Growth America/U.S. Environmental Protection Agency, Washington, D.C.

(continued)

Research has highlighted the close relationship between low-density of settlements and sprawl, as well as new needs, consumption and lifestyles impose high-impact territorial choices with high land use rates.

A topic that over time has become important in research on the effects of the sprawl is the one linked to the quantification of the collective[8] and social costs related to mobility[9] and more recently, the health care costs that the sprawl imposes. Some recent studies in fact highlight the connection between a sprawl and the health of individuals: «The study, *Relationship Between Urban Sprawl and Physical Activity, Obesity, and Morbidity*, found that people living in counties marked by sprawling development are likely to walk less and weigh more than people who live in less sprawling counties. In addition, people in more sprawling counties are more likely to suffer from hypertension (high blood pressure)»[10] as well as the need to act, regulating and limiting the development of sprawling so as to allow improvement the people's quality of life.

Research on the sprawl is also characterised by an approach which takes ever more the form of planning and is ever more linked to identifying common tools, policies and measures that are primarily aimed at limiting land use. The latter is considered the most evident criticality and the one with the larger impact since it is associated with the erosion of natural, environmental and

---

[8] On this subject, reference should be made, by way of example, to the numerous studies conducted in the United States of America on the *social* and *environmental* costs of sprawl since the 70s: Real Estate Research Corporation (1974), "The Costs of Sprawl" (US Environmental Protection Agency, Washington, DC); Ladd H. (1992), Population growth, density, and the costs of providing public services, *Urban Studies*, 29; Carruthers J., Ulfarsson G.F. (2003), "Urban sprawl and the cost of public services", *Environment and Planning B: Planning and Design*, 30. In the case of Europe and Italy, the amount of research is more restricted: Camagni R., *et al.* (2002), *I costi collettivi della città dispersa*, Alinea, Firenze; Caperchione E. (2003), "Local Government accounting system reform in Italy: A Critical Analysis", *Journal of Public Budgeting, Accounting and Financial Management*, 15(1); Hortas-Rico M., Solé-Ollé A. (2008), "Does Urban Sprawl Increase the Costs of Providing Local Public Services? Evidence From Spanish Municipalities", *Urban Studies*, 47(7); Travisi C.M., *et al.* (2009), "Impacts of urban sprawl and commuting: a modelling study for Italy", *Journal of Transport Geography*, 18(3); Fregolent L., *et al.* (2012), "La relazione tra i modelli di sviluppo urbano dispersi e i costi dei servizi pubblici: un'analisi panel", in Cappellin R., Ferlaino F., Rizzi P. (editors), *La città nell'economia della conoscenza*, Franco Angeli, Milan.

[9] McCann B. (2000), *Driven to spend. The Impact of Sprawl on Household Transportation Expenses*, Surface Trasportation Policy Project, Center for Neighborhood Technology.

[10] McCann B., Ewing R. (2003), "Measuring the Health Effects of Sprawl: A National Analysis of Physical Activity, Obesity, and Chronic Disease", Smart Growth America and Surface Transportation Policy Project, Sep., p. 1.

(continued)

landscape resources, and soil sealing.[11] Waterproofing is the result of the strong anthropization caused by the urbanisation processes and leads to the degradation of the ecosystem functions, the alteration of ecological balance, and a series of negative impacts on the environment such as a strong pressure and water resources with consequent decrease in rainfall absorption; loss of biodiversity; impact on food safety; increase in solar energy absorption due to dark asphalt or concrete surfaces, which contribute significantly along with the heat generated by air conditioning and cooling systems, and the heat produced by traffic, to the so-called 'Urban Heat Island' effect.

The evolution of the sprawl phenomenon and the quantification of its impacts have therefore pushed scholars and researchers and then also administrators and politicians to search for the measures to adopt and the possible actions to be implemented through specific and sectoral policies as well as careful planning.

The urban and regional planning may limit the sprawl also by means of infrastructural and transport regulation measures which foster a reduction in greenhouse gas emissions and allow to direct the growth and form of the urban space. In fact, density, functional *mixité*, re-compaction of the urban space, infrastructure design and the promotion of public and collective transport are the principles for developing guidelines for sustainable planning that regulates the urban space in such a way to control and reduce $CO_2$ emissions. This is how the form of the urban settlement and the planning aimed to regulate its growth that can make an important contribution to climate protection,[12] and this is why even at a European level, one of the major pushes is heading for intervention on the urban form, the compaction of the urban space, and the re-use of abandoned and dismissed areas.

---

[11] On this subject, please refer to the documents produced by the EU, in particular the Soil Thematic Strategy (COM(2006) 231); and on the measures that may be adopted to mitigate the phenomenon: European Commission (2011), *Report on best practices for limiting soil sealing and mitigating its effects*, Technical Report, 2011-050, Apr., available at: http://ec. europa.eu/environment/soil/pdf/sealing/Soil%20sealing%20-%20Final%20Report.pdf

[12] See: Intergovernmental Panel on Climate Change – IPCC (2014), "Human Settlements, Infrastructure and Spatial Planning", in Edenhofer, O., R. Pichs-Madruga, Y. Sokona, E. Farahani, S. Kadner, K. Seyboth, A. Adler, I. Baum, S. Brunner, P. Eickemeier, B. Kriemann, J. Savolainen, S. Schlömer, C. von Stechow, T. Zwickel and J.C. Minx (eds.), *Climate Change 2014: Mitigation of Climate Change, Contribution of Working Group III to the Fifth Assessment Report of the Intergovernmental Panel on Climate Change*, cap. 12, Cambridge, Cambridge University Press, United Kingdom and New York.

(continued)

### 8.1.1 Measures and Tools for Sprawl Containment

The European Union has long been committed to the promotion of a culture of sustainability which over the past two decades has helped to increase attention towards environmental issues and is now a leader with respect to the major environmental issues, from combating climate change[13] to the protection of biodiversity.

As for land and its protection – with special reference to the land use – the measures proposed by the EU collide again with the constraints stemming from the fact that land planning is a matter for the competence of the individual Member States. But other policies that affect to a greater or lesser degree, the land transformation processes and the issue of non-sustainable consumption and soil sealing, are the focus for a number of European Institutions (European Commission, European Environment Agency, and Eurostat), who initiated monitoring programs, research and awareness, although we continue to wait for a specific directive on land that is difficult to achieve.

As previously mentioned, planning is one of the strategic tools identified by the EU for a new enhancement of the city and the control of urban growth to take place in a sustainability manner, because sustainability means a transdisciplinary perspective on phenomena that requires different approaches, opinions and observations, which planning can combine and intersect in order so as to develop effective solutions. For this reason: «The majority of the EU Member States have established the principle of sustainable development in their key spatial planning regulations, referring to economic use of land resources and avoidance of unnecessary urban sprawl. However, the existence of relevant regulations does not give any insight on the effectiveness of implemented measures».[14] In addition, the actions taken by several European

---

[13] On this subject, a first assessment can be made of the planning measures adopted in the USA: "The first generation of state and local climate change plans reflects increasing consciousness of this, and these plans have begun to take important steps, such as measuring emissions. But much stronger action is needed. Instead of pursuing slow, incremental policy changes, governments at all levels must adopt a *backcasting* approach, setting goals for both mitigation and adaptation based on the best available scientific knowledge, and working backward from these targets to develop plans and programs capable of achieving them. The initiatives would then be regularly reviewed and revised to ensure progress at an appropriate rate" (Wheeler S.M. (2008), "State and Municipal Climate Change Plans. the first generation", *Journal of the American Planning Association*, vol. 74/4, pp. 481–496).

[14] Soil Thematic Strategy (COM(2006) 231); and on the measures to be adopted to mitigate the phenomenon: European Commission (2011), *Report on best practices for limiting soil sealing and mitigating its effects*, Technical Report, 2011-050, Apr., available at: http://ec.europa.eu/environment/soil/pdf/sealing/Soil%20sealing%20-%20Final%20Report.pdf.

(continued)

countries and aimed at sustainability-oriented planning are based on: «Quantitative limits for annual land take exist only in six Member States: Austria, Belgium (Flanders), Germany, Luxembourg, the Netherlands, and the United Kingdom. In all cases the limits are indicative and are used as monitoring tools",[15] for example: «In England, 10 % of the total land area, which includes country roads, is urban and, according to the Department for Communities and Local Government (2008), over 70 % of new development is taking place on this previously developed land (i.e., brownfield) at high densities to conserve greenfield land. This is a highly restrictive land use policy, constraining the supply of new houses and limiting lifestyle choice».[16] Similar rules have been adopted in other European countries: in Germany, measures were introduced for progressive control of land consumption with the goal of reaching zero consumption in 2050, by envisaging re-use of brownfield sites, requiring new urbanization to only occur in areas accessible by public transport and urban development plans to be designed in such a way to enhance compact urban centres and provide them with services, such as Active City and District Centres (2008). In France, the "Schémas de la Cohérence Territoriale" (SCOT) – large-scale plans which serve as a guide for the local plans - allow to determine perimeter of the protected natural and urbanized spaces. The SCOTs impose the principle of "extension limitée de l'urbanisation" which establishes limits to the urbanisation of non-anthropized areas and the realization of large commercial spaces.

Different regions in Italy are preparing urban regulations aimed to control land consumption: Regione Veneto is working on a draft law on the reduction of land consumption through urban regeneration; this work began at the end of 2013 but the draft law is still under discussion. Regione Toscana is engaged in reviewing its Regional planning law with the introduction of specific measures to contain land use; Regione Puglia, which has already passed a law to encourage and facilitate access of youth to agriculture and combat the abandonment and consumption of agricultural land.

In addition to planning, the most common tools for urban growth control are greenbelts and urban growth boundaries, applied in different cities around the world (London, Berlin, Portland, Beijing, Singapore, etc.) with the aim of defining and delimiting the physical boundary between the city and the countryside, with the contribution also of fiscal and regulatory measures and other indications such as the re-use of dismissed areas and buildings to encourage

---

[15] op. cit.
[16] Echenique M.H., Hargreaves A.J., Mitchell G., and Namdeo A. (2012), "Growing Cities Sustainably. Does Urban Form Really Matter?", *Journal of the American Planning Association*, vol. 78/2, pp. 121–137.

(continued)

the upgrade of that which is already in existence, and the start of urban regeneration processes in order to prevent consumption of non-anthropized land, to allow construction works to take place on free land only when all dismissed or under-used land is recovered, and when re-use of areas already compromised has been verified to be impossible. In any case the use of free land should be linked to real needs; regulating land use through regulatory restrictions and adopting different local taxation systems for the different areas and uses.

The outcomes of these measures in the various contexts of application were different, and this demonstrates that there is no solution or single measure able to minimize the sprawl but that the measures to be taken must be different and applied in different ways, but all within an unambiguous reference framework to be determined by the planning.

The adjustment and implementation of policies aimed at governing urban transformation must take place at different scales of intervention with a focus on the large-scale and/or metropolitan scale, because this allows understand, contextualize, and find solutions to phenomena that occur on a local scale.

Planning plays a key role, especially on a vast or regional scale, since the processes of urban sprawl affect large areas of land - several municipalities and provinces mutually adjacent. This scale also enables to address issues such as pollutant gas emissions, resource management, reduction of land consumption, protection of natural ecosystems, and also enables to implement welfare policies[17]; and an integrated approach to urban development policies is the only key to implementing the *European strategy for environmental sustainability*.

---

[17]Wheeler S. (2009), "Regions, Megaregions, and Sustainability", *Regional Studies*, vol. 43/6, pp. 863–876.

# Chapter 9
# Pilot Action City of Vienna – UHI-STRAT Vienna

**Doris Damyanovic, Florian Reinwald, Christiane Brandenburg, Brigitte Allex, Birgit Gantner, Ulrich Morawetz, and Jürgen Preiss**

**Abstract** The article presents the results of the pilot action "Urban Heat Islands–Strategy Plan Vienna" (UHI-STRAT Vienna). It sets out by determining what potential consistent consideration of urban climate aspects at different levels of action and decision-making has and how to implement such consideration. In a second step it looks at today's and future development of UHI and the urban climate. The report goes on to explain the three fields of action identified, i.e. awareness building, information and public relations for UHI, as well as urban infrastructure and large-scale and more detailed technical and structural measures to support strong consideration of the issue. It shows up the levels of action in planning from the master plan to the actual project and the options available in the course. Two feasibility studies reveal how UHI-relevant measures can be implemented in designated areas of the city. They make a clear distinction between measures in the development of new city quarters and measures in adapting existing ones, and they also identify two different levels of planning, the strategic master plan on the one hand and the planning of legal provisions, i.e. the land-use and building development plans on the other hand. The "Master Plan for Nordbahnstraße – Innstraße" in Vienna's 20th municipal district is used as an example to show how measures can be introduced at different stages of the master plan process. Proposed measures can be embedded in land-use and building development plans, as demonstrated in the case of the quarter surrounding the Vienna University of Technology (Karlsplatz) in the 4th municipal district. The studies were assessed as to the feasibility of the measures proposed, which involved participation of different agencies of the Vienna

D. Damyanovic • F. Reinwald • C. Brandenburg • B. Allex • B. Gantner
Department of Landscape, Spatial and Infrastructure Sciences, Institute of Landscape Planning, University of Natural Resources and Life Sciences, Vienna, Austria

U. Morawetz
Department of Economics and Social Sciences, Institute for Sustainable Economic Development, University of Natural Resources and Life Sciences, Vienna, Austria

J. Preiss (✉)
Vienna Environmental Protection Department, Municipal Department 22, Unit of Spatial Development, 1200 Wien, Dresdner Straße 45, Vienna, Austria
e-mail: juergen.preiss@wien.gv.at

© The Author(s) 2016
F. Musco (ed.), *Counteracting Urban Heat Island Effects in a Global Climate Change Scenario*, DOI 10.1007/978-3-319-10425-6_9

257

City Administration. The summary points out the project's added value for the city, indicating that the journey Vienna has taken to protect the climate while at the same time adapting to the consequences of climate change is bound for success.

## 9.1   Introduction

The pilot action "Urban Heat Islands – Strategy Plan Vienna" (UHI-STRAT Vienna) is meant to trigger discussion processes, to make the problems entailed in UHI visible for political decision-makers and the city administration, as well as to offer assistance and clearly defined solutions. The results of the pilot action UHI-STRAT Vienna are helping to put the issue of Urban Heat Islands in the focus of future urban development. Increasing heat stress and a rising temperatures during the summer months as predicted for the City of Vienna can be mitigated by putting into place the strategic and technical measures proposed. Protecting and expanding the city's green infrastructure, for example, can effectively reduce the consequences of UHI while at the same time improving people's quality of life and boosting urban biodiversity. The pilot action UHI-STRAT Vienna helps the participating administrative agencies identify the measures relevant for their work, the measures they can implement in their own areas of competence, the steering tools and levels at their disposal, and the potential of the different measures. The pilot action UHI-STRAT Vienna was developed in close coordination with representatives from the Vienna City Administration and outside experts on the basis of autonomous yet interlinked discourses (Hubo and Krott 2012).

## 9.2   A Consistent Strategy for UHI in Urban Planning

The objective of the UHI-STRAT Vienna is to integrate consistent consideration of urban climate aspects at different levels of planning. There is a wide range of tools where the (urban) climate already features strongly, along with a number of strategies and rules on how to tackle the phenomenon of Urban Heat Islands. The subject matter includes contracts under international law, such as the UN Climate Change Convention, Austrian wide approaches, such as the Austrian Strategy on Adaptation to Climate Change, regional approaches, such as the Climate Protection Programme of the City of Vienna, as well as federal and provincial laws, guidelines, planning tools and planning assistance. Mitigating the effects of UHI combined with forward-looking urban planning for the prevention of Urban Heat Islands has become a very integrative task. Different fields of action, steering levels and planning processes are either influenced by the implementation of measures or influence the latter in turn. It is important in the context to bear in mind the hierarchy of the planning tools and the chronological order the different tools are used in the course of planning.

This is why many aspects of adaptation to climate change have found their way into programmes and activities pursued by the City of Vienna, mostly in conjunction with objectives of environmental protection:

The Municipal Department for Environmental Protection has been promoting roof and façade greening for many years, measures to this end ranging from presentations, international congresses, publications and consultation with individual projects. It even has its own test area with several types of roof greening on the Department's office building. Numerous studies, expert meetings and public relations activities are dedicated to forward-looking use of rainwater (rainwater management), particularly with a view to raising the rate of evaporation. There is positive interaction with "ÖkoKauf" (EcoBuy Vienna) , a programme for sustainable procurement, and "Öko Businessplan" (EcoBusinessPlan Vienna), a cooperation initiative with the economic chamber to consult businesses on ecological measures.

The ecological criteria set out in UHI-STRAT are in line with the Environmental Department's wildlife conservation programme "Netzwerk Natur" (network nature), a topic which is also addressed in the project "Nachhaltiger Urbaner Platz" (sustainable urban space), a checklist for sustainable design of urban spaces.

"Microclimate" constitutes a separate assessment category along with other environmental goods incorporated into the "strategic preliminary assessment of environmental impacts caused by housing projects". This tool is to make different locations and projects for the creation of housing more easily comparable and comprehensible with regard to their environmental impact at the urban development level already. It is to ensure that all environmental aspects are duly considered when choosing from different planning options.

All of the above programmes and projects are brought together in the city's Smart City Strategy and the Climate Protection and Adaptation Programme (KliP II).

Implementation of UHI-STRAT Vienna must be addressed both in the different fields of action and at the various levels of planning. The UHI effect needs to be considered with measures relevant to the city as a whole, as well as with those that have a bearing on individual lots or buildings. It means acting strategically and setting specific measures within one's own competence. Aside from large-scale urban planning approaches it is also important to build public awareness and make members of different Municipal Departments and agencies with the Vienna City Administration sensitive to today's and tomorrow's challenges in tackling the UHI phenomenon.

The city administration, builders and developers, private ones too, have the right to set measures of their own accord for the purpose of reducing the UHI effect.

Circumstances and concepts for reducing or preventing the UHI effect may vary depending on the location and occasion. Each development task (e.g. planning a new city quarter, adapting and enhancing existing buildings or project-related processes) has its own set of actions and measures. Political and legal settings, as well as planning instruments provide the basis for realising urban planning and development that is sensitive to UHI. Adjustment measures to reduce Urban Heat Islands are positioned at various political and legal levels and provide the frame for UHI-STRAT Vienna.

## 9.2.1 Consolidation at the European and the National Level

The "EU Strategy on Adaptation to Climate Change" (2013) is based on the premise that climate protection measures must be paired with adaptation measures if Europe is to master the challenges of climate change. From the point of view of the EU Commission "it is cheaper to take, early, planned adaptation action than to pay the price for not adapting" (COM 2013, 2). The objective must be to raise climate resilience in Europe. The adaptation options are threefold: "gray" and "green" infrastructure approaches, as well as "flexible" structural approaches (COM 2009). Promoting functions and services within ecosystems is considered imperative as these are considered more cost-effective and sometimes more viable than simply trusting grey infrastructure (COM 2009, 6). The "Austrian Strategy on Adaptation to Climate Change" (2012) adopted by the Council of Ministers also makes it clear that along with measures to limit the global rise in temperature it also takes suiTab. and timely adaptation measures. This second pillar of climate policy constitutes a major complement to climate protection seeing as it reduces greenhouse gases. More specifically, the Austrian strategy emphasises the negative effects of heat waves on people's health and the importance of measures to reduce these (Federal Ministry of Agriculture, Forestry, Environment and Water Management 2012a, b, 5). The rise in hot days and the heat stress they create are considered tomorrow's challenges which adaptation measures are required for. Land-use planning is addressed as one of 14 main fields of activity (Federal Ministry of Agriculture, Forestry, Environment and Water Management 2012a, b, 16). "Prevention of overheating and heat islands and compensation of bioclimatic stress for people's health" is to be made possible by providing, in development plans, "green" and "blue" infrastructure for built-up areas, as well as "measures with an impact on bioclimate" (Federal Ministry of Agriculture, Forestry, Environment and Water Management 2012a, b, 117f). The strategy also calls for a 'Climate Proofing' of spatial planning and tools "to systematically consider the impact of climate change" (Federal Ministry of Agriculture, Forestry, Environment and Water Management 2012a, b, 118f).

## 9.2.2 Strategic Approach to UHI-Relevant Aspects in Vienna

The City of Vienna has taken a strategic approach to climate-sensitive action and measures to adapt to climate change.

The objectives and results of the individual aspects were combined to form the **"Smart City Vienna Framework Strategy"** adopted by the Vienna City Council in 2014. It is an umbrella strategy for the period up to 2050 to be implemented step by step, individual objectives being subject to continuous monitoring. The overriding goal is to reduce $CO_2$ emissions from currently 3.1 tons per head to approximately one ton (minus 80 % from 1990 to 2050). Unlike comparable strategies in other cities it encompasses environmental protection goals beyond that, such as reducing the share of motorised private transport from currently 28 to 15 % by 2030, or maintaining the high share of green areas of 50 %.

The current *Climate Protection Programme of the City of Vienna (KliP II)* encourages strong consideration of the UHI effect in tools of spatial planning, nature conservation plans, as well as informal tools. The "Climate Protection Programme of the City of Vienna (Klip II) – update 2010–2020" adopted by the Vienna City Council has coined as its key goal the reduction of greenhouse gases and proposes measures to adapt to and mitigate the impact of climate change. It contains a separate field of action dedicated to "mobility and urban structure". Again the focus is primarily on reducing energy consumption. The set of measures, however, clearly addresses urban planning measures that are to help reduce the UHI effect. Objectives for the field of action "urban structure and quality of life" include "pursuing integrated sets of measures to raise the quality of life in built-up urban areas (greening street space, courtyards and roofs, reducing soil sealing, upgrading green and open space,…)" (Vienna City Administration 2009, 93). Specific measures are "green paths, multiple use, activating green and open spaces already dedicated, roof greening, neighbourhood gardens and succession gardens" (Vienna City Administration 2009, 100). Regional cooperation must ensure "green and open space for the long term, linking green space (regionally) and strengthening awareness for agricultural products from the city region (Vienna City Administration 2009, 105et seq.). KliP II also for the first time stimulates Vienna's measures to adapt to global climate change.

The *Urban Development Plan 2025* (STEP 2025) in particular broaches the issues of urban climate and climate protection. Its aim is to make "climate protection and adaptation to climate change integral elements of planning, implementation and further development of city quarters and open spaces"(STEP 2025, 85). This involves, amongst others, creating open and green spaces that can contribute towards reducing the UHI effect. Specific measures include the greening of roofs and facades, as well as planting trees and avenues (STEP 2025). The chapter on open spaces in STEP 2025 has a separate focus on "adaptation to climate change". Green and open spaces in this context are granted a major role in adapting to climate change while special emphasis is placed on their positive influence on the urban climate. A network of open space is to improve the microclimate in individual city quarters. The initiative "urban green instead of air conditioning" wants to identify the areas concerned and reduce UHI.

The *Vienna Nature Conservation Act* wants to protect and "take care of nature in all its forms across the city and to ensure urban ecology functions" (Nature Conservation Act §1). Protection of green and natural areas includes urban climate aspects considering that climate is part of the landscape balance (section 3, para. 2). "All measures must be planned and implemented in such a way as not to endanger or seriously impair 1. the balance of the landscape, 2. its structure or 3. its recreational effect on human beings" Vienna Nature Conservation Act §4 para 2). Site protection as provided for in the Nature Conservation Act essentially ensures that green spaces and their role for the climate in Vienna are maintained for the long term.

The *Building Regulations for Vienna* (Vienna Urban Development, Urban Planning and Building Code) set out the principles of urban planning, land use and construction engineering. The first part lists the objectives for determining or amending land-use and building development plans. These refer to climate-relevant aspects only indirectly, e.g. the objective "to (4.) preserve or create environmental conditions that will ensure a healthy environment, in particular with a view to housing, work and leisure time " (Building Code for Vienna, §1, para 2 Z4).

Protection of the urban climate has been embedded in the strategic and legal tools to enable targeted measures for reducing the UHI effect.

## 9.3 UHI and the Urban Climate in Vienna – Status Quo and Future Developments

Building up natural permeable surfaces is considered the main culprit in the development of Urban Heat Islands (Kuttler 2011). The UHI effect is further enhanced by both a steady decrease and fragmentation of urban green spaces and the waste heat produced by industrial processes, air conditioning and motor vehicles. Construction developments also increase the surface roughness, slowing down wind speed in the course. They prevent cold air flows generated in undeveloped "cold air production sites" from entering the densely built-up city. Building developments in many cases act as an additional blockade for cold air flows from undeveloped environs to agglomeration areas. Generally speaking temperatures are expected to rise from the periphery to the city centre (see Fig. 9.1).

The isothermal map highlights the Urban Heat Islands, the outlines of the built-up area, as well as the "hot spots", such as sealed car parks or industrial areas, and "cold spots", such as parks, agricultural areas and bodies of water in Vienna.

Forecasts for climate development are subject to a certain amount of uncertainty. From today's point of view temperatures in Vienna are reckoned to increase. "The 2040ies in the eastern parts of Austria will likely see an increase in temperatures of 1.3–1.8 °C in winter, 1.8–2.5 °C in spring, 2.0–2.5 °C in summer and 2.5–.0 °C in autumn, compared to the 1980s. Heat waves will be on the rise. Between 1961 and 1990 there were an average of 5.1 heat wave days per year (also known as "Kysely days"), between 1976 and 2005 there were as many as 9.1 already, and the current forecast for the period between 2010 and 2039 in the centre of Vienna is an average 17.7 Kysely days per year, the inner districts, because of the UHI effect, being more affected by the heat stress than the periphery" (Vienna City Administration 2009, 196) (Fig. 9.3).

The Central Institute for Meteorology and Geodynamics in Vienna (ZAMG), during the project "Focus I" (Zuvela-Aloise et al. 2013), calculated high-resolution, climate simulations of future heat stress in Vienna and examined the effectiveness of adaptation strategies in urban planning aimed at reducing heat stress in densely populated areas. The simulation showed how to improve buildings and open spaces by raising the amount of green and water surfaces, as well as the level of desealing, and by exploiting the Albedo (reflection coefficient) effect on surfaces and roofs.

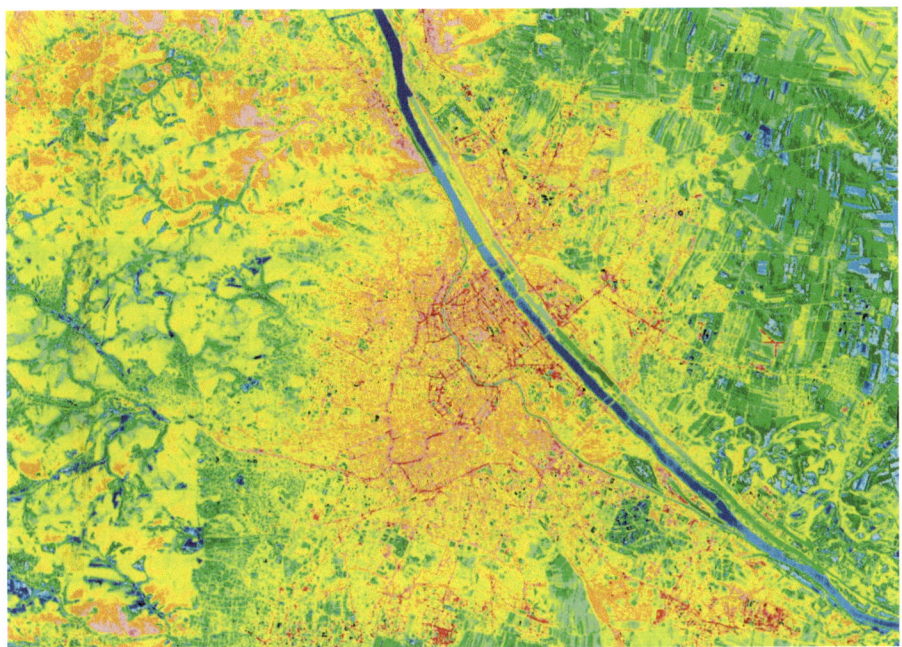

**Fig. 9.1** Thermal image of Vienna and surroundings by night. There is a noticeable difference between the urban agglomeration and the cooler rural areas (Source: City of Vienna, MA 22)

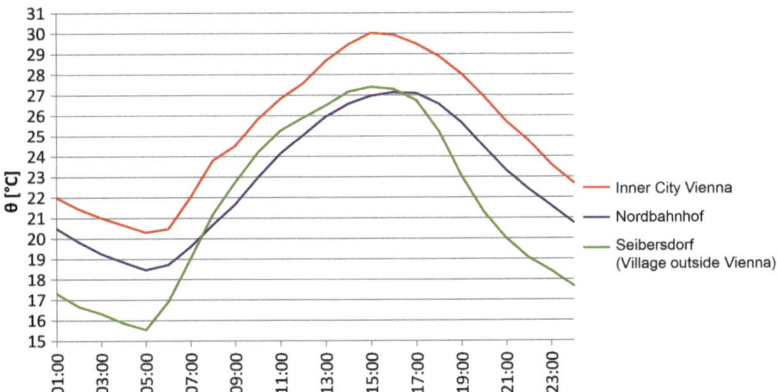

**Fig. 9.2** The Fig. shows the average hourly temperature distribution on a given day in the summer of 2012 – pictured here are two selected areas in Vienna (see Sect. 9.6.2) compared to a rural area in Seibersdorf. Results clearly reveal significant differences in the microclimate of the areas studied, with conspicuously high temperatures in the city centre (Source: Vienna University of Technology, Mahdavi et al. 2014)

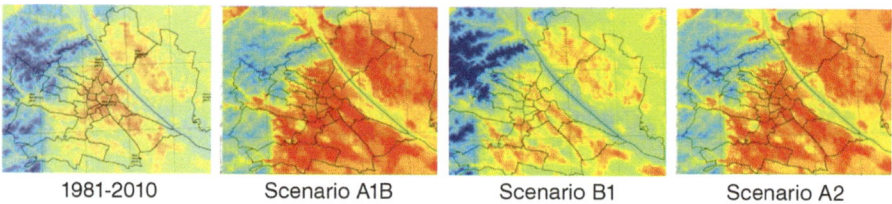

| 1981-2010 | Scenario A1B | Scenario B1 | Scenario A2 |

**Fig. 9.3** Average annual number of summer days 1981–2010 (*far left*) and possible climate scenarios 2071–2100 (Source: Zuvela-Aloise et al. 2013)

The aim was to create a scientific foundation for urban planning to build on. The MUKLIMO_3 experiments, for the purposes of urban development strategies, arrived at the conclusion that there is a great need for adaptation measures if heat stress in the city is to be reduced significantly. Targeted small-scale measures, when combined with each other, such as increasing the share of green space (+20 %), reducing building density (−10 %) and desealing (−20 %), also have a decidedly positive effect (Zuvela-Aloise et al. 2013). All of the above can compensate the impact of climate change at least to some extent. Model results also revealed that because of the topography, the prevailing atmospheric circulation (NW and SE winds) and the different urban structures adaptation measures will not be felt the same throughout the city. Raising the share of green space (+30 %), for example, has a higher cooling effect in the city centre than it does in districts on the outskirts.

## 9.4 Fields of Action for UHI-STRAT Vienna – City-Wide Actions and Actions for Individual Lots and Buildings

UHI-STRAT Vienna identifies three fields of action to enhance consideration of the UHI effect, i.e. (1) awareness building, information and public relations for UHI, (2) urban infrastructure and large-scale strategic measures (3) and more detailed technical and structural measures.

### 9.4.1 Awareness Building, Information and Public Relations

This field of action covers measures aimed at making heat a major issue for future urban planning and development and at building awareness among people and planning experts in general.

It is primarily about providing information, whereby a distinction has to be made between information for residents and visitors to the city on the one hand and information relevant for planning purposes on the other hand. Information on (impending) heat stress for the city has been available since 2010. The Vienna Health Board in cooperation with the Central Institute for Meteorology and Geodynamics provides preventive information on expected heat waves on its website and via the local media (http://www.wien.gv.at/gesundheit/sandirektion/hitzebericht.html). This site, as well as that of the Public Health Services of the City of Vienna (MA 15) (www.gesundheitsdienst.wien.at), also explains about what to do in the event of a heat wave.

Implementing the UHI-STRAT also means building awareness and competences for the UHI issue and its impacts among the departments responsible for planning and projects at the City Administration. Spatial research and research projects on climate change are already generating information relevant for planning to some extent.

### 9.4.2   Urban Structure, Large-Scale Strategic Measures

When implementing measures a distinction is made between long-term strategic measures and the more specific technical and structural ones, the difference being the scale – from the city as a whole down to individual buildings and open spaces – and the time horizon. Building an interconnected network of open spaces to generate and distribute cold air and expanding the tree population in the city are strategic measures with a long-term effect.

### 9.4.3   Specific Technical and Structural Measures

This field of action describes different approaches for implementing the strategic goals of UHI-STRAT Vienna, as well as large-scale strategic measures for the planning and project stages. The 24 specific technical and structural measures are divided into five different areas, (1) green and open spaces (incl. streets), (2) water bodies in the city, (3) shading, (4) mobility and (5) buildings. The measures prepared take into account suitable courses of action with both existing and planned new structures.

## 9.5   Level of Action – From Master Plan to Project

Bearing in mind the overarching significance of Urban Heat Islands, the environmental and climate policy approaches for the protection of the (urban) climate and the rules and regulations associated with them the following main levels of action were identified for UHI-STRAT Vienna: (1) master plans and urban development guidelines; (2) strategic environmental assessment (SUP) and environmental impact assessment (UVP); (3) land use and building development plan; (4) planning and development of public green and open space; (5) developer competitions, housing initiatives and public housing construction; (6) planning and development of public utility buildings and (7) subsidising measures.

When implementing measures it is important to take into account the hierarchy of planning levels and the chronological order different tools are employed in during the planning process. Interfaces with the various tools call for integrated planning and harmonisation across departments and agencies if the measures employed against the UHI phenomenon are to be successful.

### 9.5.1 Master Plans and Urban Planning Mission Statements

Urban development mission statements and master plans have a major bearing on subsequent steps of planning and development in city quarters (MA 21B, 2010). They harmonise public and private interests and create the foundation for further planning. Urban development structures, building density and distribution of open spaces are determined right here. While this planning level is not legally binding it is usually confirmed by a City Council decision to be used as a guiding principle for further development.

Urban development master plans, as a rule, are developed through a number of processes, e.g. citizen participation and competitions, and take into account the challenges planning entails, such as planning of new buildings, the development of former railway locations or branches of industry. Major subject matter and strategies are incorporated into this planning level to weigh up (partly) contradictory urban development objectives, such as densification vs. expansion of open space.

### 9.5.2 Strategic Environmental Assessment and Environmental Impact Assessment

Major projects require various testing methods, more specifically the environmental impact assessment (UVP) and strategic environmental assessment (SUP). UVP is used for the approval of specific projects that have a major impact on the environment while SUP is implemented as early as the planning stage to set the course for decisions relevant to the environment. Both assessment methods investigate the impact of projects on the following protected goods: human beings, animals, plants and their habitat, soil, water, air and climate, landscape, material goods and cultural assets, as well as the interactions between them. Climate already ranges high with the assessment methods and projects are currently run to find our whether, how and to what extent climate change is considered in these methods.

### 9.5.3 Land Use and Building Development Planning

Land use and building development planning sets out legally binding conditions for all subsequent planning and development processes. Here is where building types, building heights and their orographic alignment are decided. Special Conditions also determine a number of UHI measures at this stage. Aside from building alignments and size, rules may be defined to determine the amount of green space on a given parcel of land, as well as the size and location if windows. Details on roof and façade greening may also be provided at this point.

### 9.5.4 Planning and Development of Public Green and Open Space

Planning and development of public streets, squares, green and open spaces is vital for the implementation of UHI-reducing measures, because here is where qualities are determined for the long term. Major emphasis is placed on incorporating UHI-sensitive criteria into design competitions. Internal guidelines and checklists, some of which contain climate-sensitive aspects, facilitate implementation of measures at this level.

### 9.5.5 Developer Competitions, Housing Initiatives and Public Housing Construction

Approximately 60 % of households in Vienna live in subsidised apartments (Kolbitsch and Stalf-Lenhardt 2008). This level of action is therefore relevant for many parts of the city. Developer competitions have proven successful in Vienna since 1995. The competitions help to promote quality in subsidised apartments. Four main criteria are used to assess the quality of drafts: architecture, economy, ecology and social sustainability. In addition there are "theme" competitions for low-energy and passive houses or car-free housing developments. Competitions to date have considered microclimate for the design of open spaces and have also included the vision of "climate neutral cities" (e.g. Aspern Urban Lakeside).

The housing initiative launched in 2011 has contributed to ensuring quality based on a two-tier cooperative planning process. Both programmes have always emphasised climate protection but have not paid much attention to adaptations to climate change. Evaluation of these instruments (Liske 2008) shows that new and quite specific topics can be integrated into urban development at this level and turned into pilot projects for other projects to copy.

## 9.5.6    Planning and Development of Public Utility Buildings

Being a "model" in its own sphere of competence allows the City of Vienna to influence commercial developers and participants in competitions. This applies to all Viennese kindergartens, schools and campuses (Vienna Model where different school levels, from kindergarten to secondary schools, share the same building), as well as administrative buildings and other city-owned buildings. The "Space Book" (Municipal Department 34 – Building and Facility Management) and the "Criteria for Energy-conscious Building for Service Buildings in Vienna" (Municipal Department 20 – Energy Planning) define quality standards for the purpose. These guidelines contain a number of UHI-relevant aspects and measures, such as effective sun protection, reducing the externally induced cooling energy provided for in the building code or avoiding large glass constructions to prevent overheating.

## 9.5.7    Subsidising Measures

Subsidies are a way of influencing private persons and institutions. Municipal Department 42 (Parks and Gardens) has been subsidising roof greening, courtyard and vertical greening successfully since 2003. Subsidies for roof greening are calculated on the basis of the thickness of the rooting substrate. The example shows how subsidies can promote quality-assuring aspects and measures for the reduction of UHI.

Strategic urban developmet

Master Plans and urban design concepts

Land use and development plans

Green and open space planning

Building planning and design

**Fig. 9.4** Planning levels in the city relevant for the reduction of the UHI effect (source: from top to bottom: Stadtentwicklung Wien, Magistratsabteilung 18 – Stadtentwicklung und Stadtplanung, 2014, STUDIOVLAY; Stadtentwicklung Wien; Büro tilia; Jürgen Preiss, MA 22)

## 9.6   Feasibility Studies

The feasibility studies described below want to demonstrate how UHI-relevant measures can be put into practice using two selected areas in the city as examples (Fig. 9.5). They make a clear distinction between measures in the development of new city districts and measures in adapting existing ones, and also identify two different levels of planning, the strategic master plan on the one hand and the planning of legal provisions, i.e. the land-use and building development plans on the other hand. The "Masterplan Nordbahnstraße – Innstraße" in Vienna's 20th municipal district is used as an example to show how measures can be introduced at different stages of the master plan process. Proposed measures can be embedded in land-use and building development plans, as demonstrated in the case of the quarter surrounding Vienna University of Technology (Karlsplatz) in the 4th municipal district. Workshops were held with different agencies at the Vienna City Administration to assess how the UHI catalogue of measures can feasibly be implemented at these planning levels. The Institute for Building Physics and Building Ecology at Vienna University of Technology simulated measures for both selected areas (e.g. tree planting, roof greening) to find out what impact these measures have on air temperature (Figs. 9.10 and 9.12).

The results of a survey carried out for the case study UHI STRAT Vienna are presented here to set the scene for the description of the feasibility studies. The

**Fig. 9.5** Location of the two pilot areas in the city (Source: Vienna GIS)

survey reflects people's attitude towards heat in the city, their behaviour during heat waves, as well as their assessment of the measures employed to reduce the UHI effect.

## 9.6.1  People's Attitude Towards Heat in the City

385 answers were collected during this postal survey among people in Vienna to assess their perception and attitudes towards heat in the city.

The survey was done in August 2013. Questionnaires were sent to 3792 house-holds in Vienna, which approximately 10 % of the addressees replied to. 27 blocks of flats were picked out randomly from different areas in Vienna, some more densely built-up than others, and the responses were weighted to arrive at as representative as possible a sample. Almost everyone in Vienna has witnessed at least one heat wave already. Three quarters consider this a negative experience. Heat is felt particularly strongly in the streets and in people's homes (Fig. 9.6).

People in their homes try to adjust to the heat and find ways to reduce its effect. Most frequently cited measures to fight heat are: open windows during the night (88 %), make sure to take in more liquids (86 %), keep blinds and curtains closed (80 %). A negligible number of people considered leaving the city or working fewer hours an option during the last heat wave. Only 6 % of the respondents used air-conditioning in their homes. Approximately half of the respondents used fans.

Most frequently perceived public measures against the heat are air-conditioning in public transport (64 %), drinking fountains in the city (59 %) and trees in the streets (51 %). Respondents have hardly noticed measures, such as brightening of street surfaces, shading of pavements or greening of rail or tram tracks.

A vast majority (86 %) believes that trees are a suiTab. measure for reducing heat stress in the city. An even greater number agrees that trees have a positive effect on

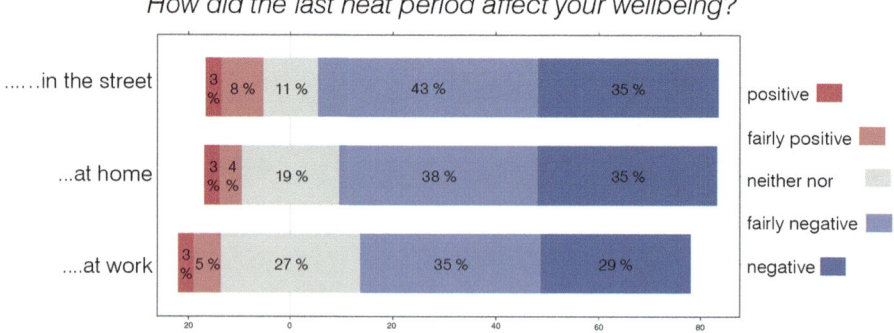

**Fig. 9.6** Responses to: how did the last heat period affect your wellbeing? (Source: INWE)

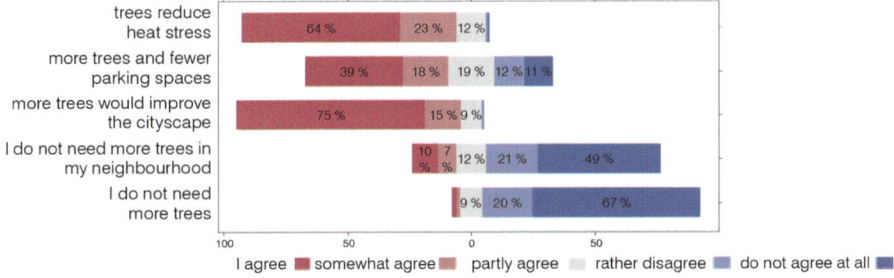

**Fig. 9.7** Responses to: attitude towards measures against heat development, weighted values (Source: INWE)

the streetscape. Most respondents would like to see more trees in their neighbourhood (70 %) and across the city (87 %). A majority of 54 % endorses the claim for "more trees and fewer parking spaces in my district". Only 24 % are not in favour of this measure (Fig. 9.7).

## *9.6.2 Feasibility Study for the Urban Development Master Plan*

Planning and realisation of large-scale urban expansion and urban development projects can take years and even decades which is why it so important to find ways of reducing heat in city quarters at this strategic level. Urban development master plans and guidelines require intensive planning processes, usually in combination with participation processes. To implement UHI sensitive planning and development means to involve experts early on in the process, ideally when preparing the strategic objectives for the master plan.

### 9.6.2.1 UHI-Relevant Links in the Master Plan Process

The following description of ways to incorporate UHI-relevant issues into the different stages of developing an urban development master plan or guideline is based on the study "planning as a process" commissioned by Municipal Department 21B – District Planning and Land Use – to collect experience with master plan processes in Vienna and internationally. Master plan development is characterised by the four stages of "opening, setting the programme, consolidating and implementing" (MA 21B 2010), during which UHI-relevant issues may be introduced and put into practice.

During the opening stage political and planning requirements, as well as the various expectations with regard to future development are identified. Process structures and participants are determined at this point so it is imperative to include persons knowledgeable in climate-sensitive urban planning. This stage also deter-mines what basic information, plans, expert reports and studies will be required for the process. The master plan process has to specify what basic information on cli-mate conditions in a city quarter must be obtained (e.g. main wind directions, sig-nificance of the area as a cold air production site, link with major cold air corridors etc.).

Setting the programme for actual planning usually means drawing a rough urban development guideline to give the project direction. The interests of politics, inves-tors, landowners and representatives from the administration are translated into functional and structural specifications for the development of an area.

Structural and urban development criteria to prevent heating in future city quar-ters may be introduced at this stage. The objectives, challenges and general frame-work defined here provide the setting for further development. Analysing the planning area also reveals links that UHI-relevant aspects can be attached to. This means, amongst others, assessing the availability of green and open space in neigh-bouring quarters, wind corridors, air flows and water permeability of the soil.

The most important step towards incorporating UHI-relevant issues and mea-sures (see below) at this stage is to define the requirements for preparing qualifica-tion processes and urban development competitions. The actual urban development qualification process rounds off this phase.

The "consolidation" phase in the planning and development process is about turning the competition results into specific guidelines, preparing feasible concepts and developing detailed implementation projects. By transferring the requirements to the land-use plan and preparing the environmental assessments and environmen-tal impact assessments as needed UHI-relevant strategic objectives and clearly defined measures are introduced to the process. Issues, such as the effect of planned construction on the microclimate, must be dealt with in detail at this stage.

The phase is completed by an interface with the legally binding land-use and building development planning "Not every urban development aspect in the master plan requires binding regulations. By the same token it would be negligent to waive binding and reliable regulations in favour of informal agreements " (MA 21B 2010, 51). Ways of embedding UHI-relevant measures in the land-use and building devel-opment plan are described extensively in the second example.

The implementation stage is about developing individual projects for the social and technical infrastructure provided by the public authorities, about implementing public space and building development. Technical and planning measures to reduce the UHI effect are put into practice at this stage (Fig. 9.8).

### 9.6.2.2 UHI Measures in the Master Plan for "Nordbahnstraße – Innstraße"

"Nordbahnstraße – Innstraße" is located on the premises of the former Nordbahnhof (a railway station in the 20th municipal district) developed gradually over the course of the past few years. Employees of the Vienna City Administration acted out a

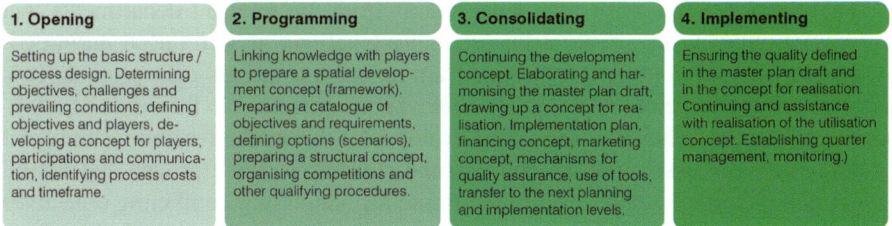

| 1. Opening | 2. Programming | 3. Consolidating | 4. Implementing |
|---|---|---|---|
| Setting up the basic structure / process design. Determining objectives, challenges and prevailing conditions, defining objectives and players, developing a concept for players, participations and communication, identifying process costs and timeframe. | Linking knowledge with players to prepare a spatial development concept (framework). Preparing a catalogue of objectives and requirements, defining options (scenarios), preparing a structural concept, organising competitions and other qualifying procedures. | Continuing the development concept. Elaborating and harmonising the master plan draft, drawing up a concept for realisation. Implementation plan, financing concept, marketing concept, mechanisms for quality assurance, use of tools, transfer to the next planning and implementation levels. | Ensuring the quality defined in the master plan draft and in the concept for realisation. Continuing and assistance with realisation of the utilisation concept. Establishing quarter management, monitoring.) |

**Fig. 9.8** Stages of the master plan process and links to implementation of the measures (Source: MA 21B 2010)

**Fig. 9.9** Feasibility Study "Nordbahnstraße – Innstraße" – section from an aerial view (*left*) and measures discussed (*right*) for the winner in the urban development competition (Source: City of Vienna)

scenario to transfer competition results into an urban development mission statement that encompasses UHI-relevant measures (Fig. 9.9).

To spark the discussion objectives were defined for the "competition inviting urban development ideas for Nordbahnstraße – Innstraße". Additional competition documentation defined quality objectives along with "hard" project requirements, such as gross floor area, the mix of residential areas/offices/retail/commerce in percent, as well as social infrastructure. Following spatial analyses and information campaigns for the public the "general conditions and objectives for the competition inviting urban development ideas" (MA 21A 2011) were drawn up. It was during this early planning stage that the first UHI-relevant goals and criteria were drafted. A number of solutions mentioned in the collection of measures were strategically positioned at this point already. The objectives for the urban development competition reveal modalities of how these measures may have a bearing on subsequent implementation stages. One of the requirements, for instance, was to create a system of green and open spaces with a high quality of use for everybody, another was to link the new city quarter with the surrounding main green and open spaces. Other requirements included minimising the degree of soil sealing, as well as considering and integrating urban climate aspects (sun/shade/wind/humidity) in competition submissions across the board (MA 21A 2011).

There was general agreement among staff from the relevant departments that most UHI-reducing measures at this planning level can be introduced during the phases of opening and setting the programme for the master plan process. It is important if not imperative to coin UHI-relevant propositions in the urban development guidelines already. Participatory development during the feasibility study and cross-agency discussions about chances and restrictions have proven successful. These negotiating processes can set the frame for addressing conflicting objectives and challenges and thus support the process of weighing up individual objectives. Attention also needs to be paid to bringing on board the "implementers", e.g. Vienna Public Transport for matters relating to designing and placing bus or tram stops, or coordinating green and open spaces across construction sites to minimise overheating in a quarter. Listed below are the points and issues that can and ought to be addressed and finalised during this early stage of urban planning and development: (1) What impact will the planned project have on climate? (2) Which measures for reducing the UHI effect can be implemented in the urban development scenario proposed? (3) Who is responsible for implementation? (4) Which tools will be employed and which planning processes applied to implement the measures? Which challenges does or may implementation pose?

### 9.6.2.3 Modelling Measures and Their Impact with the Example of "Nordbahnstraße – Innstraße"

The Department of Building Physics and Building Ecology at Vienna University of Technology (Mahdavi et al. 2014) was commissioned to simulate the impact of the master plan on microclimate based on the results of the winner in the competition inviting ideas for development of the former brownfield Nordbahnstraße – Innstraße.

As soon as the buildings were simulated the mean night air temperature in the area under investigation was seen to rise. This may be explained by a reduced sky view factor, an increase in thermal mass in the area and an increase in the long-wave radiation emitted as a result. In the daytime, however, a significant reduction in mean air temperature was noted (see Madhavi et al. 2014) (Fig. 9.10).

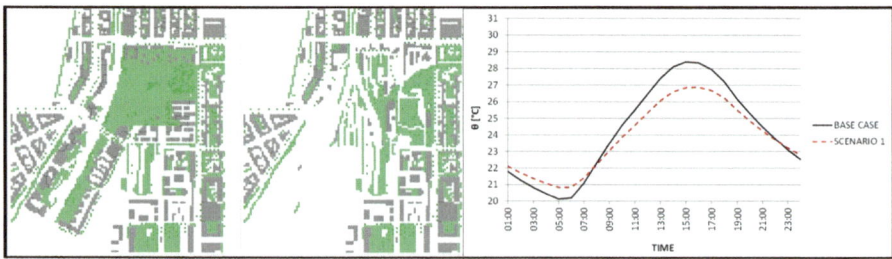

**Fig. 9.10** ENVI-met model before and after building development plus a comparison of average hourly temperatures in the course of a reference day before and after building development (Source: TU Vienna, Mahdavi et al. 2014)

## 9.6.3    Feasibility Study at the Level of Land-Use and Building Development Planning

Permissible utilisation and development on individual sites are made legally binding for owners and developers during land-use and building development planning. As a rule sites are zoned for building purposes, as green areas or as traffic areas. Aside from this classification the land-use and building development plan also defines the building categories, building methods and building regulations, height and cross-section of traffic areas (§5 Vienna Building Code), as well as additional definitions and "Special Conditions".

### 9.6.3.1    UHI-Relevant Links in the Land-Use and Building Development Plan

In principle a distinction has to be made between new developments and structural improvements to existing buildings. Of course, there is more scope for implementing UHI-relevant measures with new developments. However, there is a wide range of regulations that can be implemented for future projects in existing buildings as well. Where major changes are made to existing buildings these must be adapted to the land-use and building development planning valid at the time. Again, this proves the long-term strategic significance of this planning level.

Essentially there are two areas where UHI-relevant topics and measures can be incorporated: in the drawn and in the written part. The drawn part sets out rules for escape routes, conditions for use or building classes, i.e. rules to do with the urban structure and the shape of buildings, as well as measures aimed specifically at reducing the UHI effect, such as various greening measures, or at mitigating the impact, such as requirements for shaded pathways and arcades. "Special Conditions" (BB) in the written part of the building development plan contain specifications for the defined area, offering additional suggestions for integrating measures to reduce the UHI effect. This includes, in particular, targets for garden design, roof greening, façade greening, desealing, greening of courtyards and tree planting.

Measures in the drawn part may range from directions of the streets to the geometry of a building. Streets heat up more in the course of a day than their environment. It is recommended that street layout and adjacent buildings with a shading effect on the streets are considered at this level. The width of streets is connected to the height of buildings with relevant regulations set out in the Vienna Building Code (§75 para. 4). These regulations are generally applicable with the exception of protected zones or areas designated "urban development hotspots". There is little point in narrowing the cross sections of streets as this would necessitate a reduction in building heights to avoid difficulties with lighting and exposure to light. Wider cross sections combined with green infrastructure can help to reduce the UHI effect. Depending on the direction of a street (E-W, N-S) measures, such as planting rows of trees or utilisation of surfaces may have more or less of an impact. Alignment of streets must take into account the main wind direction so as not to hinder the exchange of air. The height of buildings, their position in relation to each other and the shade they subsequently produce must be coordinated separately for each location. There is no rule of thumb here as the local wind situation, topography and supply of green space vary widely. For complex urban development situations or where climate challenges, such as strong winds, prevail microclimate simulation with different building scenarios is recommended. The drawn part can set the scene for "public pathways" and "arcades" for sun protection along major pedestrian axes. Measures may also be specified in the written part, i.e. the special conditions, as demonstrated in the second, inner-city example below.

### 9.6.3.2    UHI for the Land-Use and Development Plan for Karlsplatz and Surroundings

The second example is located in the area surrounding Vienna University of Technology in the 4th municipal district of Vienna. An analysis was made as to how to incorporate requirements when revising the land-use plan to make sure that new constructions with and renovations of existing buildings take into account the phenomenon of UHI. Most of the area was developed during the Gründerzeit (in the late nineteenth century) with an utilisation mix of apartments, offices and commerce and is comparable to many quarters in the city centre of Vienna.

Special Conditions are particularly suitable for determining how UHI-relevant measures can be implemented in areas already developed. The Fig. 9.11 shows the potentials staff from the Vienna City Administration gathered during an experimental game based on the requirements set out in the Special Conditions for land-use and building development planning. The Special Conditions proposed are concerned primarily with tree planting, roof and façade greening, landscape design of surfaces, as well as with requirements that have a bearing on the level of soil sealing, both in public and in private areas. Qualities, such as substrate thickness with roof

**Fig. 9.11** Feasibility Study "Karlsplatz" – section from an aerial view (*left*) and the measures discussed (*right*) (Source: City of Vienna)

greenings, or accessibility of roof gardens may also be defined in the Special Conditions. Other issues addressed may include taking the necessary steps to enable tree planting along streets and in public squares, determining the permissible percentage of sealing on a plot of land to reduce the level of soil sealing in park areas or specifying whether arcades are to be built in the area (Fig. 9.11).

### 9.6.3.3 Modelling Measures and Their Impact with the Example of Karlsplatz and Surroundings

Three adaptation measures were modelled for assessment of the city centre. The scenarios include: a base case without measures, (1) tree planting, (2) roof greening and (3) a combination of tree planting and roof greening. The figures below show the difference in climate conditions between the current building stock and the simulated implementation of individual measures on a reference day. The models were built by the Department of Building Physics and Building Ecology, Vienna University of Technology using ENVI-met 4.0 (Mahdavi et al. 2014). Clearly visible are the differences in air temperature between the current situation and after the simulated impact of the measures selected.

Results reveal that adaptation measures have the potential to reduce air temperature in the research areas on hot summer days. As expected different adaptation measures also have different levels of impact. Roof greening in the city centre has no noticeable effect on air temperature in the open spaces of streets (scenario 2), while trees do (scenario 1). The combination of the two selected measures proved particularly effective (scenario 3). Looking at the time patterns showed that differences in air temperature are more distinct in the evening and during the night (see Fig. 9.12).

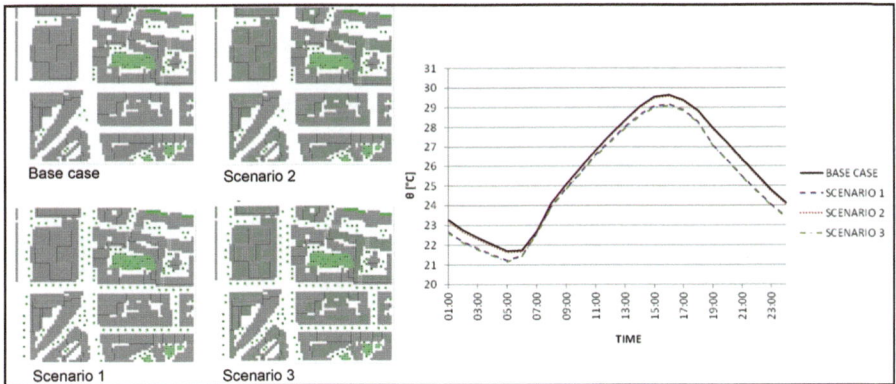

**Fig. 9.12** The research area in the ENVI-met model and after implementation of selected adaptation measures, as well as average hourly temperature on a reference day, shown for the current building stock and for the three adaptation scenarios (Source: Vienna University of Technology, Mahdavi et al. 2014)

**Fig. 9.13** Cover of the UHI-STRAT Vienna; Download (in German): https://www.wien.gv.at/umweltschutz/raum/pdf/uhi-strategieplan.pdf

## 9.7   Conclusion

The pilot action UHI-STRAT Vienna shows how measures for reducing the UHI effect can be implemented in urban planning and urban development in Vienna. There is in fact a wide range of tools to trigger action at various levels of planning and to make urban planning climate-sensitive, from strategy planning to development and completion. Close cooperation with the administrative agencies relevant for planning confirmed that existing tools of urban planning, formal and informal, are quite capable of reducing the UHI effect. Many examples revealed during the project process are proof that urban climate is an issue already for many administrative agencies in their day-to-day business. The examples can help to make sure that Vienna will continue its successful venture of protecting the climate while at the same time adapting to the impact of climate change. This must be considered at an early stage at the strategic level and then broken down to the various levels of planning and finally development.

"Green" measures proved especially effective for Vienna. A growing city where densification of built-up areas is necessary to keep distances short can employ these measures to create green and recreational areas for residents, while at the same time reducing the UHI effect. There are strong synergies between measures to reduce the UHI effect and other strategies pursued by the City of Vienna, e.g. reducing (leisure time) traffic, promoting biodiversity, improving water retention and establishing a network of open space. UHI-STRAT Vienna provides the setting for the implementation of these measures Fig. 9.13.

## References

COM. (2009). Weissbuch - Anpassung an den Klimawandel: Ein europäischer Aktionsrahmen, Brüssel.
COM. (2013). Mitteilung der Kommission an das Europäische Parlament, den Rat, Den Europäischen Wirtschafts- und Sozialausschuss und den Ausschuss der Regionen - Eine EU-Strategie zur Anpassung an den Klimawandel, Brüssel.
Federal Ministry of Agriculture, Forestry, Environment and Water Management (Eds.). (2012a). Die österreichische Strategie zur Anpassung an den Klimawandel. Teil 2 – Aktionsplan. Handlungsempfehlungen für die Umsetzung (Austrian strategy on adaptation to climate change, part 2 – action plan. Recommendations for implementation) Vienna.

Federal Ministry of Agriculture, Forestry, Environment and Water Management (Eds.). (2012b). Austrian strategy on adaptation to climate change, Part 2 – action plan. Recommendations for implementation. Available online: http://www.bmlfuw.gv.at/dms/lmat/umwelt/klimaschutz/ klimapolitik_national/anpassungsstrategie/strategieaussendung/Anpassungsstrategie_ Aktionsplan_23-10-2012_MR.pdf

Hubo, C. & Krott M. (2012). Erfolgsfaktoren im Naturschutz und ihre Förderung durch Programme am Beispiel chance.natur. In GAIA 21/2 (2012), 135–142 (success factors in nature conservation and programmes to promote these using the example of chance.natur).

Kolbitsch A., & Stalf-Lenhardt, M. (2008). Studie über Wirtschaftlichkeitsparameter und einen ökonomischen Planungsfaktor für geförderte Wohnbauprojekte in Wien. Vienna (German) (study on feasibility parameters and an economic planning factor for subsidised housing projects in Vienna).

Kuttler W. (2011). Klimawandel im urbanen Bereich. Teil 2, Maßnahmen. *Environmental Sciences Europe, 23*(21) (climate change in urban areas, part 2, measures).

Liske H. (2008). Der "Bauträgerwettbewerb" als Instrument des geförderten sozialen Wohnbaus in Wien – verfahrenstechnische und inhaltliche Evaluierung. Vienna ("developers competitions as a tool for subsidised social housing in Vienna.")

Mahdavi A., Kiesel K., Vuckovic M. (2014). ENVI-met Simulation of Karlsplatz and Nordbahnhof, TU Vienna.

Municipal Department 21 B. (2010). Planung als Prozess- Gestaltung dialogorientierter Planungs- und Umsetzungsprozesse. Werkstattbericht Nr. 109. Vienna (German) (planning as a process – designing dialogue-focused planning and implementation processes).

Municipal Department 21A. (2011). Städtebaulicher Ideenwettbewerb, Zielsetzungen Freiraum / Gender Mainstreaming.

STEP 2025. (2014). Stadtentwicklungsplan 2025. Vienna (German) (urban development plan 2025).

Vienna City Administration 2009, MDKLI. (2009). Klimaschutzprogramm der Stadt Wien. Fortschreibung 2010–2020.

Zuvela-Aloise et al. (2013). Future of Climatic Urban Heat Stress Impacts – Adaption and mitigation of the climate change impact on urban heat stress based on model runs derived with an urban climate model. http://www.zamg.ac.at/cms/de/forschung/klima/stadtklima/focus-i

# Chapter 10
# Pilot Actions in European Cities – Stuttgart

**Rayk Rinke, Rainer Kapp, Ulrich Reuter, Christine Ketterer, Joachim Fallmann, Andreas Matzarakis, and Stefan Emeis**

**Abstract**  The field of urban climatology has a long tradition in Stuttgart. It exists as discipline in Stuttgart since 1938. Stuttgart was the first city to establish its own Department of Climatology to research ways of improving the flow of fresh air into the city and to reduce thermal stress in most populated city districts. The specialist department of Urban Climatology, within the Environmental Protection Office, deals with tasks relating to environmental meteorology within the scope of air pollution control and also relating to urban and global climate protection. So in Stuttgart the urban heat island phenomenon (UHI) is studied for several decades, leading to a high level understanding of the UHI and the problems which it causes. The UHI causes an increase in air temperatures and thermal stress, that are identified as most negative impacts on human health and urban living. In the view of global climate change and the predicted temperature rise for the Stuttgart region of 1.5–2 K in this century, the negative impacts of UHI on human health and urban living will become more problematic in the future. According to the results of climate models the frequency of very hot days is expected to jump by nearly 30 % at the end of the century. The rising temperatures due to the global climate change in combination with the temperature shift as a result of the UHI will intensify the heat stress in

R. Rinke (✉) • R. Kapp • U. Reuter
Section of Urban Climatology, Office for Environmental Protection, Municipality,
Gaisburgstraße 4, 70812 Stuttgart, Germany
e-mail: Rayk.Rinke@stuttgart.de

C. Ketterer
Albert-Ludwigs-University of Freiburg, Werthmannstr. 10, D-79085, Freiburg, Germany

iMA Richter & Roeckle, Eisenbahnstrasse 43, 79098, Freiburg, Germany

J. Fallmann
UK Met Office, Exeter

A. Matzarakis
Albert-Ludwigs-University of Freiburg, Werthmannstr. 10, D-79085, Freiburg, Germany

Research Center Human Biometeorology, German Meteorological Service,
Stefan-Meier-Str. 10, D-70104, Freiburg, Germany

S. Emeis
Head of Research Group "Regional Coupling of Ecosystem-Atmosphere", Karlsruhe Institute
of Technology (KIT), Institute of Meteorology and Climate Research, Atmospheric
Environmental Research (IMK-IFU), Garmisch-Partenkirchen, Germany

© The Author(s) 2016                                                                 281
F. Musco (ed.), *Counteracting Urban Heat Island Effects in a Global Climate
Change Scenario*, DOI 10.1007/978-3-319-10425-6_10

urban areas, that leads to a significant increasing risk to human health, in particular to the very young and elderly. Not least due its importance for the human health and the quality of urban life in Stuttgart, the UHI is focussed by urban planners and is noticed by the future development of the city.

Within the pilot action study in Stuttgart several measure for reducing the UHI and the impacts on urban living and human health are analysed by the use of micro-scale and macro-scale simulations. With the help of these analysis realisable measure are selected. The most useful measures are implemented into a development outline plan for the redevelopment of the city district Stuttgart-West by the municipal urban planners.

**Keywords**  Climate change • Urban climatology • Urban heat island (UHI) • Urban planning • Restructuring • Air ventilation • Thermal stress • Development outline plan • Green roofs • Heat warning system

## 10.1   The City of Stuttgart

In the following chapter a short overview of the urban and climatic conditions in Stuttgart is given. Especially the basics of the complex topographic situation and the city structure, that influence strongly the urban climate in Stuttgart is described. Also the urban heat island phenomenon in Stuttgart is presented. More information can be found at the website: http://www.stadtklima-stuttgart.de

### 10.1.1   Urban and Climatic Situation in Stuttgart

Stuttgart is the capital of the state of Baden-Wuerttemberg located in south-western Germany. As the sixth-largest city in Germany, Stuttgart has a population of about 590.000 and is the centre of a densely populated area, the Greater Stuttgart Region, with a population of 2.6 million. Stuttgart covers an area of 207 km² thereof 49 % are settlements. The population density is about 5410 person/km².

Stuttgart's area is characterised by a complex topographic situation with local distinctions (Fig. 10.1). It is one of the greenest cities in Germany. The land use distribution of Stuttgart is shown in Fig. 10.2. Greenery in the form of vineyards, forests, parks, etc. is prevalent throughout the city. In Stuttgart 39 % of the surface area has been listed as protected green belt land or nature conservation area; a record in the whole of Germany. Despite this greenery populated, industrial and commercial areas are densely built-up. The city's location, building and topographical characteristics have a negative impact on urban climate and cause an intense urban heat island (City of Stuttgart 2010).

The city is located in a river valley (the Stuttgart basin), nestling between vineyards and thick woodland. Stuttgart's centre is situated close by, but not on the River Neckar in a Keuper sink. The city area is spread across a variety of hills and valleys.

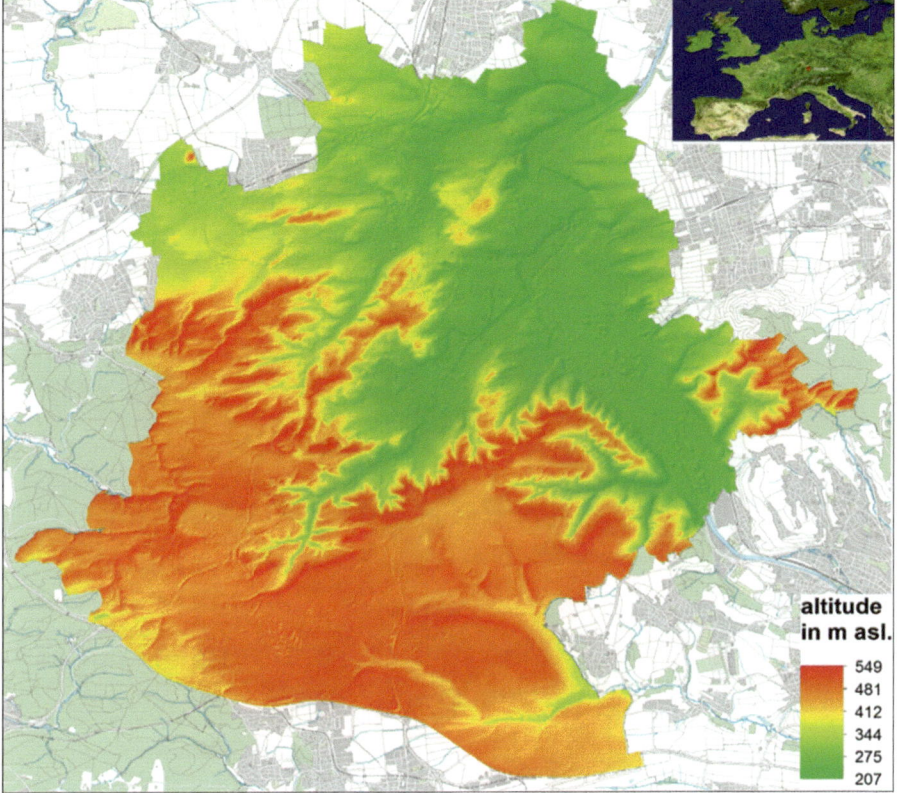

**Fig. 10.1** Topographic map of Stuttgart's city area and Stuttgart's location within Europe

Steep hill slopes surround the city centre on three sides. The elevation ranges from 207 m above sea level by the Neckar River to 549 m on Bernhartshöhe hill. The complex terrain has a significant influence on all climatic elements like radiation, air temperature and wind, resulting in large climatic distinctions within the city area. Stuttgart's overall climate is mild with an average annual temperature of about 10 °C in the Stuttgart basin (city centre) and about 8 °C in the more elevated outskirts situated about 400 m asl. Figure 10.3 shows the annual mean temperature distribution in the city area). Besides the Upper Rhine Valley, Greater Stuttgart is one of the warmest regions in Germany. The month of July is the hottest month with an average temperature of 18.8 °C, while temperature in January averages 1.3 °C.

A major element of Stuttgart's climate is the light wind, that causes a lack of adequate air exchange. The light wind results not only of the city's position between two bights of the Keuper plains. The whole Neckar Valley is known for low wind speeds and very frequent lulls. This is the result of small air pressure differences common to Southern Germany and of Stuttgart's sheltered position between the Black Forest, the Swabian Alb, the Schurwald and the Swabian-Franconian Forest. Due to orographic conditions, it is impossible to indicate a consistent wind rose for the whole of Stuttgart. The sheltered position between the surrounding mountain

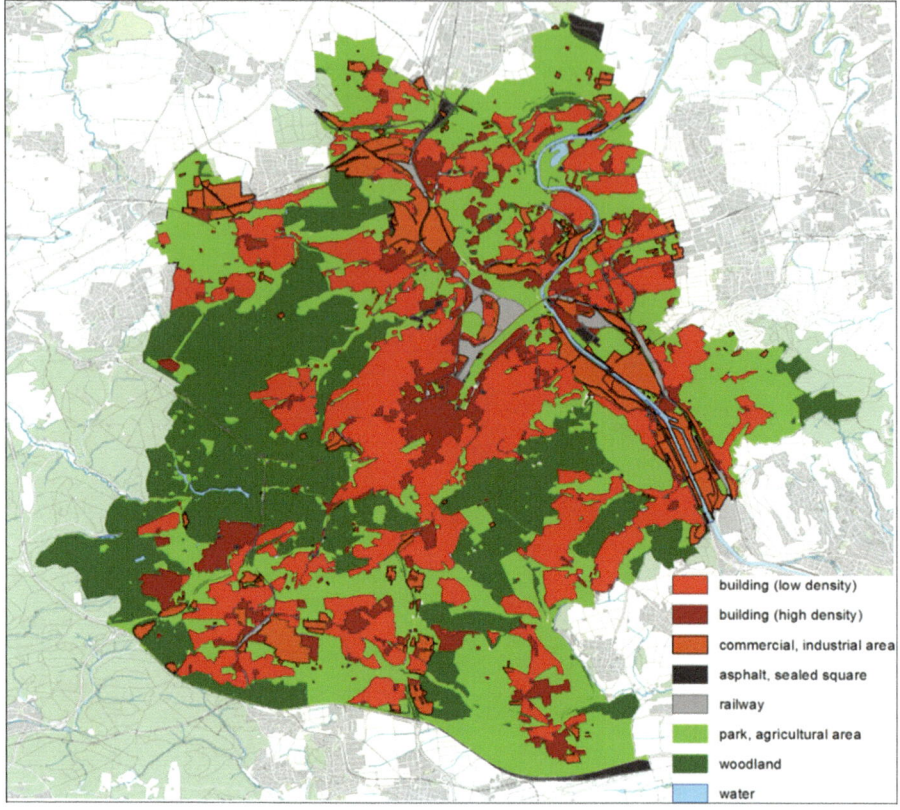

**Fig. 10.2** Land use map of Stuttgart

ranges leads to a frequent development of local wind systems, especially at the slopes and in the valleys. In addition, over large green areas in the surrounding and the city area especially at the higher altitudes, in the nighttime cold air is produced, that generates cold air streams. Even if these winds have no high wind speeds, they play a significant role for the ventilation and local fresh or cold air supply in some city districts. Preserving these local winds and streams is an important objective in the urban planning process in Stuttgart with focus on environmental and urban climate protection since decades. It becomes apparent that primarily cold air flows effectively reduce UHI caused thermal stress in nighttime.

### 10.1.2   UHI in Stuttgart

The lack of adequate air exchange in combination with high building density and a huge amount of sealed surfaces, especially in most populated and industrial city districts, facilitate the development of an intensive UHI. Quantifying the intensity of

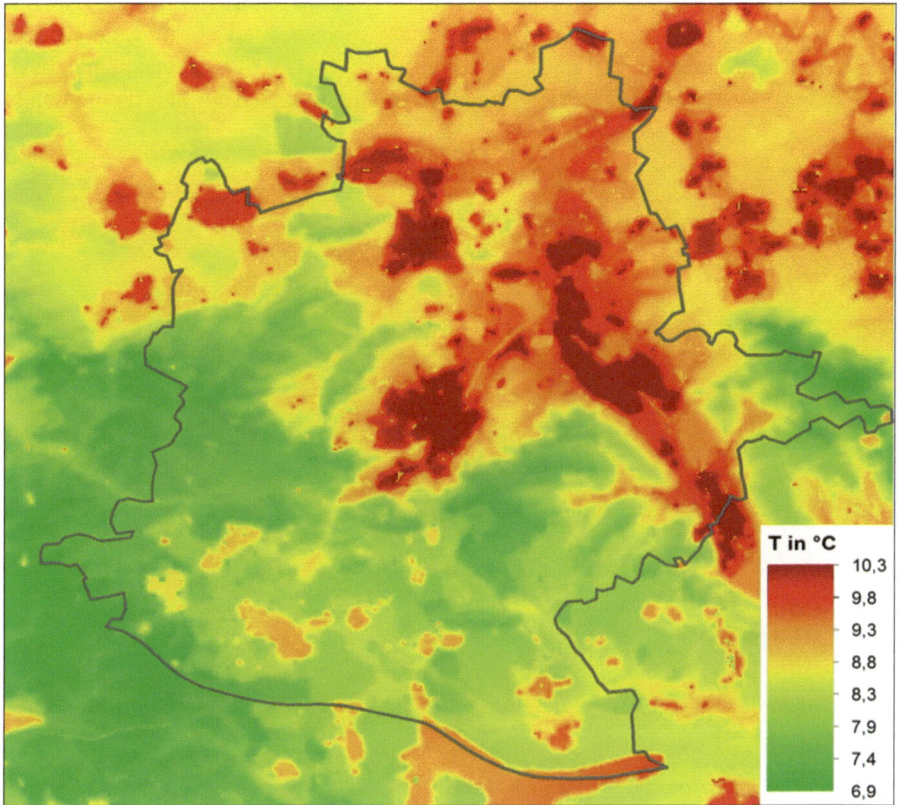

**Fig. 10.3** Annual mean temperature in Stuttgart. The city area is marked with a *grey line*

UHI in Stuttgart is complicate, because of the differences in altitude, which influence the air temperature and overlay the temperature shift due to the UHI. However, the overall temperature raise due to the UHI phenomenon ($UHI_{Ta}$) in Stuttgart is identified within several studies by 1–2 K in annual mean, but locally the $UHI_{Ta}$ intensity can reach more than 5 K. The UHI phenomenon modifies the climate in Stuttgart. For example in the surrounding of Stuttgart, the Filder region, there are 28–32 summer days (days with more than 25 °C daily maximum temperature) as compared with 40–47 summer days in the Innercity region and the Stuttgart bight. The UHI turns Stuttgart's inner city into a region with high heat stress (about 32 days, Fig. 10.4) and only occasional cold stress.

For the longterm characterisation of UHI meteorological values such as temperature, solar radiation and humidity are measured continuously at about ten sites in the city area, operated by German Weather Service (DWD), municipality Stuttgart (MS), University Hohenheim (UnH) and the environmental protection agency of Baden-Württemberg (LUBW). In this study the UHI is analysed using air temperature and also the thermal index Physiologically Equivalent Temperature PET (Mayer

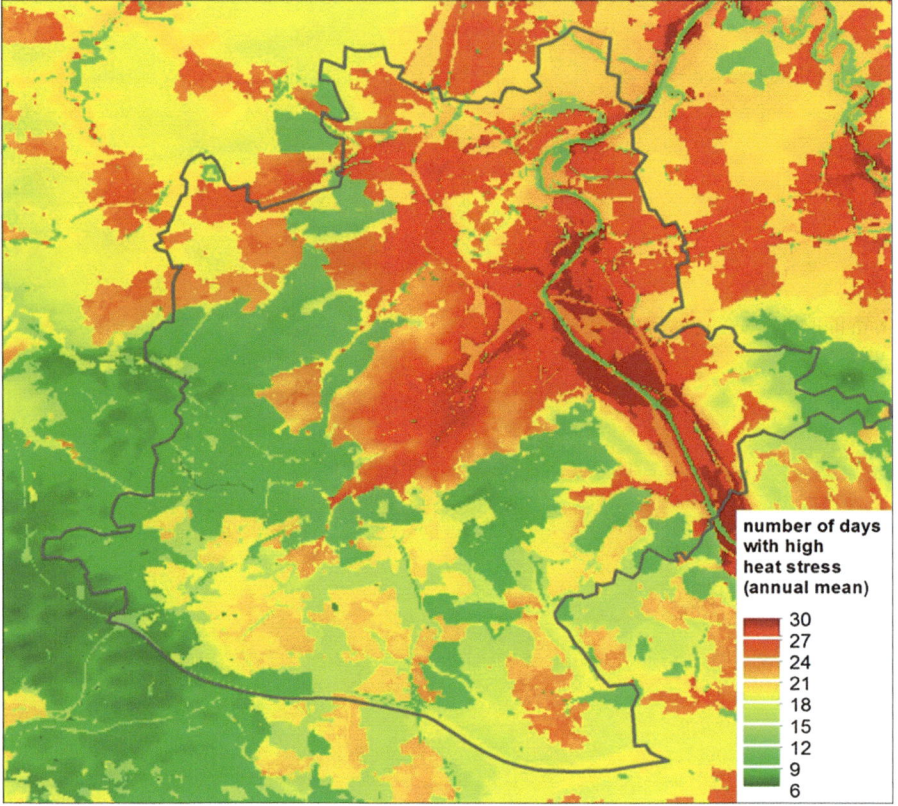

number of days
with high
heat stress
(annual mean)

30
27
24
21
18
15
12
9
6

**Fig. 10.4** Annual number of days with high heat stress in Stuttgart. The city area is marked with a *grey line*

and Höppe 1987; Höppe 1993, 1999; Matzarakis et al. 1999) measured at the sites Schwabenzentrum (MS), Schnarrenberg (DWD), airport station Echterdingen (DWD) and at University Hohenheim (UnH). Average annual $UHI_{Ta}$ intensity (based on air temperature) at Schwabenzentrum is 2 K and at Schnarrenberg 1.6 K. In the Neckar valley, the mean $UHI_{Ta}$ is 0.9 K. At the suburb Hohenheim, the $UHI_{Ta}$ of 0.3 K is not pronounced as the suburb is surrounded by agricultural areas and has a higher elevation as the rural reference station Echterdingen. The urban-rural differences in PET are higher with 4.1 K (3.1 K) between Schnarrenberg (Schwabenzentrum) and Echterdingen.

During summer, the $UHI_{PET}$ ($UHI_{Ta}$) is by 15.2 % (8.1 %) higher than 6 °C in the city center. However, a UHI between 0 and 6 °C is most frequent at the other measuring sites. The UHI effect is stronger and more frequent during summer than during winter, increasing the already existing heat load. The minimum $UHI_{Ta}$ occurs in the late morning, whereupon the rural air temperature is often higher than the urban, especially during warm seasons. The $UHI_{Ta}$ peaks at 6:00 p.m. in the winter and 9:00 p.m. during spring, summer and autumn at Schwabenzentrum. At Neckartal, the amplitude of the diurnal cycle is weaker and $UHI_{Ta}$ is maximal in the early morning. The monthly maximum UHI $_{Ta}$ occurs in winter in the city center due to anthropogenic heat production. However, considering hourly averages, the maximum UHI is experienced in summer. It can be observed that air temperature differences are largest at nighttime, but the PET differences are highest at daytime. The urban heat island intensity was compared to the air pressure as well as flow patterns. The UHI is more pronounced during periods with anticyclonic weather situation (Ketterer and Matzarakis 2014a).

## 10.2   Pilot Action Study Stuttgart

In this chapter the pilot action study in Stuttgart is presented. For the pilot action area the city district Stuttgart-West is chosen. The area was selected in view of a problematic climatic situation with high thermal stress caused by the structure of building. An important point for the selection of the area is an initiative launched by the municipal urban planners for restructuring the district in the next decades. Expected changes in urban living, the predicted increase in urban population, within the district existing brownfields and also the poor climatic situation are facts for a necessary restructuring of the district. The objective of the pilot action study is to find out realisable options for improving the local climatic situation in the district mainly due to better the air ventilation and the reduction of thermal stress. The results of these study should be integrated into a development outline plan of the district, which is under development by the municipal urban planners. In Stuttgart development outline plans are an established pathbreaking helpful urban planning tool for the sustainable future development and restructuring of single city districts weighting residental, economical, public, natural, environmental and climatical aspects. Reducing the negative impacts of the UHI on urban living will be a topic of the development outline plan. For the implementation into the development outline plan, first the UHI intensity and hotspots of high thermal stress in the district have to be known. As a second the effectivity of measures must be analysed to set up the most valuable ones. Within the pilot action study, the UHI intensity and its impacts on urban living is analysed using meteorological measurements and micro- and macro-scale simulation tools. Micro-scale simulation tools are also used to verify the local effectivity of thermal stress reducing measures. To estimate the potential of measures for a city wide reduction of the UHI intensity, macro-scale simulation tools are used.

### 10.2.1   Pilot Action Area

The pilot action area Stuttgart-West (valley floor) is located in a small valley close to the city centre in the western inner-city region (Fig. 10.5). The area is surrounded by steep hills at three sites (South, West and North). Stuttgart-West is the most densely populated district in Stuttgart and has a population of about 33,000 and a

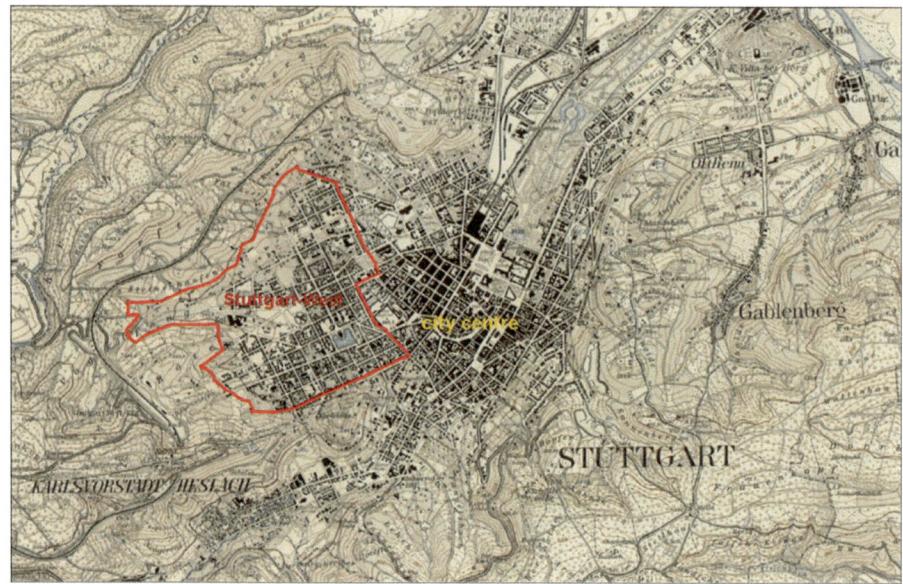

**Fig. 10.5** City map of Stuttgart with the location of the pilot action area Stuttgart West (*red marked area*)

**Fig. 10.6** Airviews of Stuttgart-West, which illustrate the typical building structure

population density of 18,370 person/km$^2$. About 10 % of the inhabitants are younger than 15 years and about 15 % are older than 65 years, that means 25 % of the inhabitants are in early danger by thermal stress. The area is characterised by a high building density with predominant residental buildings. A high number of historical buildings, that have to be preserved, limit the redeployment of the district. Green areas and places for the recreation of the inhabitants are sparse available. The

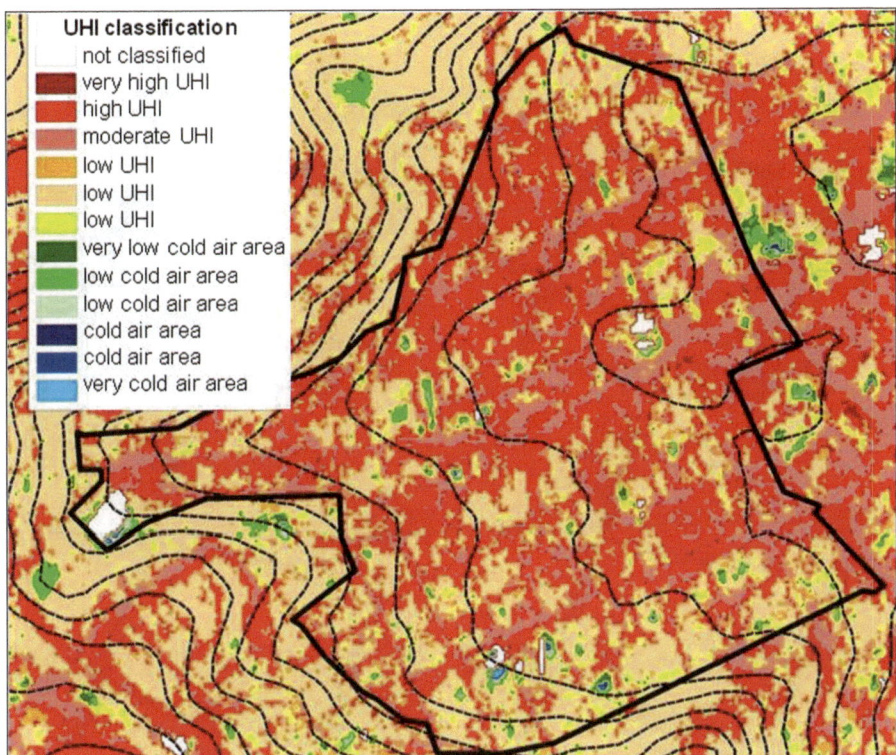

**UHI classification**

- not classified
- very high UHI
- high UHI
- moderate UHI
- low UHI
- low UHI
- low UHI
- very low cold air area
- low cold air area
- low cold air area
- cold air area
- cold air area
- very cold air area

**Fig. 10.7** UHI classification in Stuttgart-West (*left*) and number of days with high thermal stress in Stuttgart-West (annual mean, *right*)

typical building structure in the district are blocks with additional buildings in the inner areas of the blocks (Fig. 10.6). These characteristics causes poor ventilation of the district and a high UHI intensity with increased thermal stress (Fig. 10.7). In addition the air pollution is on a high level. Due to these atmospheric conditions the pilot action area is less attractive for living with potential risks for human health.

The average $UHI_{Ta}$ intensity is about 2 K, but can be many times higher on local hotspots depending on daytime and season. On hotspots an $UHI_{Ta}$ intensity of more than 6 K is measured frequently. At the surrounding hill slopes local wind systems arise and at nighttime cold air flows are induced at the hill slopes. Because of the high building density in the valley floor, these local streams are blocked and mostly don't reach the inner district area. The pilot action area is the most thermally stressed area in Stuttgart. Based on case studies an optimized building structure for the pilot action area to reduce the thermal stress is developed.

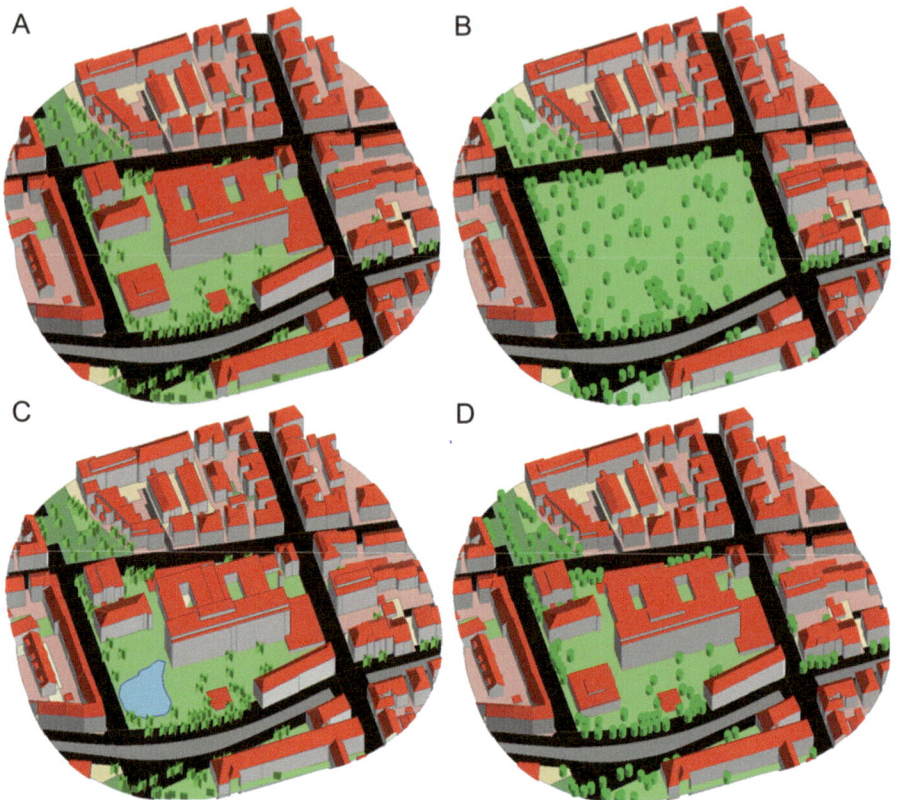

**Fig. 10.8** Different scenarios for the Olga Hospital site as input for the micro-climate simulations. *Panel A* depicts the current state of the Olga Hospital (also with green roofs for every building with flat roofs), *Panel B* the park scenario, in *Panel C* one building is replaced by a small pond (shallow water) and in *Panel C* the number of trees along the streets was increased

## 10.2.2 Case Study Olga Hospital (Stuttgart-West)

### 10.2.2.1 Quantification of Mitigation and Adaptation Possibilities

For the case study the area of the Olga Hospital in Stuttgart-West is chosen. The Olga Hospital is a hospital, that is not longer in use and should be redesigned into a residental area in the next years. The human thermal comfort conditions of the case study area and different scenarios (Figs. 10.8 and 10.9) were analyzed using micro-scale models RayMan (Matzarakis et al. 2007, 2010) and ENVI-met 3.5 (Bruse and Fleer 1998; Huttner 2012). The input parameters for the ENVI-met simulations are based on the measurements of the 24th July 2010 (wind speed 2.6 ms$^{-1}$, wind direction: 250°, no clouds, shortwave radiation adjustment factor 0.83, relative humidity and potential temperature were forced). So, these case studies are representative for hot summer days with a high amount of solar radiation. The average

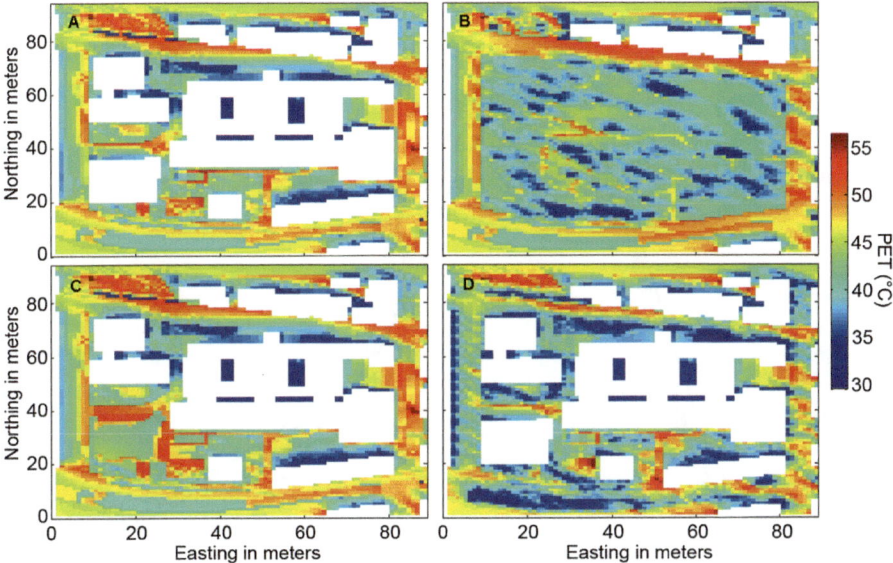

**Fig. 10.9** Physiologically Equivalent Temperature (PET) of different scenarios (see Fig. 10.8) of the Olga Hospital area. The basic meteorological variables were simulated by ENVI-met 3.5 and PET was calculated by TIC-ENVI-met (Ketterer and Matzarakis 2014b). Finally, the data were averaged from 10 a.m. to 4 p.m. for the height of 1.5 m above ground

mean radiant temperature in the whole area is 57 °C; ranging from 23.8 °C to 75.6 °C. The lowest mean radiant temperature was calculated in the shadow of trees in green areas. The mean radiant temperature is at least 3 K higher in the shadow of buildings and 45 K higher in sealed courtyards. The mean radiant temperature has the greatest influence on PET in the daytime on a sunny summer day. PET rises up to 58 °C above sealed surfaces with low albedo, high solar irradiation and low wind speed. In green areas, PET ranges between 18 and 28 °C in shaded, but does not exceed 35 °C in unshaded areas. Streets which are parallel to the wind direction (e. g. Bebel and Bismarck Street), featured lower PET ($\Delta$PET $\leq 10$ K) than other streets (Senefelder Street). The difference in PET between sealed and non-sealed areas is at least 10 K (Ketterer et al. 2013).

#### 10.2.2.2   Micro-scale Simulations

##### 10.2.2.2.1   Surface Types

Thermal conditions over green areas, paved and water surfaces are quantified using ENVI-met simulations. PET rises up to 58 °C above paved, unshaded surfaces with low albedo. Above green areas, PET does not exceed 35 °C in unshaded areas and 25 °C in shaded areas. The difference between paved and green areas is at least

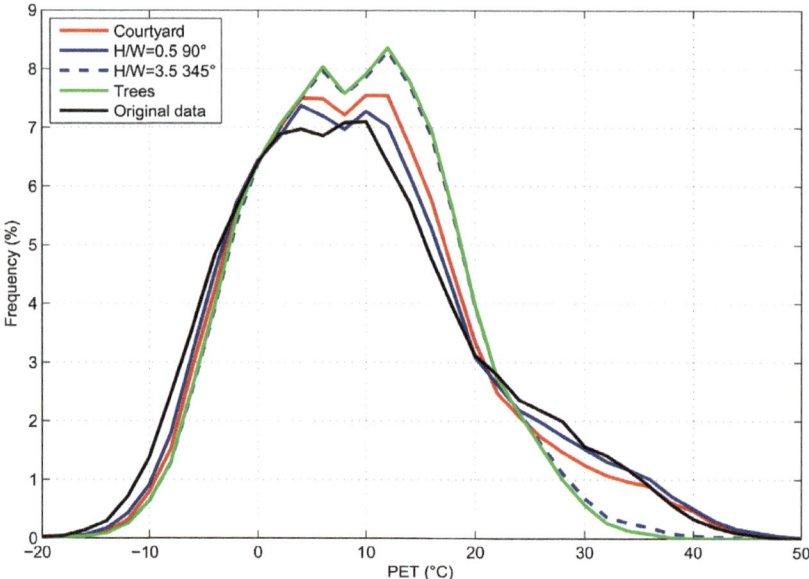

**Fig. 10.10** Frequency distribution of the Physiologically Equivalent Temperature (PET) for following different urban morphology: courtyards, street canyon with aspect (H/W) ratio of 0.5 and 3.5 and rotation of 90° (E–W) and 345° (NNE–SSW), under a group of trees and the original data of the measuring station Schwabenzentrum (city center) for the period 2000–2010

10 K. Considering the assessment scale of Matzarakis and Mayer (1996), the thermal stress can be reduced from strong heat stress above paved surfaces to light heat stress above green surfaces (Fig. 10.9).

The installation of a small pond has no significant impact on the spatial average of the studied Olga Hospital area, but a local impact on the air temperature. Air temperature is decreased due to the smaller Bowen ratio and enhanced latent heat flux. Additionally, water has a very high specific heat capacity. However, small and shallow water surfaces heat up relatively fast, so that they can have a warming effect during evenings and nights in mid-summer as well as in early autumn.

ENVI-met simulations of the current scenario with the Olga Hospital and for a scenario with a park were done for a calm, hot summer day and compared for 14 LST. The specially averaged PET value decreases by 2.6 K in the park scenario. The wind speed increases due to lower roughness in the lee (east) side of the park and decreases PET, too. The PET value was decreased by maximum 7.0 K an on average by 1.7 K in the street east of the park. On the streets in the north and south of the park, PET is 0.4 and 0.8 K lower than in the current state. A park with a continuous green area is 1–20 K PET colder then green areas on the built-up area. The more trees in a park, the cooler PET on a hot summer day and the smaller the diurnal amplitude of the temperature. However, the air temperature differences between different scenarios are below 2 °C (Ketterer et al. 2013; Ketterer and Matzarakis 2014b).

## 10.2.2.2.2   Trees

PET in 1.5 m height was found to be around 10 K lower under trees compared to green areas and 25 K lower than over asphalt (Fig. 10.9). Therefore, shading by trees could reduce the frequency of daytime heat stress significantly (Fig. 10.10).

The increasing number of trees in the Olga Hospital area has no significant impact on the averaged air temperature during moderate warm conditions. However, during hot summer days it could reduce the air temperature in this area by 3.0 K (spatial average).

## 10.2.2.2.3   Green Roofs

The effect of green roofs was quantified using ENVI-met and by changing all roofs of the hospital scenario into green roofs. The effect of green roofs on the local thermal conditions experienced by humans on street level are on a very low level ($\Delta$PET<0.06 K). The local air temperature differences on street level are even lower. However, green roofs significantly reduce the warming of urban roof surfaces in daytime. Inside green roofs the accumulation of heat is decreased, resulting in a lesser heat emission in nighttime. A large-scale revegetate of roofs is an effectively measure for the mitigation of UHI intensity especially in nightime.

## 10.2.2.2.4   Urban Morphology

The urban morphology is analyzed using RayMan Pro (Matzarakis et al. 2007, 2010). The morphology of street canyons influences solar access and radiation and therefore thermal comfort. The importance of solar access for city dwellers depends on the climate zone. While south of the Alps sun is considered as harmful, solar access is favored in northern cities such as Stuttgart. East-west oriented street canyons do not have solar access during winter months due to the low zenith angle of the sun. But during summer, the street canyon and especially the northern façade is illuminated during the whole day. Accordingly high is the frequency of heat stress in this E-W oriented street canyon. A N-S oriented street canyon is accessed by sunshine during the midday hours throughout the year (Ketterer and Matzarakis 2014b).

The daily maximum value of PET could be reduced by 10 K due to a changing H/W ratio from 0.5 to 3.5 and an orientation of 120° on a hot summer day. Throughout the year, the frequency of heat stress can be reduced by 477 h (4.3%). Additionally, the occurrence of thermal comfort conditions could be increased by 10%. However, a change in H/W ratio from 0.5 to 1 (2.5) could already reduce the frequency of heat stress by 192 (333) hours per year (Fig. 10.10).

10.2.2.2.5    Courtyards

The micro-climate of courtyards was studied using the micro-climatic models RayMan and ENVI-met. The human thermal comfort conditions in two courtyards (Schlossstrasse and Senefelderstrasse) in the Olga Hospital area were compared to the conditions in street canyons (Breitscheidstrasse, Senefelderstrasse) and on a green area (Elisabethenstrasse – Hasenberstraße) over 11 years. Therefore, the micro-scale RayMan model employing fish-eye photos was used to describe long term conditions. In the courtyards, the frequency of heat stress (PET > 29.1 °C) and thermal comfort is between 45.5 % and 51.6 % from May to September. Whereas the frequency of thermal comfort is between 3.6 % and 13.8 % higher in east-west and NNW – SSE oriented street canyons. Additionally, PET is also higher at night-time than during daytime due to the smaller sky view factor in courtyards. Multiple reflections can also increase PET in courtyards. Another factor is the low wind speed in these sheltered locations, triggering a further increase in PET. ENVI-met simulations for a hot summer day show that PET is up to 25 K higher over a paved courtyard compared to a park area covered with plants and grass.

## 10.2.3    Macro-scale Simulations

Specific urban planning strategies, like green roofs or facades and highly reflective materials are able to reduce the UHI. Taha (1997) demonstrated that increasing the albedo by 0.15 can reduce peak summertime temperatures for the urban area of Los Angeles by up to 1.5 °C. During the DESIREX Campaign 2008, Salamanca et al. (2012) stated that a higher albedo leads to about 5 % reduction in energy consumption through air conditioning during summertime periods for the area of Madrid. The regional energy saving effect of high-albedo roofs can also be found in Akbari et al. (1997) and on a more global perspective in Akbari et al. (2009).

    In the course of the project UHI – Development and application of mitigation and adaptation strategies and measures for counteracting the global "UHI phenomenon" (3CE292P3) – CENTRAL Europe. (2011–2014), these kinds of scenarios are conducted for the urban area of Stuttgart. Due to its geographical location in a valley, the weak mountain – valley circulation leads to increasing potential for natural heat trapping in the urban region. Modelling work of the environmental agency of Stuttgart shows, that the area with more than 30 days/year heat stress is anticipated to increase from 6 % (1971–2000) up to 57 % (2071–2100). This reflects the calculations of the Intergovernmental Panel on Climate Change (IPCC) on global climate change.

    The Karlsruhe Institute of Technology (KIT) conducts simulations using the numerical mesoscale Weather Research and Forecasting Model WRF Skamarock et al. (2005) on regional scale, coupled to urban parameterization schemes (Kusaka et al. 2001; Martilli et al. 2002). The results reflect the effects of certain urban planning strategies on near surface air temperature and on UHI intensity.

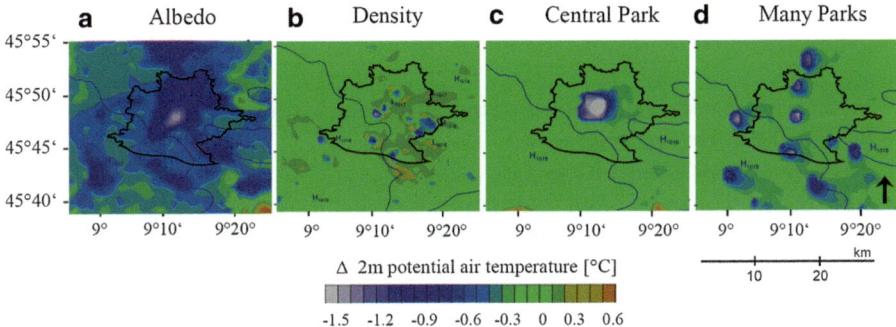

**Fig. 10.11** Difference in potential 2 m air temperature for the four scenarios: (**a**) changed albedo for roofs and walls, (**b**) modified proportion street width/building height and the two urban greening scenarios with one big park (**c**) and a number of smaller parks (**d**); projected time is August 13 2003 8:00 p.m

**Table 10.1** Impact on UHI formation expressed as the difference between mean urban and mean rural temperature, August 13 2003 8 p.m

| Scenario | Control | Albedo | Many parks | Big park | Density | delta [°C] Albedo | Many parks | Big park | Density |
|---|---|---|---|---|---|---|---|---|---|
| T mean urban [°C] | 33.1 | 31.5 | 32.5 | 32.3 | 32.4 | −1.60 | −0.60 | −0.80 | −0.70 |
| T max [°C] | 34.3 | 31.9 | 33.5 | 33.3 | 33 | −2.40 | −0.80 | −1.00 | −1.30 |
| **UHI; delta Θ** | **2.52** | **0.84** | **1.47** | **1.19** | **1.32** | **−1.68** | **−1.05** | **−1.33** | **−1.2** |

The control run indicates 'real' conditions

Four case studies were applied representing different mitigation measures.

1. Increase of the reflectivity of roof and wall surfaces in the urban area ('Albedo')
2. Decreasing the building density by 20 % by increasing the Sky View Factor ('Density')
3. Replacing urban surface by natural vegetation in the city center ('Central Park')
4. Replacement of single urban areas scattered around the city area ('Many Parks')

The difference in 2 m temperature between scenario- and base case ('reality') run reflects the efficiency of the mitigation procedure (Fig. 10.11). To refer to an extreme case scenario, a period during the European Heat Wave 2003 (August 11th–18th 2003) was chosen, where summertime temperatures exceeded the annual average.

Table 10.1 presents the mean and maximum urban temperature as well as the UHI, calculated from mean urban and mean rural temperature with regard to Aug 13 2003 8:00 p.m. The difference between Control run and Scenario is calculated for both temperature and UHI.

Referring to Table 10.1, a changing of the albedo of wall and roof surfaces has the strongest effect on temperature, causing a decrease of the UHI intensity by nearly 2 °C.

Both vegetation scenarios show a decrease of about 1–1.3 °C. With 1.2 temperature reduction, the effect is similar for the 'Density' case.

Because of insufficient observation data in the rural surrounding, it is difficult to retrieve the UHI intensity from measurements. The difference between air temperatures observed at 'Stuttgart Schwabenzentrum' (37.4 °C) and at Stuttgart Hohenheim in the near surrounding (33.1 °C) accounts for 4.3 °C, for August 13 18:00 UTC. Assuming a height dependent temperature decrease of 1 °C per 100 m between urban (250 m NN) and rural (400 m NN) location, the adapted observed UHI amounts to 3 °C. The second parameter in Table 10.1 describes the mean temperature for the whole modelling period for one single urban grid cell in the centre of the city, whereas all other parameters treat aerial statistics for one temporal snap-shot.

The above findings describe the climatological and meteorological background for the forthcoming mitigation and adaptation actions to reduce the urban heat islands and its impacts. Besides the modeling results it is now of great importance to find the best urban planning strategy for the specific urban area of interest, considering a mixture of different mitigation strategies. However, due to the coarse resolution it is difficult to directly apply the measures proposed by that study. Rather, these kinds of modelling studies can be used as a decision support and provide meteorological boundary conditions for high resolution street scale models.

## 10.3 Transfering the Findings of the Pilot Action Study to Urban Planning Process in Stuttgart

The findings of the studies within the pilot action Stuttgart, have shown, that several measure can be effectively reduce the UHI intensity and the impacts on urban living and human health. However, reducing UHI intensity in a city, which is growing over the centuries, requires deep changes in the city and building structure. Changing a city to improve the urban climate is a hard and challenging transaction, which needs a sustainable future-oriented urban planning. Most of the measures for reducing UHI in a city are only effective by large-scale implementation, but changing a city due to urban planning are mostly concentrated on single buildings or small areas like existing brownfields. Additional difficulties are a low awareness for the problems caused by UHI by the public and political boards, existing national and internation strategies for the future development of urban areas, which potential forces the UHI intensity (for example the European sustainable development strategy, which supports the development of more compact and more densed cities), a low number of free available areas to set up measures like parks, contrary interests of public, industry and economy and the ownership structure. Also, in Germany a legal basis for the consideration of UHI related aspects in the urban planning process is currently not available.

### 10.3.1   Legal Basis for the Consideration of UHI Related Aspects Within the Urban Planning

To date in Germany no independent "Urban Climate Protection Act" exists in its own right. Also planning measures for Urban Heat Island specific requirements are not directly regulated. Instead, these concerns have been integrated into the structure of existing environmental legislation. This is due to the circumstance that many of the classical disciplines of environmental protection or rather ecology simultaneously exercises positive repercussions for climate change and that a firm foothold can be provided for climate protection within the framework of existing legislation. Examples of this include the Federal Building Code (BauGB), the Federal Nature Conservation Act (BNatSchG) and a variety of regulations issued by the Federal Immission Control Act (BImSchG). Also rulings given by a series of laws and regulations such as the Energy Saving Act (EEG) and the Energy Saving Ordinance make a specific contribution to global climate protection, as well as the "Greenhouse Gas Emissions Trading Act" (TEHG).

The German Building Code is the most powerful act in Germany for the municipal administration to arrange measures for urban planning in respect to environmental, nature, urban and global climate protection. The code offers differentiated possibilities for urban development that is urban and global climatically just.

### 10.3.2   Development Outline Plan

The development outline plan (DOP) constitutes a non-formalized level of spatial planning. It is not codified by the Federal Building Code and non-obligatory. In practice, however, the DOP proved to be a valuable and flexible tool to steer urban development within built up areas. It is an essential function of the DOP to define the municipality's development and planning goals for those parts of the city that show tendencies of urban change. In practice, the planning intentions for public spaces and streets can be described more precisely than those for private building sites. This is why the DOP often also functions similarly to a local design plan. The DOP is not subject to legal regulations.

In Stuttgart development outline plans are used to set up the urban planning strategy for the sustainable future development and restructuring of single city districts weighting all aspects of urban living, economy and nature, environment and climate protection.

For the pilot action area Stuttgart-West a development outline plan is under progress by the municipal urban planning department. Within the development outline plan Stuttgart-West the strategy for the future development of the city district is ascertained. One aspect of this strategy is to improve the climatic situation and to reduce the negative impacts of the UHI inside the district. Based on the analysis simulation results, done within the pilot action study, hotspots of high thermal

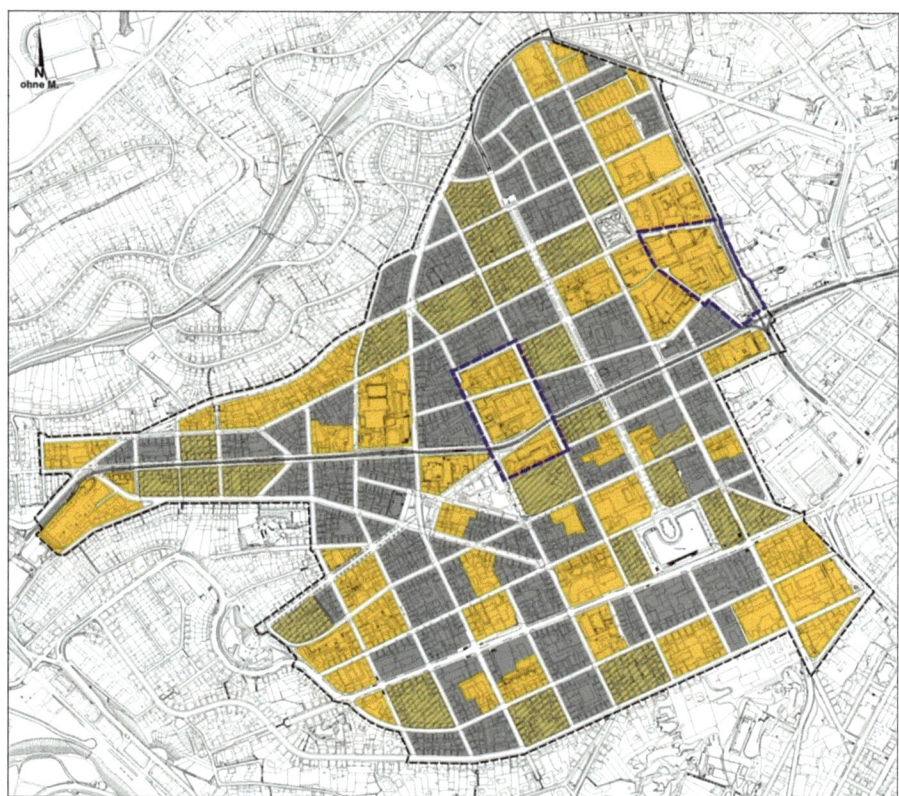

**Fig. 10.12** Map based on the analysis of areas with high thermal stress. The *grey shaded blocks* are characterised by high thermal stress mainly due to a prevented ventilation of the inner area and due to high building density and unavailable greening. The *grey blocks* should be redesigned according the development outline plan. The purple surrounded blocks are currently under reconstruction

stressed areas in the district are indentified. Also effective and realisable measure for reducing the thermal stress on the hotspots are choosen for the implementation into the development strategy and the development outline plan for the district.

Hotspots, which are high thermal stressed are located mostly in the inner region of the district (Fig. 10.12). These areas are characterised by a high building density with additional buildings inside the blocks, a high degree of sealing and a lack of greening. The high building density prevents a ventilation of the block inner area, which forces the accumulation of heat. For these blocks an optimised building structure (Fig. 10.13) for the potential reconstruction is developed based on the micro-scale simulation done in the pilot action study (Olga Hospital, see Chap. 2.2). These optimised building structure supports a better ventilation of the block and the thermal stress is reduced compared to the existing building structure due the greening of the block in form of greened courtyards, green roofs and green facades or the use of cool materials for roofs and facades. However, the optimised building structure is

**Fig. 10.13** Optimised building structure, which is offered to reduce the thermal stress inside the block. These building structure is part of the development outline plan is not obligatory for potential reconstruction of existing blocks, but gives a reference to improve the local climatic situation. Gaps between the buildings facilitate the ventilation of the inner area. The courtyard of the block (*green marked area*) is greened. Buildings inside the blocks (*red-green marked*) must be equipped with green roofs and the height of these buildings is limited. Buildings around the block (*orange marked*) should be equipped with green roofs, but it is not compulsory. Facades which are orientated to the south (*dark green marked*) have to be greened or designed with cool materials

**Fig. 10.14** Optimised design of street canyons in Stuttgart-West. *Green lines* marking street canyons, where the thermal stress is reduced due to facade greening of the use of cool materials for facades. *Green circles* marking possible positions for trees

**Fig. 10.15** Possible creation of green connections (*green lines*) in Stuttgart-West

not obligatory for the redevelopment of a block, but gives a reference to improve the climatic situation. If the suggested building structure is absolutely the optimum has to be checked from case to case under consideration of the ambient conditions. But setting up green roofs on new buildings can only be prevented for a comprehensible reason. Green roofs on new buildings are a standard in Stuttgart.

Beside the optimised building structure also suggestions for the design of public spaces are offered (Figs. 10.14 and 10.15). The design of public spaces to reduce the thermal stress is relatively easy realisable, there no ownership structure must be respected. The measures offered for the design of public spaces mainly strive the improvement of the sojourn quality, due the reduction of thermal stress in street canyons, squares and parks. To reduce the thermal stress in street canyons, especially possible tree positions inside street canyons, which are orientated from east to west are identified and implemented into the development outline plan. Also for building facades along east to west orientated street canyons a design with cool materials and/or a facade greening is offered. The greening of existing brownfields mainly due to parks are also part of the development strategy of the district. The creation of connections between existing green areas (Fig. 10.14) are an utmost concern, which is considered within the strategy for the improvement of the climatic situation inside the district.

Also the ventilation of the district due to local wind systems and cold air streams is analysed (Fig. 10.16). To improve the ventilation of the district, reducing the

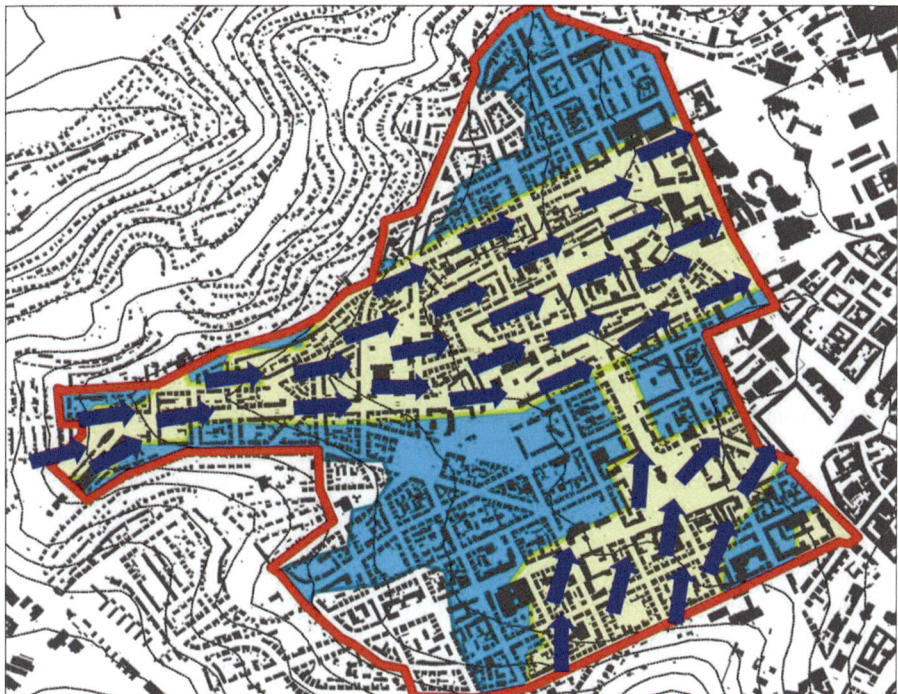

**Fig. 10.16** Major cold air streams in Stuttgart-West. The *yellow shaded areas* mark areas with restriction to buildings to reduce the obstruction of existing streams

obstruction of existing streams, due to the enlargement of the major stream axes (mostly street canyons) is implemented into to development outline plan for Stuttgart-West. For the enlargement of the stream axes in the concerned areas, a maximum building high, an optimal building axes orientation and the reinstatement of buildings are offered.

## 10.4   Stuttgart's Heat Warning System HITWIS

The increase in the number of days exposed to high temperatures and high humidity will result in heat stress for the population. This can pose a danger, particularly for the elderly and those in poor health. In the summer of 2003, the extreme conditions accounted for an estimated 2000 additional deaths. So it is necessary to inform and to warn respectively. Additionally a customized behaviour must be advised different target groups.

To improve the more or less existing heat warning system in Stuttgart HITWIS was constituted a working group "Heat waves/heat stress" including some parts of the municipality (health care, urban climate, social welfare office etc.). This working group is well connected to external institutions like ambulance services, housing societies, social services etc..

The following measures are recommended and have been partly realised:

- Supply of a leaflet including recommendations for a more adapted behaviour in a heat case.
- Development of a special heat app(lication) (f.e. inclusive a drink reminder) for mobile devices running on different platforms.
- Distribution of the web-based heatwarning of the german weather service for Stuttgart.
- Publication of special thematic website within the Internet presence of the Municipality of Stuttgart.
- Organisation and operation of a heat phone to warn and inform elder and lone people.
- Composing of a heat city map including "Cooling Zones", water posts etc..
- Public relations in different media (sensibilisation, reminding), high visibility events with small gifts, promotional articles f.e. folding fans
- Instructions for different target groups f.e. families, sportsmanlike people etc.
- using of electronic advertising panels displaying prepared warnings and hints before and during heat waves.

# References

Akbari, H., Bretz, S., Kurn, D. M., & Hanford, J. (1997). Peak power and cooling energy savings of high-albedo roofs. *Energy and Buildings, 25*(2), 117–126.

Akbari, H., Menon, S., & Rosenfeld, A. (2009). Global cooling: Increasing world-wide urban albedos to offset CO2. *Climatic Change, 94*(3–4), 275–286.

Bruse, M., & Fleer, H. (1998). Simulating surface–plant–air interactions inside urban environments with a three dimensional numerical model: The challenge of awareness in developing societies. *Environmental Modelling and Software, 13*, 373–384.

Höppe, P. (1993). Heat balance modelling. *Experientia, 49*, 741–746.

Höppe, P. (1999). The physiological equivalent temperature – A universal index for the biometeorological assessment of the thermal environment. *International Journal of Biometeorology, 43*, 71–75.

Huttner, S. (2012). *Further development and application of the 3D microclimate simulation ENVImet*. Johannes Gutenberg-Universität Mainz.

Ketterer, C., Ghasemi, I., Reuter, U., Rinke, R., Kapp, R., Bertram, A., & Matzarakis, A. (2013). Veränderung des thermischen Bioklimas durch stadtplanerische Umgestaltung. *Gefahrstoffe – Reinhaltung der Luft, 73*, 323–329.

Ketterer, C., & Matzarakis, A. (2014a). Human-biometeorological assessment of heat stress reduction by replanning measures in Stuttgart, Germany. *Landscape and Urban Planning, 122*, 78–88.

Ketterer, C., & Matzarakis, A. (2014b). Human-biometeorological assessment of the urban heat island in a city with complex topography – The case of Stuttgart, Germany. *Urban Climate, 10*, 573–584.

Kusaka, H., Kondo, H., Kikegawa, Y., & Kimura, F. (2001). A simple single-layer urban canopy model for atmospheric models: Comparison with multi-layer and slab models. Kluwer Academic Publishers; *Boundary-Layer Meteorology, 101*(3), 329–358.

Martilli, A., Clappier, A., & Rotach, M. (2002). An urban surface exchange parameterisation for mesoscale models. Kluwer Academic Publishers; *Boundary-Layer Meteorology, 104*(2), 261–304.

Matzarakis, A., & Mayer, H. (1996). Another kind of environmental stress: Thermal stress. *WHO Collaborating Centre for Air Quality Management and Air Pollution Control Newsletters, 18*, 7–10.

Matzarakis, A., Mayer, H., & Iziomon, M. G. (1999). Applications of a universal thermal index: Physiological equivalent temperature. *International Journal of Biometeorology, 43*, 76–84.

Matzarakis, A., Rutz, F., & Mayer, H. (2007). Modelling radiation fluxes in simple and complex environments – Application of the RayMan model. *International Journal of Biometeorology, 51*, 323–334.

Matzarakis, A., Rutz, F., & Mayer, H. (2010). Modelling radiation fluxes in simple and complex environments: Basics of the RayMan model. *International Journal of Biometeorology, 54*, 131–139.

Mayer, H., & Höppe, P. R. (1987). Thermal comfort of man in different urban environments. *Theoretical and Applied Climatology, 38*, 43–49.

Reuter, U., Hoffmann, U., & Kapp, R. (2010): City of Stuttgart. (2010). *Climate change – challenge facing urban climatology*. Publication series No. 3/2010. Available online at: http://www.stadtklima-stuttgart.de

Salamanca, F., Martilli, A., & Yagüe, C. (2012). A numerical study of the urban heat island over Madrid during the DESIREX (2008) campaign with WRF and an evaluation of simple mitigation strategies. *International Journal of Climatology, 32*(15), 2372–2386.

Skamarock, W. C., Klemp, J. B., Dudhia, J., Gill, D.O., Barker, D. M., Wang, W., & Powers, J. G. (2005). A description of the advanced research WRF version 2. Available online at: http://oai.dtic.mil/oai/oai?verb=getRecord&metadataPrefix=html&identifier=ADA487419

Taha, H. (1997). Urban climates and heat islands: Albedo, evaporation, and anthropogenic heat. *Energy and Buildings, 25*, 99–103. Available online at: http://www.javeriana.edu.co/arquidis/educacion_continua/documents/Urbanclimates.pdf

# Chapter 11
# Urban Heat Island and Bioclimatic Comfort in Warsaw

Krzysztof Błażejczyk, Magdalena Kuchcik, Wojciech Dudek, Beata Kręcisz, Anna Błażejczyk, Paweł Milewski, Jakub Szmyd, and Cezary Pałczyński

**Abstract** This chapter will introduce the UHI phenomena in Warsaw, in particular after a the definition of the pilot area, experimental microclimatic measurements were made in two housing estates, Koło and Włodarzewska, located at a similar distance from the city centre and from the city limits but different in terms of building periods and materials. A specific analysis of vegetation is provided to put in relationship UHI effects and allergenic factors. The case is completed by some solutions in terms of mitigation and adaptation to reduce urban warming impact.

**Keywords** Microclimate • Urban spatial organization • Green areas • Mitigation

## 11.1   Introduction – UHI as an Effect of Spatial Organization of the City

Warsaw is the largest city in Poland. Its area of almost 515 km$^2$ has significant differentiation of land use. Currently about 248 km$^2$ is built-up area (48 %). Within this the greatest part (about 57 km$^2$) is covered by industry, trade units and transport systems. Forests make up about 15 % of the city. Urban parks and other recreational green areas cover 10 %. 12 % of the city territory is used as arable land, for crops and pasture. The category "heterogeneous agricultural areas" includes sparsely built areas and allotment gardens – 11.3 % (Table 11.1). With 1.7 million residents and

K. Błażejczyk • M. Kuchcik (✉) • P. Milewski • J. Szmyd
Geoecology and Climatology Department, Institute of Geography and Spatial Organization, Polish Academy of Sciences, Twarda 51/55, 00-818 Warszawa, Poland
e-mail: mkuchcik@twarda.pan.pl

W. Dudek • B. Kręcisz • C. Pałczyński
Nofer Institute of Occupational Medicine, Łódź, Poland

A. Błażejczyk
Bioklimatologia. Laboratory of Bioclimatology and Environmental Ergonomics, Warszawa, Poland

© The Author(s) 2016
F. Musco (ed.), *Counteracting Urban Heat Island Effects in a Global Climate Change Scenario*, DOI 10.1007/978-3-319-10425-6_11

**Table 11.1** The land use types in Warsaw, according to Corine Land Cover 2006 (EEA 2007)

| Land use | Area [km²] | Area [%] |
|---|---|---|
| Urban fabric | 191.0 | 37.1 |
| Industrial. commercial and transport units | 56.9 | 11.1 |
| Construction sites | 3.3 | 0.6 |
| Artificial. non-agricultural vegetated areas | 53.4 | 10.4 |
| Arable land | 40.0 | 7.8 |
| Permanent crops | 0.5 | 0.1 |
| Pastures | 21.2 | 4.1 |
| Heterogeneous agricultural areas | 58.3 | 11.3 |
| Forests | 77.8 | 15.1 |
| Scrub and/or herbaceous vegetation | 0.6 | 0.1 |
| Open spaces with little or no vegetation | 1.6 | 0.3 |
| Inland wetlands | 0.2 | 0.0 |
| Inland waters | 9.9 | 1.9 |
| Total | 514.6 | 100.0 |

over 3.2 million residents of the greater agglomeration area, it has become the 10th most populous city in the European Union (Eurostat 2014). After the Second World War, the area of Warsaw increased gradually; small villages and rural areas as well as natural forests were included into the city.

During the last 20 years, many fields, pastures and meadows were adapted for residential districts.

The recent tendency in city development is to build dense settled residential districts (both, small single family buildings and 4–6 floor blocks) as well as to insert new buildings into free spaces in the city centre (which was dramatically destroyed during the II World War). At the administration level of Warsaw there is not one single vision for city development. For the whole city there is only a general overview of investment intentions (Studium… 2010).

The shape of UHI in Warsaw resembles a diamond and reflects the distribution of the densest built area. The mean yearly intensity of the UHI-index (difference of the minimum daily temperature for the considered site to the value of the minimum daily temperature for Warszawa-Okęcie station) reaches over 2 °C in the city centre. On the outskirts and in the forest area in south-east Warsaw, the UHI-index is from 0.5 to 1.0 °C lower than at the airport station. During spring and summer, the intensity of UHI is comparable to the average. The most intensive UHI is to be observed in autumn. The very centre of the city is warmer in the night by 2.5 °C comparing to Okęcie station. The lowest UHI-index occurs in winter – only 1.5 °C, but then the spatial extent of UHI is greater than in other seasons. This situation is associated

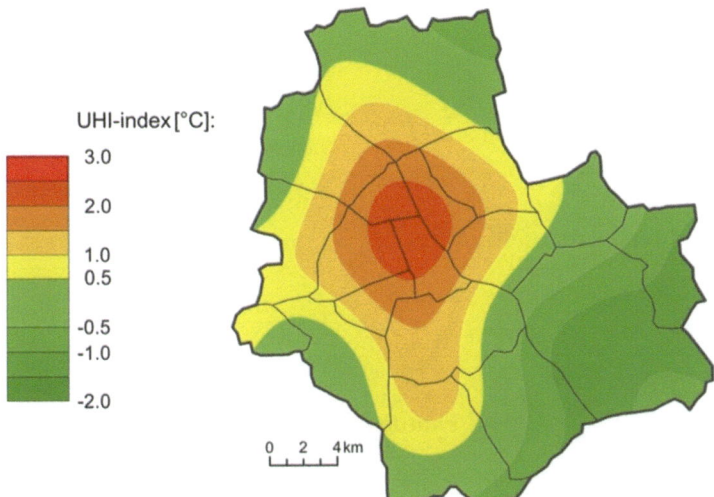

UHI-index [°C]:

3.0
2.0
1.0
0.5
-0.5
-1.0
-2.0

0   2   4 km

**Fig. 11.1** Spatial distribution of UHI-index in Warsaw, mean values for the years 2011–2012

with the usage of house heating stoves in places not connected to central heating plants (Fig. 11.1).

Thus, the general idea of pilot studies in Warsaw was: (1) to verify how the varying structures of space organization and land cover of selected residential districts influence UHI intensity and perceptible thermal conditions, and (2) how architectural solutions (e.g. planting additional lawns and trees, organizing green roofs) can minimise UHI and affect perceptible thermal conditions.

An additional aspect of the pilot studies was to assess the allergenic potential of plant cover (trees and bushes) growing inside the studied residential districts. It allows the validation of the health impact of vegetation and consequently, to give recommendations regarding plant composition which would be more friendly for the local population.

## 11.2   Pilot Areas Methodology

The pilot studies were designed on the basis of a network of microclimatic measurements in Warsaw and its surroundings, working since 2006 as part of climate research carried out in IGSO PAS (Kuchcik et al. 2008; Błażejczyk et al. 2013b).

To cope with the aims of research, three small areas in Warsaw were chosen: Twarda (in the centre of the city) as well as Koło and Włodarzewska housing estates (in the western part of the city). As a reference site, representing outside rural conditions, the station situated in the Botanical Garden in Powsin was chosen (Fig. 11.2). For each area a detailed inventory of the greenery, type of surfaces, heights of buildings and horizon limitations was made.

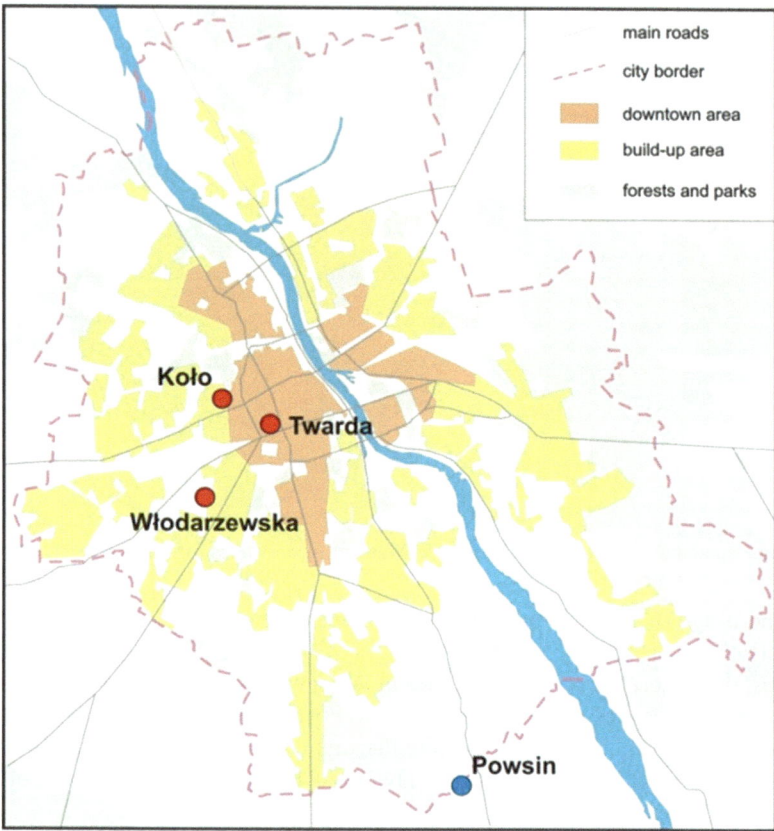

**Fig. 11.2** Location of the pilot study areas (Twarda, Koło, Włodarzewska) and peripheral reference station (Powsin)

The Koło estate was established about 50 years ago (in the 1960s). The 4–5 floor buildings are built in low density and the majority are built from clay bricks. Few parking places are located inside the estate. Wide spaces between buildings are covered by lawns and tall, mature, deciduous trees. The RBVA (Ratio of Biologically Vital Areas), i.e. the ratio of areas covered by vegetation or open water (not sealed areas) in the plot size (according to Szulczewska et al. 2014) is 54.3 % and FAR (Floor Area Ratio)[1] is 0.8 (Fig. 11.3a).

The Włodarzewska estate was built about 15 years ago (1995–2000) and is surrounded by many open spaces and a park, but arranged in a way which effectively precludes the entrance of air from the outside. It is characterized by compact development. The 4–5 floor blocks are very densely built up. They are constructed mostly from concrete and include underground car parks. Parking places are also organized along communication roads inside the estate. Many small flowerbeds and lawns

[1] Floor Area Ratio, is calculated as the area of all building contours (Barea) multiplied by the number of floors (fn) and divided by the total area of the plot (Tarea), FAR = Barea·fn/Tarea.

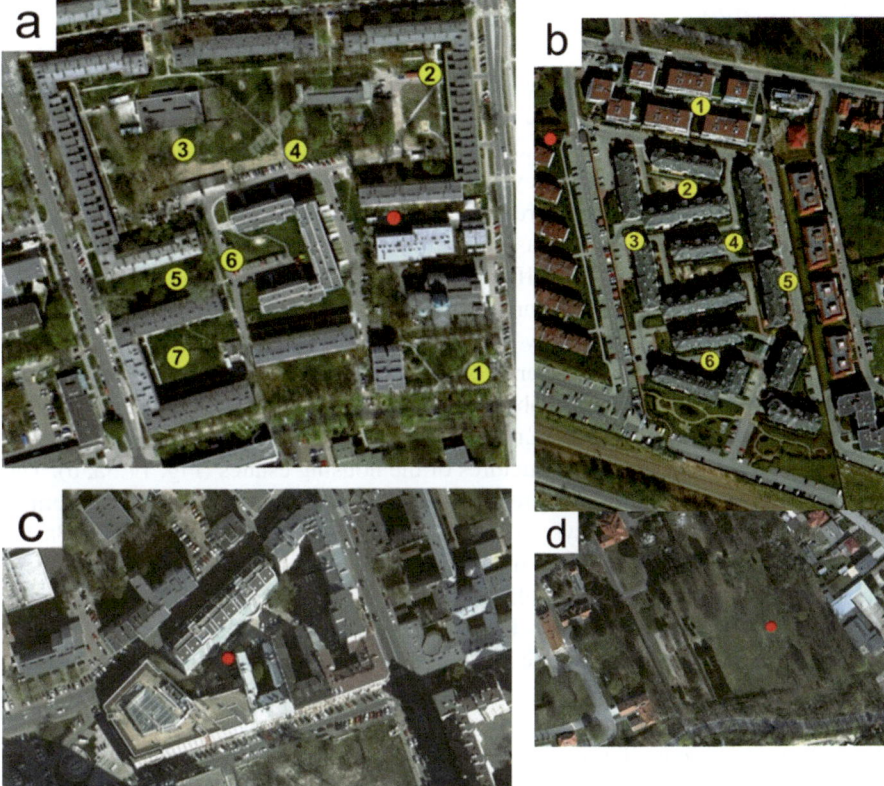

**Fig. 11.3** Aerial view of pilot areas: Koło housing estate (**a**), Włodarzewska housing estate (**b**), Twarda district (**c**) and Powsin reference station (**d**); *yellow points* indicate sites of microclimatic measurements, the *red points* indicate sites of permanent microclimatic measurements

with coniferous shrubs grow between the buildings. Only a few young deciduous trees grow there. RBVA is 40.7 % and FAR is 1.25 (Fig. 11.3b).

The Twarda district is located in the city centre. It consists of a mixture of 80-year old buildings (from clay bricks), which were not destroyed during II World War as well as of newly constructed compartments (from concrete). 6–7 floor buildings predominate. Vegetation cover is very poor and only few trees and lawns can be found at the bottom of deep courtyards. Almost all the space between buildings is used as parking places. The area is surrounded by streets without any vegetation. The RBVA for this area is only 4 % and the FAR index is 2.74 (Fig. 11.3c).

The characteristics of UHI and perceptible thermal conditions in each of the selected studied areas were compared with air temperature (Ta) and Universal Thermal Climate Index (UTCI) values observed at the peripheral reference station in Powsin. The station is situated in the Botanical Garden and represents open area conditions with ground covered by grass. Horizon shading is about 10 % and the station is exposed to sunbeams almost the whole day (Fig. 11.3d).

## 11.3   Methods

The pilot studies in Warsaw have been composed of two steps. In the first step, experimental microclimatic measurements were made in two housing estates, Koło and Włodarzewska. They are located a similar distance from the city centre and from the city limits. However, they differ in type, density and age of buildings as well as in composition of green areas and the percentage of biologically vital area. The aim of the microclimatic measurements was to assess influence of local space organization on differences of UHI and perceptible thermal conditions. The results of the measurements were compared both with the Powsin reference station and with the city centre represented by Twarda district.

During the microclimatic research on 21 and 22 May 2013, the air temperature and humidity, wind speed and global solar radiation were measured in two periods of the day: early morning (5–7 a.m.) and at midday (12 p.m.–2 p.m.). The posts were situated in various micro structures of the housing estates (Fig. 11.3a, b). The perceptible thermal conditions were assessed with the use of the Universal Thermal Climate Index UTCI (Błażejczyk 2011; Bröde et al. 2012; Błażejczyk et al. 2013a, b). The index assesses heat stress in man caused by a complex outdoor environment (air temperature and humidity, solar radiation and wind speed). The spatial variability (in early morning and midday hours) of air temperature and UTCI and differences between microclimatic posts and the reference peripheral station were analysed.

As a result of the greenery inventory, all plant species have been divided into four classes according their allergenic potential. The classification was evaluated according to the Polish Society of Allergology guidelines for diagnosis and management of allergic diseases supported additionally by local allergists' experience.

Class 3 – great allergenicity, frequently sensitizing species (Alder, Birch, Hazel)
Class 2 – moderate allergenicity, rare sensitizing species (Poplar, Elm, Willow, Beech, Oak, Plane, Ash, Linden)
Class 1 – slight allergenicity, very rare sensitizing species or isolated case reports only (Acacia, Hornbeam, Maple, Elder, Spruce, Pine, Jasmine, Ambrosia, Olive)
Class 0 – no allergenicity, species with no or unknown allergenic potential (female cultivars of dioeciously plants from higher classes were also included in this class due to no pollen production).

If a plant species was not mentioned in any guidelines or local allergists' statement, the EBSCO scientific journal database was used to determine the potential allergenicity of the plant. If, during the last 15 years, there had been three or more cases or scientific reports published on the possibility of respiratory tract allergy induction, the plant was classified as Class 1; otherwise it was classified as Class 0.

Microclimatic measurements became the basis for the second, simulation step of the pilot studies. In this step a few scenarios of possible changes in land cover in Włodarzewska and Twarda, which could reduce UHI, were considered. The simulations of air temperature for 4 days a year representing spring, summer, autumn and winter were made with the use of ENVI-Met software in the Vienna University of Technology, Department of Building Physics and Building Ecology, Institute of Architectural Sciences.

## 11.4  The Role of Urban Vegetation in Reduction of UHI – The Results of Microclimatic Research

The variation of air temperature (Ta) and heat stress (UTCI) inside residential districts was examined on two sunny days, 21–22 May 2013, in two housing estates, Koło and Włodarzewska, which differ in the provision of green areas as well as in the arrangement of buildings (Fig. 11.3a, b). On the measuring days, the wind was weak (<2 m·s$^{-1}$), the mornings had clear sky, *Cumulus* clouds were created during the day and a short 10-min shower occurred on 22 May. Global solar radiation in Koło was more differentiated inside the estate than in Włodarzewska, which significantly influenced the calculated UTCI values. The air temperatures in the housing estates and in the reference station (Powsin) fluctuated between 8 and 15 °C in the mornings and 20 and 23 °C during the middle of the day.

The differences in air temperature in various micro structures inside both housing estates ranged from 2.5 to 2.8 °C in the early morning to 3.3 to 3.5 °C at noon, reaching higher values in Włodarzewska. At midday, the spatial differences in air temperature increased due to an increase in solar radiation (up to about 600–800 W·m$^{-2}$). The warmest air was above artificial surfaces (asphalt, concrete), at well insulated sites and the coldest air was found over natural, shaded surfaces (lawns under trees). The wind tunnel effect was clearly seen at post 2 (close to the tunnel under the block of flats) and 5 (between two buildings) in Koło, and at post 5 (on a street along a long block of flats) in Włodarzewska.

The spread of UTCI values inside the analysed estates was 3 times greater than that of air temperature, though they mostly fall within one heat stress category, named "no thermal stress". In the Koło estate, the differences in UTCI values between posts at individual points in time reached 10.5 °C in the morning. The coldest was recorded on a vast lawn in the centre of the estate, surrounded by high trees and buildings (post 3), and the UTCI values indicated "slight cold stress" there. The warmest was a calm site under a canopy of high trees (post 1), where, at midday, the differences in UTCI reached 10.4 °C. The coldest was a shaded site under a tree canopy (post 1) and the warmest – an asphalt surfaced parking area (post 6), where moderate heat stress was noted.

In the Włodarzewska estate, the differences in UTCI values were much smaller and they reached only 6.7 °C in the morning. The coldest were pavements next to a block of flats, the windiest in the estate (post 5), and the warmest – a calm and sunny square between buildings (post 6). At midday, the spread of UTCI values was up to 6.8 °C. The warmest was a small lawn with young trees, squeezed between blocks of flats (post 4).

**Fig. 11.4** The Ratio of
Biologically Vital Areas
(RBVA) and the Floor
Area Ratio (FAR) of the
pilot areas

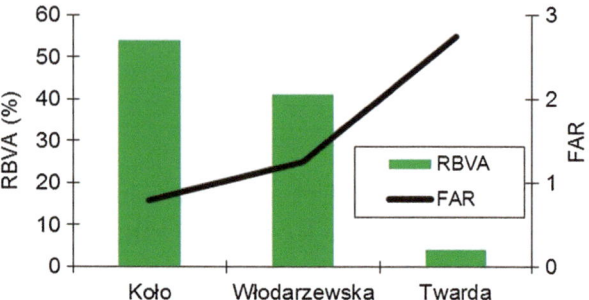

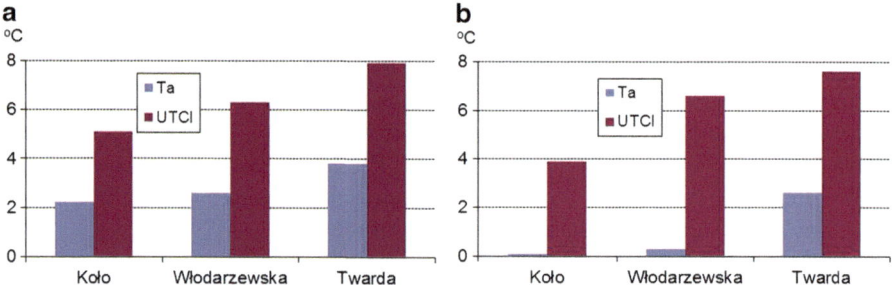

**Fig. 11.5** The difference of the air temperature (Ta) and heat stress index (UTCI) between pilot areas and the Powsin reference point (**a** – morning hours, **b** – midday hours)

Concluding, green areas, including both lawns, bushes and trees, play an important role in creating urban heat stress. The results of experimental research revealed a bigger variation of Ta and UTCI inside a housing estate with higher RBVA and lower FAR (Fig. 11.4). Significant differences were found when Ta and UTCI values observed inside pilot areas were compared with the peripheral part of Warsaw (Powsin). In early morning (which represents the classical definition of UHI), the greatest Ta differences were observed in the city center and the smallest – in the Koło area. Similar relations of Ta were found for the midday hours, though Koło was characterized by the same temperature as the peripheral station and Włodarzewska was little warmer. The differences between the studied areas in comparison to the peripheral reference station (Powsin) were even greater when considering the perceptible thermal conditions represented by UTCI. The Koło estate has UTCI values that are, on average, 4–5 °C higher than at the periphery station, though in the Twarda district (city center), UTCI is 7.5–8 °C higher. The reason for such good thermal conditions in Koło is the presence of well-developed vegetation with predominantly high, leafy trees which cause many places in the Koło estate to be as cool as is observed in the botanical garden (Fig. 11.5). The UTCI is very sensitive to even small changes in the meteorological variables induced by different urban structures on a very detailed scale. The comparison of two housing estates with different structures shows that low density settlements with a great portion of biologically vital surfaces and trees can create relatively mild biothermal conditions.

## 11.4.1    Allergenic Potential of Vegetation

In the ecophysiographic description of three pilot study areas (in this case vegetation was also analysed for the southern part of the Włodarzewska estate where microclimatic conditions were not considered), a total of 97 different plant species have been described (see Table 11.2). Table 11.3 contains the general evidence of allergenicity in classes of plants in the studied pilot areas.

It is generally recognised that the Urban Heat Island (UHI) phenomenon has a detrimental impact on the health of populations that live under its influence. This phenomenon can act as an amplifier to heat wave events, mainly due to a lack of night-time thermal body regeneration. It can cause thermal stress through heat accumulation during consecutive days. Such conditions affect a certain population of city dwellers known to be at risk of developing heat-related illnesses. This population comprises subjects with chronic pulmonary and cardio- or cerebrovascular disorders, elderly people, young children and the disabled (Basu 2009). It is indisputable that phenomenon such as UHI need to be counteracted by launching various mitigation and adaptation strategies. One must remember that some of those strategies involve introduction of a new plant species into an urban area. It can mitigate the UHI phenomenon quite efficiently, but simultaneously can give rise to another major health problem. Improper plant choice may cause people who are susceptible to seasonal airborne allergens to develop symptoms of asthma, rhino conjunctivitis or urticaria/dermatitis. It is essential for mitigation and adaptation strategies to select appropriate plant species that do not aggravate the symptoms of airborne allergies. Although the definite impact of the UHI phenomenon on the allergenic activity of plants has never been described or proven before, there is some evidence supporting the hypothesis that such phenomena can alter plant physiology, causing them to be more allergy-aggressive. It has been proved in many studies, that in warmer climate plants can produce larger amounts of pollens when compared to those in cooler regions. An increase in carbon dioxide level has a similar impact (Cecchi et al. 2010). Factors typical for urbanized UHI areas such as: elevated ambient temperature, elevated carbon dioxide levels, increase in concentration of anthropic pollutants, i.e. sulphur dioxide, nitrogen dioxide, carbon monoxide, ozone and airborne particulate matter (PM) affect plant physiology causing an increase in allergen production. Pollen grains released in such an environment contain more allergen proteins on their external surfaces than they do in cooler settings (Todea et al. 2013; Beck et al. 2013). Typical pollen grain diameters range from 15 to 40 μm. Such a diameter allows pollen grains to reach only the upper region of the respiratory tract to trigger rhinoconjunctival symptoms. Only particles smaller than 10 μm can penetrate the respiratory tract down to its deeper structures to provoke asthma seizures. Pollen grains are extremely resistant to fragmentation, however, allergens can easily be transferred from pollen onto smaller particles ($PM_{10}$, $PM_{2.5}$) and, in this way, easily reach every compartment of the respiratory tract. Therefore, increased plant allergenic activity and air pollutants can act synergistically and thus dramatically reduce the quality of life of subjects susceptible to airborne allergens.

**Table 11.2** Plant species recognised in three pilot study areas (Latin names in alphabetical order)

| | | | |
|---|---|---|---|
| *Abies concolor* | *Cotoneaster horizontalis* | *Mahonia aquifolium* | *Rhus typhina* |
| *Abies koreana* | *Cotoneaster lucidus* | *Malus purpurea* | *Ribes alpinum* |
| *Acer campestre* | *Crataegus monogyna* | *Malus spMorus alba* | *Robinia pseudoacacia* |
| *Acer negundo* | *Crataegus xmedia* | *Parthenocissus quinquefolia* | *Rosa sp.* |
| *Acer platanoides* | *Daphne mezereum* | *Philadelphus sp.* | *Salix alba* |
| *Acer platanoides* | *Deutzia scabra* | *Physocarpus opulifolius* | *Salix babylonica* |
| *Acer pseudpolatanus* | *Elaeagnus angustifolia* | *Picea abies* | *Salix caprea* |
| *Acer saccharinum* | *Euonymus fortunei* | *Picea pungens* | *Salix caprea 'Klimanrock'* |
| *Aesculus hippocastanum* | *Fagus sylvatica* | *Pinus mugo* | *Salix fragilis* |
| *Alnus glutinosa* | *Forsythia xintermedia* | *Pinus sylvestris* | *Sambucus nigra* |
| *Berberis thunbergii* | *Fraxinus excelsior* | *Platanus xhispanica* | *Sorbus aria* |
| *Betula pendula* | *Hedera helix* | *Populus alba* | *Sorbus aucuparia* |
| *Buxus sempervirens* | *Hydrangea sp.* | *Populus nigra* | *Spiraea 'Grefsheim'* |
| *Caragana arborescens* | *Ilex aquifolium* | *Populus simonii* | *Spiraea japonica* |
| *Carpinus betulus* | *Juglans nigra* | *Potentilla fruticosa* | *Spiraea xvanhouttei* |
| *Catapla bignonioides* | *Juniperus 'Blue Carpet* | *Prunus avium* | *Symphoricarpos albus* |
| *Cercidiphyllum japonicum* | *Juniperus sabina* | *Prunus cerasifera* | *Syringa vulgaris* |
| *Chaenomeles japonica* | *Juniperus sp.* | *Prunus domestica subsp. syriaca* | *Tamarix sp.* |
| *Chamaecyparis pisifera* | *Juniperus virginiana* | *Pyracantha coccinea* | *Taxus baccata* |
| *Chamaecyparis sp.* | *Juniperus xmedia* | *Pyrus pyraster* | *Thuja occidentalis* |
| *Cornus alba* | *Larix decidua* | *Quercus robur* | *Tilia cordata* |
| *Corylus colurna* | *Ligustrum vulgare* | *Quercus rubra* | *Tilia platyphyllos* |
| *Cotinus coggygria* | *Lonicera maackii* | *Reynoutria sachalinensis* | *Viburnum opulus* |
| *Cotoneaster dammeri* | *Lonicera xylosteum* | | *Weigela florida* |
| | *Magnolia sp.* | | *Wisteria sp.* |

**Table 11.3**  Numbers of tree and shrub specimens with different allergenicity in compared housing estates

| Pilot area | Allergenicity | | | |
|---|---|---|---|---|
| | No (class 0) | Slight (class 1) | Moderate (class 3) | Great (class 4) |
| Koło (469) | 134 (28.6%) | 273 (58.2%) | 54 (11.5%) | 8 (1.7%) |
| Class 2&3 together | | | 62 (13.2%) | |
| Włodarzewska (619) | 302 (48.8%) | 276 (44.6%) | 15 (2.4%) | 26 (4 2%) |
| Class 2&3 together | | | 41 (6.6%) | |
| Włodarzewska-south (101) | 49 (48.5%) | 39 (38.6%) | 9 (8.9%) | 4 (4%) |
| Class 2&3 together | | | 13 (12.9%) | |

It is also hypothesized that combinations of those factors, in addition to triggering respiratory tract allergy symptoms, can also promote allergisation in the portion of the population so far unaffected by allergies, by facilitation of allergen penetration into the respiratory tract (Lovasi et al. 2013). The deeper the allergen can reach into the respiratory tract, the greater area of mucosa is affected and the longer the allergen stays inside the organism, the more severe allergy symptoms can be triggered as well as easier allergisation. Air pollution itself also facilitates allergisation. It can irritate respiratory tract mucosa, thus causing it to be more easily penetrated by allergen proteins. Another factor that can affect people susceptible to seasonal airborne allergens is the elongation of the pollen season (Bielory et al. 2012). In a warmer climate, plants start to pollinate earlier and continue to release pollens for longer periods. It causes the anti-allergic pharmacology therapy schedule and specific immunotherapy calendar to require additional modification (early implementation, prolonged administration). All facts described above indicate that suitable plant selection is essential for successful implementation of various greenery-related UHI adaptation and mitigation strategies. It is also important to remember that all plant allergenicity assessment is local-specific. The set of allergy patterns is different for various geographic regions. The best example is the allergy to olive trees, which is known to be a frequent allergy in the Mediterranean region but not in north of Europe. Therefore, for performing such an assessment, the cooperation of local urbanists, botanists and allergists is needed to develop a suiTab. model of plant cover for UHI mitigation and adaptation strategies.

The prevalence of plant species considered to be a recognizable hazard to people with seasonal airborne allergies (Class 2 and 3) ranged from 6.6 to 13.3%. The prevalence of Class 3 plants alone, known to cause the greatest allergological risk, range from 1.7% (Koło area) up to 4.2% (Włodarzewska area). Although Class 2 and Class 3 plants are almost evenly scattered throughout the pilot areas, there are two spots of Class 3 plant compaction close together. First, the spot at the south east corner of the Koło area contains six Class 3 plants. The second spot, at the north east corner of the Włodarzewska area, contains seven Class 3 plants. These spots are recommended for immediate remodeling. This limited intervention will allow the reduction of Class 3 prevalence to 0.4% in Koło and 3.4% in the Włodarzewska area.

316 K. Błażejczyk et al.

## 11.5 Possibility of Reducing UHI Intensity

For the calculations of the scenarios of the possible UHI reduction, two areas were chosen – Włodarzewska and Twarda (very city centre). Koło was eliminated because there was nothing to change and the existing thermal conditions were friendly for humans.

Looking for the benefits from trees and the non-dense locations of buildings, a reduction in the number of buildings (pulling down 2 of them) and an increase in the number of lawns and trees (in the places of 2 new empty parcels) (Fig. 11.6) was proposed for the Włodarzewska estate. For the Twarda district, the possible effects of two scenarios were verified. In the 1st scenario, planting an additional lawn area and deciduous trees inside court yards and along streets was proposed. In the 2nd scenario all roofs were additionally covered by vegetation (Fig. 11.7).

Four characteristics of the urban heat island for 4 days of the year were analysed: maximum UHI, i.e. the highest daily urban to rural air temperature difference (dTa), minimum UHI, i.e. minimum daily dTa value, average dTa value for night hours (9 p.m.–7 a.m.) and average dTa value for day-time hours (10 a.m.–4 p.m.).

Simulations of air temperature for the Włodarzewska housing estate showed that the replacement of some buildings by lawns and trees could lead to a significant

**Fig. 11.6** Land cover of Włodarzewska housing estate: (**a**) – the present state, (**b**) – changes in land cover proposed in scenario 1 (replacing 2 buildings by lawns and trees)

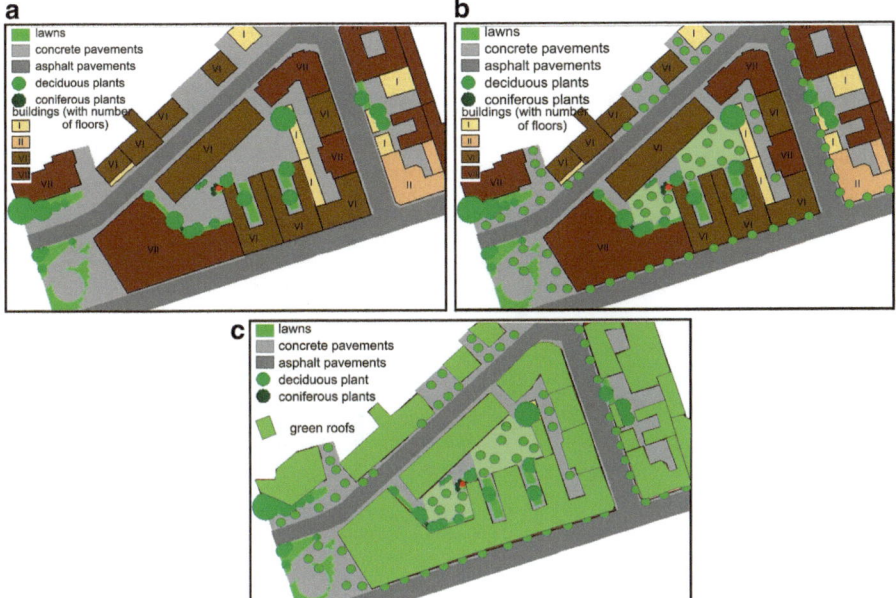

**Fig. 11.7** Land cover of the Twarda district; (**a**) – present state, (**b**) – changes in land cover proposed in scenario 1 (additional trees and permeable surfaces), (**c**) – changes in land use proposed in scenario 2 (additional trees, permeable surfaces and green roofs)

reduction of UHI (Fig. 11.8). Maximum and minimum UHI values in spring, summer and autumn could be lowered by about 0.5 °C, and in winter up to 1 °C. The proposed land use changes would also cause a reduction of average UHI values, greater at night than during day-time hours.

For the Twarda district, simulations have verified that the 1st scenario of land cover changes (additional lawns and trees) would result in no significant changes of UHI intensity. Some lower values could occur, especially in summer, but in autumn, values of minimum UHI and midday hours UHI can be higher than in the present state. A more positive effect was obtained when the 2nd scenario (additional green roofs) was applied. Lower values of UHI occurred in spring and summer but in autumn and winter no changes were observed.

## 11.5.1   Assessment of Mitigation Measures

Simulations showed that increasing the number of lawns and trees as well as implementing green roofs in Warsaw city centre cause only small changes in the intensity of the urban heat island. To make a significant improvement, some radical action is needed, involving the reorganization of the existing built-up area. In places where it is possible some buildings should be pulled down and replaced by lawns and trees.

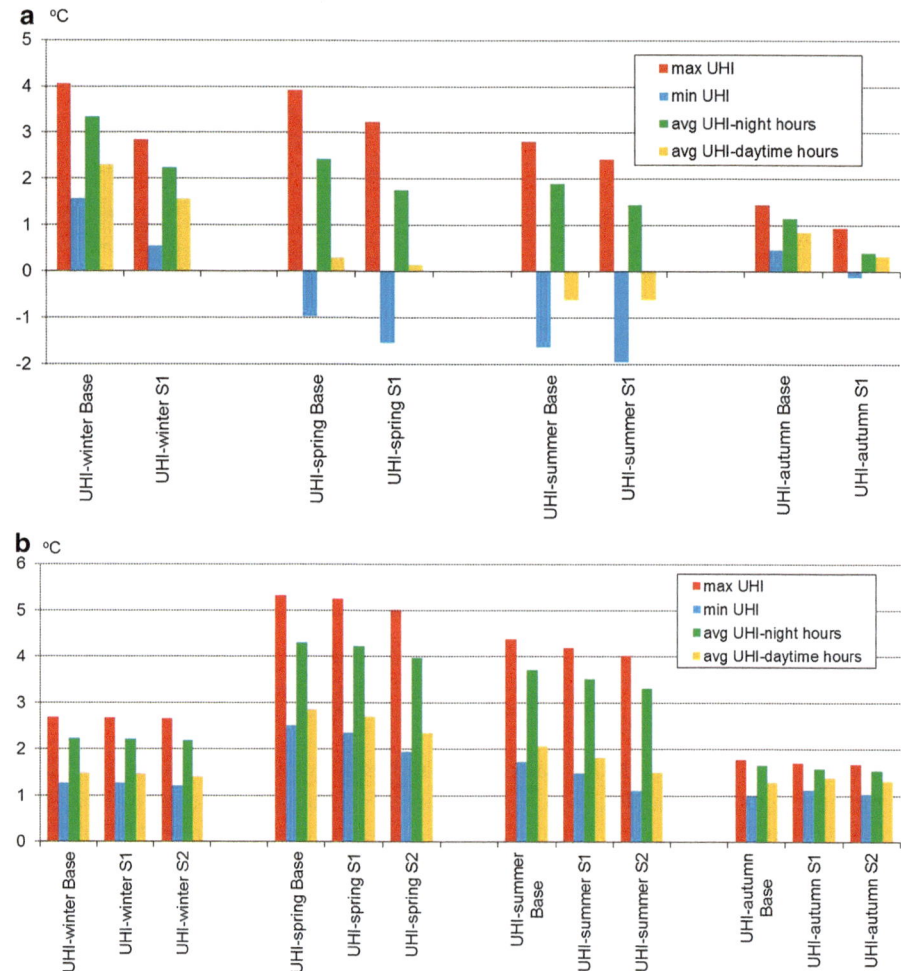

**Fig. 11.8** Simulations of UHI-index (°C) made by ENVI-Met software for 4 days of a year representing winter, spring, summer and autumn for Włodarzewska housing estate (**a**) and Twarda district (**b**)

It is also advisable to forbid compaction of downtown and to take special care of present green areas.

There are many indications when designing new housing estates which could clearly mitigate UHI effects: applying a building layout which does not close the interiors of the housing estates but open them to the neighbouring green areas, using openwork metallic fencing was much more conducive to ventilation than high brick walls, which did not allow infiltration of the air from the outside.

Mitigation of the urban heat island in areas located at a greater distance from the city centre is easier. Small measures, e.g. planting deciduous trees, planting flowerbeds or reducing artificial surfaces can clearly mitigate UHI. Summarising, the following mitigation actions should be undertaken on all planning levels: (1) protection within the city structure of open spaces which can intensify air movement and remove heat, (2) protection of all existing green areas inside the city which can reduce air temperature and heat stress as well as allow to the human organism to recuperate during hot episodes, (3) when planning new estates and compartments the urbanists and decision makers must remember about green areas and sufficient open space between buildings, (4) planting tree belts along streets to reduce surface heating, (5) planting trees on squares and play areas for children and the elderly, (6) increase the number of green roofs, terraces, balconies and facades, especially in the city centre.

However, when planning new plants, the appropriate composition of species must be considered. All plants of great allergernicity (Class 3) should be removed as soon as possible and plants of moderate allergenicity (Class 2) are recommended to be removed during the soonest area greenery remodelling and replaced with low- or non-allergenic species (Class 0 or Class 1).

Examples of plant species suitable for greenery related UHI mitigation and adaptation strategies in Warsaw:

*Trees* Maple, Hornbeam, Elder, Horse chestnut, Rowan, Acacia, Cherry plum, Mountain pine, Spruce, Fir, European larch, female cultivars of Poplar, Ash and Willow

*Shrubs and Bushes* Hedge cotoneaster, Boxwood, White dogwood, English dogwood, Border forsythia, Old-fashioned weigela, Staghorn sumac, Japanese quince, Red barberry, Common hawthorn.

*Climbing Plants (suitable for Green Facades)* Common ivy, Five-leaved ivy, Russian vine.

## 11.6  Adaptation Strategies

It is not possible to abolish the Urban Heat Island, and society must adapt to its occurrence. In 2013, the Polish Ministry of Environment published a strategic adaptation agenda for climate change in Poland to the year 2020 (Ministerstwo Środowiska 2013). There are several links to Urban Heat Island phenomenon when considering adaptation of cities to climate change. Several adaptation actions were also defined in the area of human health. Some of them refer to increased air temperature as well as to the increasing risk of allergies.

Taking into consideration action proposed in the adaptation agenda and the results obtained within UHI project, the following adaptation activities are recommended in Warsaw and should be implemented by city authorities: (1) education of all groups of society (city officers and decision makers, planners, architects, teachers and general population) about UHI phenomenon, its influence on quality of life and mitigation possibilities, (2) incorporation of UHI monitoring and information systems about UHI intensity and extension, (3) installing air conditioning in public buildings, (4) supporting air conditioning in private buildings and apartments, (5) incorporating a warning system of cardiovascular disorders, asthma and allergenic risks, (6) incorporating a system supporting elderly and disable people during episodes of extreme heat waves and intensive Urban Heat Island.

# References

Basu, R. (2009). High ambient temperature and mortality: A review of epidemiologic studies from 2001 to 2008. *Environmental Health: A Global Access Science Source, 8*, 40. Retrieved May 21, 2014, from http://www.ncbi.nlm.nih.gov/pmc/articles/PMC2759912/

Beck, I., Jochner, S., Gilles, S., McIntyre, M., Buters, J. T. M., Schmidt-Weber, C., Behrendt, H., Ring, J., Menzel, A., & Traidl-Hoffmann, C. (2013). High environmental ozone levels lead to enhanced allergenicity of birch pollen. *PLoS One, 8*(11), 1–7.

Bielory, L., Lyons, K., & Goldberg, R. (2012). Climate change and allergic disease. *Current Allergy and Asthma Reports, 12*(6), 485–494.

Błażejczyk, K. (2011). Mapping of UTCI in local scale (the case of Warsaw). *Prace i Studia Geograficzne WGSR UW, 47*, 275–283.

Błażejczyk, K., Jendritzky, G., Bröde, P., Fiala, D., Havenith, G., Epstein, Y., Psikuta, A., Kampmann, B., & Tinz, B. (2013a). An introduction to the Universal Thermal Climate Index (UTCI). *Geographia Polonica, 86*(1), 5–10.

Błażejczyk, K., Kuchcik, M., Błażejczyk, A., Milewski, P., & Szmyd, J. (2013b). Assessment of urban thermal stress by UTCI – Experimental and modelling studies: An example from Poland. *Die Erde, 144*(3), 105–116. doi:10.12854/erde-144-8.

Bröde, P., Fiala, D., Błażejczyk, K., Holmér, I., Jendritzky, G., Kampmann, B., Tinz, B., & Havenith, G. (2012). Deriving the operational procedure for the Universal Thermal Climate Index (UTCI). *International Journal of Biometeorology, 56*(3), 481–494.

Cecchi, L., D'Amato, G., Ayres, J. G., Galan, C., Forastiere, F., Forsberg, B., Gerritsen, J., Nunes, C., Behrendt, H., Akdis, C., Dahl, R., & Annesi-Maesano, I. (2010). Projections of the effects of climate change on allergic asthma: The contribution of aerobiology. *Allergy, 65*(9), 1073–1081.

Eurostat (2014). http://ec.europa.eu/eurostat/web/cities. Retrieved June 30, 2014.

Kuchcik, M., Baranowski, J., Adamczyk, A. B., & Błażejczyk, K. (2008). The network of microclimatic measures in Warsaw agglomeration. In K. Kłysik, J. Wibig, & K. Fortuniak (Eds.), *Klimat i bioklimat miast* (pp. 123–128). Łódź: Wydawnictwo Uniwersytetu Łódzkiego.

Lovasi, G. S., O'Neil-Dunne, J. P. M., Lu, J. W., Sheehan, D., Perzanowski, M. S., MacFaden, S. W., King, K. L., Matte, T., Miller, R. L., Hoepner, L. A., Perera, F. P., & Rundle, A. (2013). Urban tree canopy and asthma, wheeze, rhinitis and allergic sensitization to tree pollen in a New York City birth cohort. *Environmental Health Perspectives, 121*(4), 494–500.

Ministerstwo Środowiska (2013). *Strategiczny plan adaptacji dla sektorów i obszarów wrażliwych na zmiany klimatu do roku 2020 z perspektywą do roku 2030.* [Ministry of Environment (2013). *Strategic adaptation agenda for sectors sensitive to climate change to the year 2020 with the perspective to the year 2030*]. Retrieved May 21, 2014, from https://www.mos.gov.pl/g2/big/2 013_10/0f31c35e8e490e9d496780f98d95defc.pdf

Studium uwarunkowań i kierunków zagospodarowania przestrzennego Miasta Stołecznego Warszawy (2010). [Strategic conceptions of conditions and directions of spatial development of Warsaw (2010)]. Council of Warsaw, Legal act No XCII/2689/2010, October 7, 2010. Retrieved May 21, 2014, from http://bip.warszawa.pl/NR/exeres/65234DA5-353F-4DAB-B0F6-8A7BCF587DA3,frameless.htm

Szulczewska, B., Giedych, R., Borowski, J., Kuchcik, M., Sikorski, P., Mazurkiewicz, A., & Stańczyk, T. (2014). How much green is needed for a vital neighbourhood? In search for empirical evidence. *Land Use Policy, 38*, 330–345.

Todea, D. A., Suatean, I., Coman, A. C., & Rosca, L. E. (2013). The effect of climate change and air pollution on allergenic potential of pollens. *Notulae Botanicae Horti Agrobotanici Cluj-Napoca, 41*(2), 646–650.

# Chapter 12
# Urban Heat Island in the Ljubljana City

Blaž Komac, Rok Ciglič, Alenka Loose, Miha Pavšek, Svetlana Čermelj,
Krištof Oštir, Žiga Kokalj, and Maja Topole

**Abstract** Ljubljana made the first climate mapping of the city in 2000, putting the differences in terms of UHI effects between the inner part of the city and the surrounding areas are quite evident. An efficient planning of city infrastructure takes into account increasing average temperatures in a city, and therefore provides solution to reduce and monitor the effect of its urban heat island. The chapter presents the urban heat island in the Ljubljana region, temperature modelling based on satellite data and UHI project pilot actions in the Ljubljana city.

**Keywords** Urban heat island • Urban heat island atlas • Satellite termography measurements • Pilot actions • Ljubljana • Slovenia

## 12.1 Introduction

Urban heat island is an urban area where temperatures are higher than in the surrounding area. This is mainly due to the accumulation of heat in buildings, concrete and asphalt land during the day. This leads to a greater night long wave radiation from the surface and thus to a smaller cooling than the surroundings (Žiberna 2006). The city also has lower albedo, greater diffuse solar radiation due to pollution, reduced evapotranspiration and lower relative humidity which influences quality of life (Ravbar et al. 2005; Tiran 2016).

B. Komac (✉) • R. Ciglič • M. Pavšek • M. Topole
Anton Melik Geographical Institute ZRC SAZU, Novi trg 2, 1000, Ljubljana, Slovenia
e-mail: blaz.komac@zrc-sazu.si

A. Loose
Energy Manager of the City of Ljubljana, City of Ljubljana,
Adamič-Lundrovo nabrežje 2, 1000, Ljubljana, Slovenia

S. Čermelj
Department of Environmental Protection, City of Ljubljana,
Zarnikova 3, 1000, Ljubljana, Slovenia

K. Oštir • Ž. Kokalj
Institute of Anthropological and Spatial Studies ZRC SAZU,
Novi trg 2, 1000, Ljubljana, Slovenia

© The Author(s) 2016                                                        323
F. Musco (ed.), *Counteracting Urban Heat Island Effects in a Global Climate
Change Scenario*, DOI 10.1007/978-3-319-10425-6_12

Climate Chart of Ljubljana as the basis for
spatial planning and urban development in Ljubljana (1:25,000)

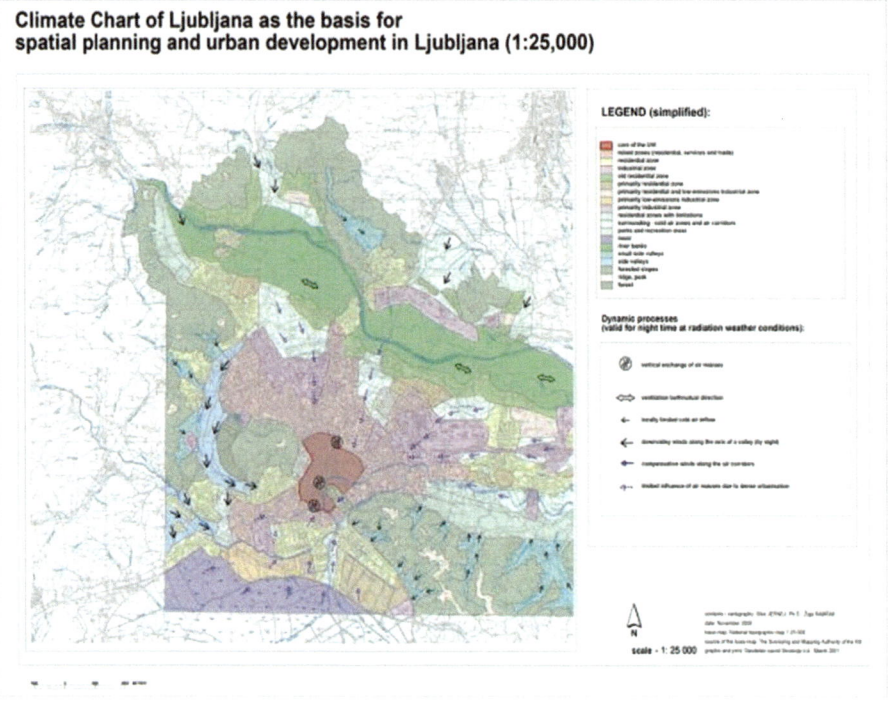

**Fig. 12.1** Climate chart of Ljubljana showing the urban heat island area

In Ljubljana, urban heat island was first detected when creating The climate map in 2000 (Fig. 12.1). Area with above-average temperatures was lying between the Rožnik and Golovec hills (Rejec Brancelj et al. 2005; Smrekar et al. 2016). In the context of climate map guidelines for spatial planning were prepared, which preserved the green wedges and fresh air corridors for the city centre (Internet 3).

During the UHI project we found that the development of the city and in particular the construction of many shopping centres with large parking places increased the intensity of urban heat island phenomenon. The project was designed primarily for mitigating as well as adapting to this phenomenon and thus climate change.

## 12.2 Meteorological Measurements

During the project the urban heat island index was calculated for the Ljubljana Bežigrad (46 °07′ N, 14°52′ E, 299 m a.s.l.) and Brnik (46 °22′ N, 14°22′ E, 364 m a.s.l.) meteorological stations (Internet 1). The Ljubljana Bežigrad weather station is within the inner circle of the city. The area is representative for a typical urban Ljubljana setting, 1.5 km away from the city centre. The weather station is a synoptic one and thus equipped with many for the UHI-recognition meteorological

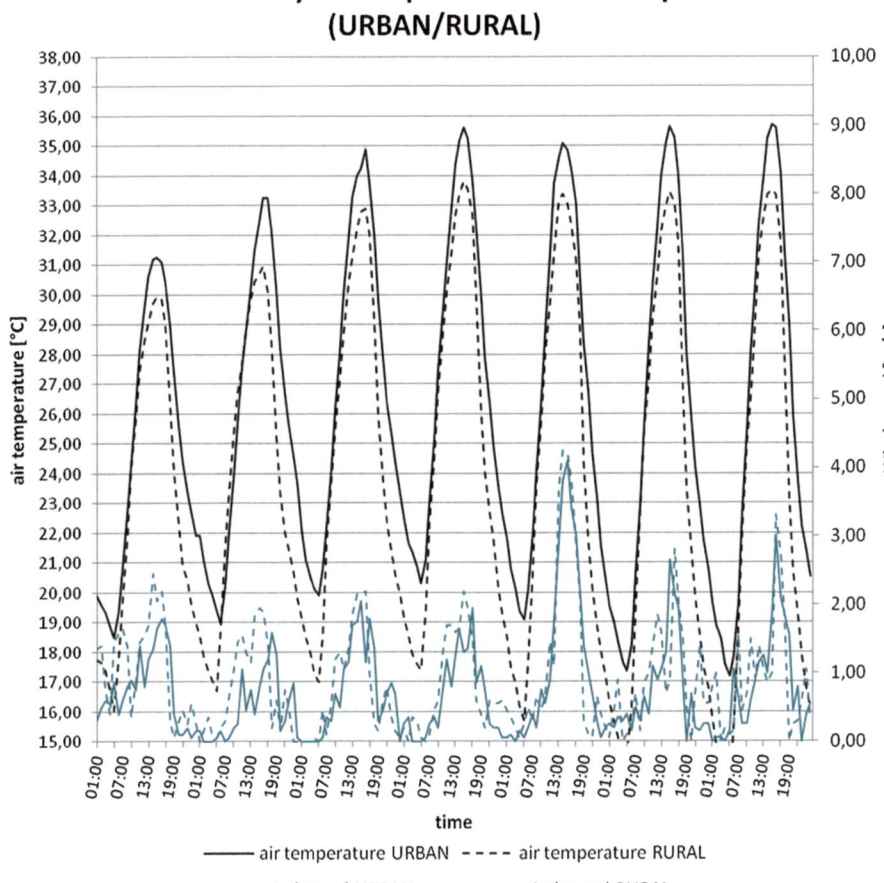

**Fig. 12.2**   Mean hourly air temperature and wind speed data in urban and rural station

devices (Internet 1). The Brnik weather station is in a complete rural area and stands on the SE edge of the Jože Pučnik Airport runway, over 100 m NW from it and more than 1 km away from the nearest bigger building. First larger forest area is over the airport's runway more than 500 m away. The surrounding terrain is in general flat, open and grassy. Weather station is a climate one and the measurements are taken automatically. The air distance between both weather stations is something more than 17 km; the difference in altitude is 65 m (Fig. 12.2).

The UHI intensity-cycle was calculated for the period July 20. to July 26. 2011. Its intensity was quite similar every day. The lowest intensity was present in the morning of the second day, when the heating at the rural station has started earlier than at the urban station and it was significantly more rapid. Before a midday the urban station became warmer, but in the evening rural station was cooling significantly faster. That caused that UHI intensity has reached the third highest

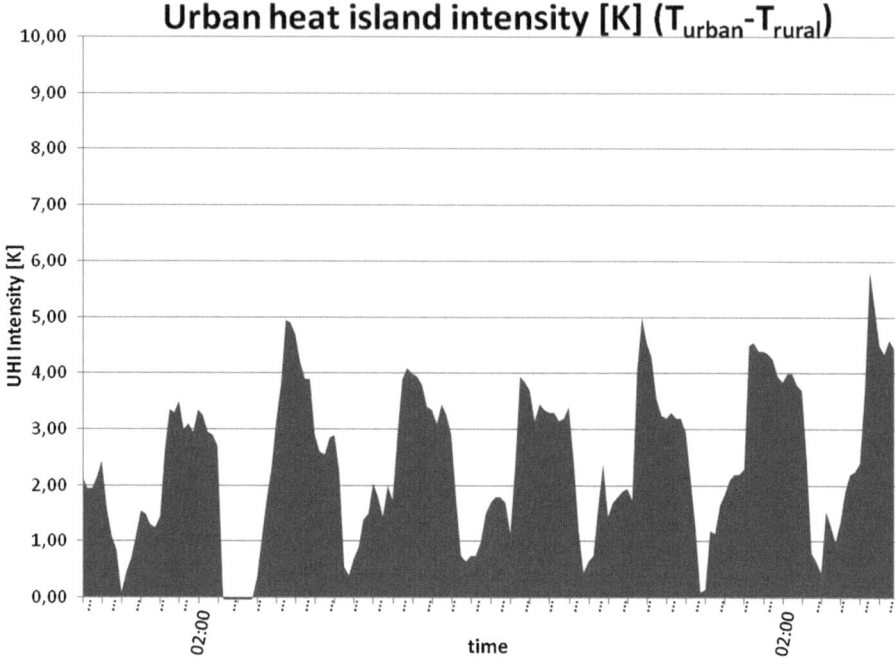

**Fig. 12.3** Mean urban heat island intensity data

value in the selected period (4.95 K). A little higher air humidity compared to following days in the period was probably a cause for thin-fog formation near the middle of the night, which has decreased cooling rate at the urban site. In the whole period wind was varying from calm in the night/early morning time to the top speed around 2–3 m/s in the afternoon on the both stations. In the beginning of selected period the prevailing wind was weak mid-atmosphere NE winds, which usually doesn't reach the valley's bottom. That is why none of the stations shows a signal of constant NE winds. The measured wind variations were thus thermally driven by a daily heating and nocturnal cooling, because in the selected period the sky was clear all the time. The biggest anomaly of the wind in the selected period occurred during the day on the 24th of August, when the wind direction in the middle altitudes has changed to SW, which descends over the Dinaric Mountains to Ljubljana basin and has some characteristics of the foehn wind. Also the air humidity was the lowest in that time at the both stations. A weak SW wind was blowing till the end of a selected period and very low humidity has caused the lowest night-time temperatures especially at the rural station and thus the UHI intensity has raised for 0.5–2 K (Figs. 12.3 and 12.4).

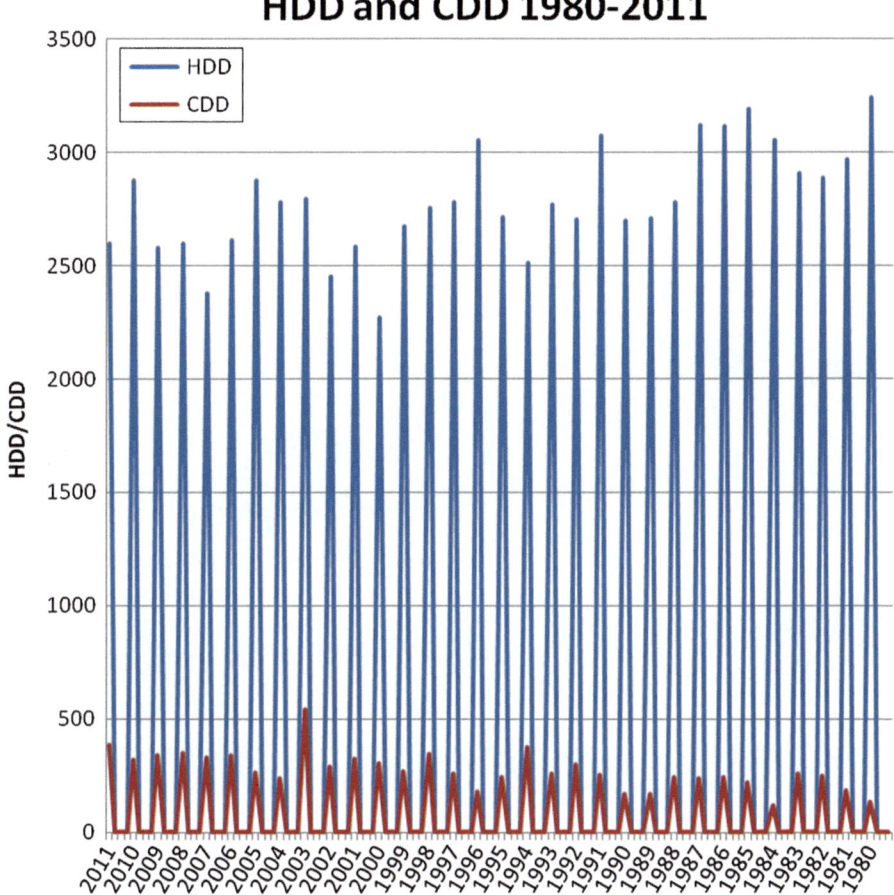

**Fig. 12.4** Mean heat degree days (HDD) and cold degree days (CDD) in the period 1980–2011 for Ljubljana

## 12.3   Satellite Thermography Measurements

Efficient planning of city infrastructure and its activities has to take into account increasing average temperatures in a city, and therefore consider and monitor the effect of its urban heat island (Zakšek and Oštir 2012).

Land surface temperature can be monitored with satellite systems (Colombi et al. 2007). To assess the extent and state of urban heat island of Ljubljana we used Landsat 8 imagery that has adequate spatial resolution for a detailed analysis of temperature variations (Bechtel et al. 2012). Images of Ljubljana and its vicinity are acquired twice every 16 days. Land surface temperatures were calculated from two cloudless images per season in the period between April 15th 2013 and March 8th 2014.

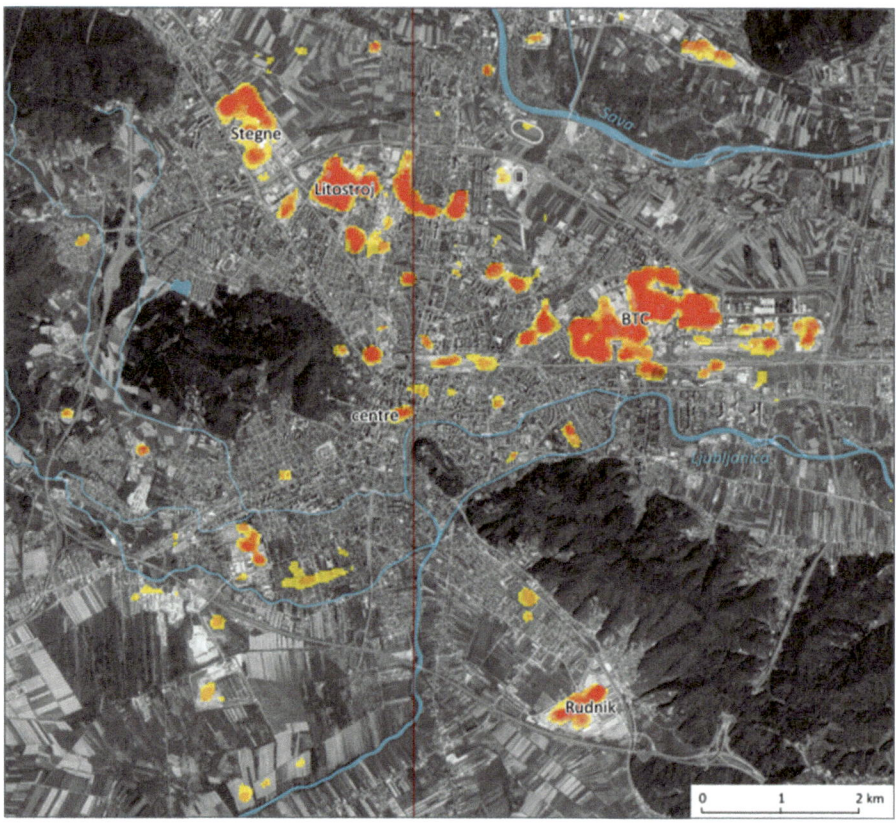

**Fig. 12.5** The frequency of occurrence of the top two percent of the highest temperatures in build-up areas

The area of interest was divided into three zones, with the first being inside the city ring road, the second a 3 km band from the first, and the third a 2.5 km band from the second. The zones have been further divided according to land use; we have grouped the classes into: built up, water, agricultural land, and forest. Build up and water areas smaller than the resolution of the thermal band (1 ha) were excluded from the analysis. The first zone was further partitioned to city districts, and the second and third zones to areas of settlements.

Comparison of temperatures included build up areas of smaller cities close to Ljubljana (Kranj, Domžale, Kamnik, Grosuplje, and Vrhnika) and some of the city's more interesting regions (the centre, industrial areas, business-shopping centres, areas of individual housing, and parks). The impact of water and trees on the surrounding surface temperature was assessed with profiling – across the whole city and through specific areas of interest.

We have found that Ljubljana exhibits a distinct urban heat island. Furthermore, some of its districts are constantly warmer than others, and specific areas are definitive hot spots (Figs. 12.5 and 12.6).

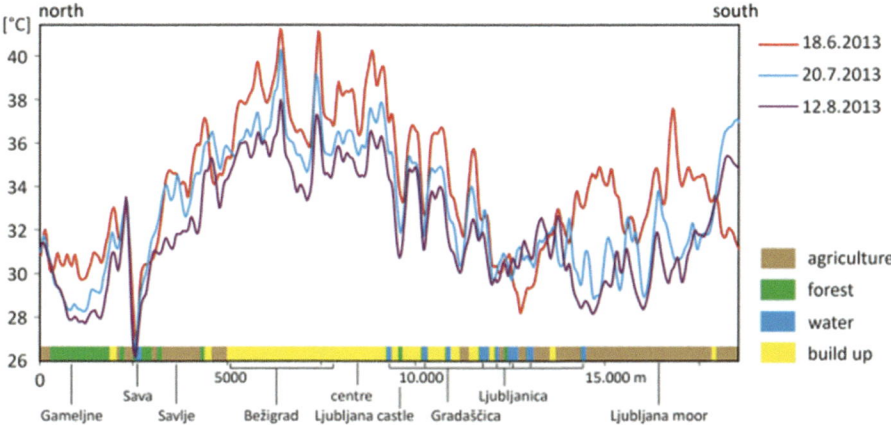

**Fig. 12.6** Profiles of summer land surface temperatures across Ljubljana, with indicated land use and prominent orientation points. The location of the profile is marked as a *red line* on Fig. 12.5 but it extends beyond the boundaries of the figure

## 12.4   Urban Heat Island Atlas

The Urban heat island atlas is an internet application for visualization of spatial data, related to urban heat island in the area of Central European countries (Ciglič and Komac 2015). The database was elaborated in GIS environment using ArcGIS Desktop and published online using ArcGIS Server. This project result is intended to support decision-makers in the field of spatial planning and is therefore free-accessible to everyone at http://zalozba.zrc-sazu.si/p/1352.

The atlas presents urban heat island influencing factors including elevation, normalized difference vegetation index, land use (Corine land cover and Urban atlas data), human activity shown by night scene image (detected by VIIRS sensor data), air temperature at 2 m and land surface temperatures (Komac and Ciglič 2014).

Viewer can analyse the data by making profiles across the temperature layers for April and August (at 2 p.m.). In this way, temperatures in the city and the surroundings can be compared (see the example of Ljubljana in Fig. 12.7).

Various local and regional datasets provided by the project partners are also presented in the atlas.

The atlas was prepared by the Anton Melik geographical institute ZRC SAZU and Hungarian meteorological service.

## 12.5   Mitigation of the Urban Heat Island in the Ljubljana City Area

The purpose of the project was to establish the existence of a heat island, which was confirmed by data from meteorological stations and satellite thermography, and then use this knowledge to mitigate the effects of heat islands.

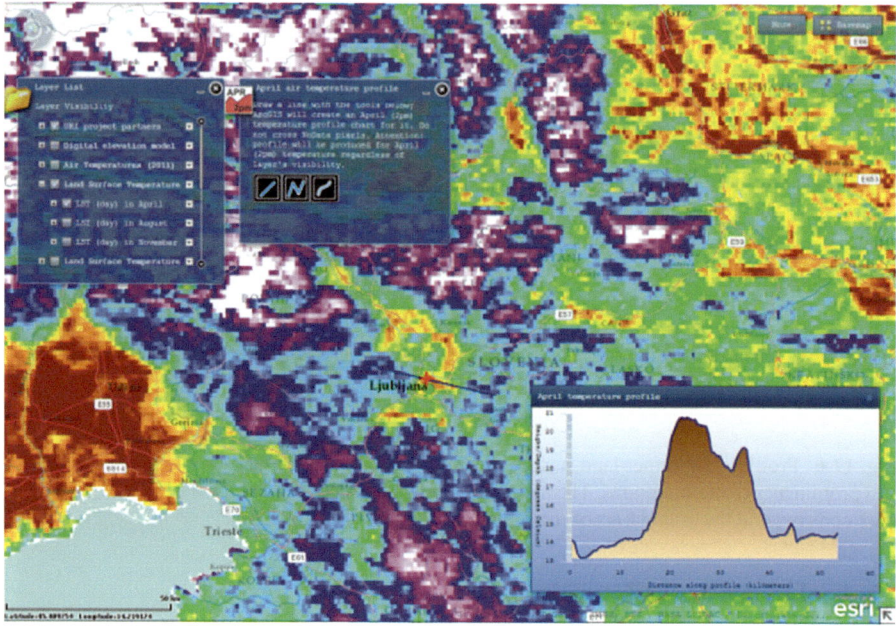

**Fig. 12.7** Print screen of the Urban heat island atlas. Its user friendly interface enables users to select between different layers, make profiles across April (*as below*) and August temperatures in Central Europe and zoom to UHI partner data

As urban island is affected by several factors such as green areas in the city, location and design of the land use, types of building and roofing materials, layout of buildings, and their energy efficiency, including spatial development of the city, we examined some of the characteristics of the city of Ljubljana. We carried out some pilot actions to draw attention to the importance of taking into account the effects of the urban heat island in the planning of land use and activities in the city. In 2010 the Municipal spatial plan (Internet 2) was adopted defining the planned land use with special focus on urban land uses and future development of green areas (Fig. 12.8).

In order to minimize the effects of the urban heat island the basic strategic objective of the Ljubljana city is to create an interconnected and transparent network of open public spaces of high quality. During several last decades an action called »Ljubljana – My Town« has been under way. It stimulates the facades' renovation all over the important public areas or representative areas.

One of the most important measures is preserving the city green areas and protecting the city forests as a part of green wedges. A specific feature of the city is its green ring, a way along barbed wire, which encircled the city during the WW2 being a popular recreational area. The focus area is the axis along the river of Ljubljanica and green areas along the Sava river, the Gruber's canal, the Špica bank, and the Gradaščica and Glinščica brooks.

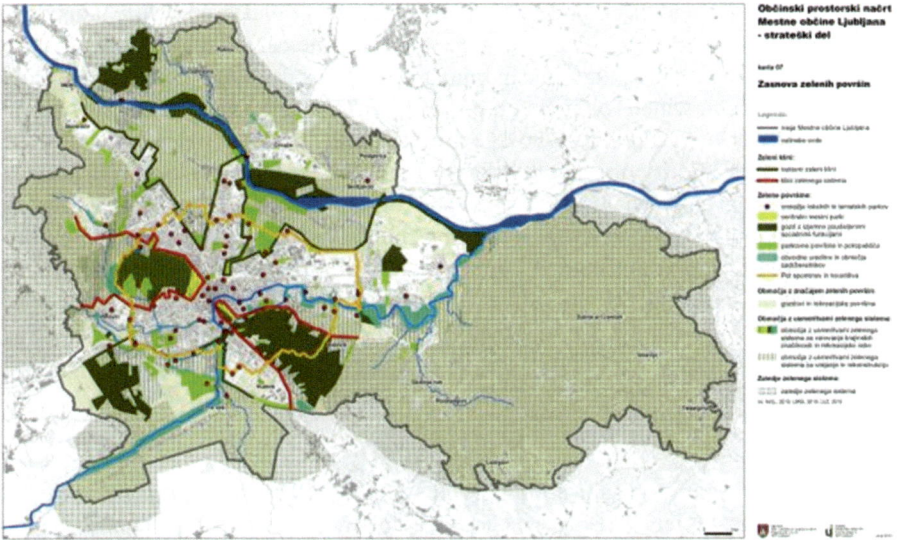

**Fig. 12.8** Future development of green areas in the City of Ljubljana (Internet 3)

Related to green areas a special attention is given to open public areas network that connect the surrounding of the city with its squares, streets, riverbanks and walkways, and also serve as an air-corridor network. They create micro-climatic and mesoclimatic conditions, enable ventilation of the city. The stretches of waterways also constitute a system of open public spaces. Adequate directions for protection and management are defined for such areas.

The extent of pedestrian areas increased lately due to the traffic limitation and complete stop in inner city centre. Still, poor airing is presented in the Ajdovščina area. The city planning measures regain pedestrian areas that would make various social interactions possible. Urban ventilation is therefore an important factor of urban heat island mitigation, and connecting natural areas in landscape parks with the city.

In Ljubljana garden-plot areas were spontaneously developed and distributed across the entire flat area of Ljubljana. Due to a urban lifestyle and a policy change by the city authorities, the area of garden plots fell considerably at the end of the twentieth and the beginning of the twenty-first centuries. Until recently garden plots were spread on 200 locations and covered 1.3 km² (Jamnik et al. 2009) Due to pesticide residues and heavy metals in the soil and produce, as well as groundwater contamination, plot gardening poses a threat to public health and the environment, and the 2008 Ljubljana Zoning Implementation Plan reduced the total area of plot-gardening areas by nearly half.

Another way of minimizing the effects of urban heat island is related to building and roofing materials. The majority of old buildings in the centre of the Ljubljana city are built of bricks and stone. In the broader city area concrete buildings were built in the 1970' and 1980', while iron was used to build some new high buildings.

Most of the stone and brick buildings in the old city are plastered, while stone was used primarily for portals, window frames and in some cases, pet mats. Mitigation is difficult due to the fact that the Ljubljana centre was declared a cultural and historical monument, which requires full protection of the cultural, aesthetic, historical and natural values (Decree … 1986). Common roofing material in Ljubljana is tiles. Tiles cover the majority of older buildings in the centre of the city. Some atriums of the buildings have been covered by glass and only few buildings and winter gardens have been equipped by green roofs until now, built mostly by incentives of individuals (Črnuče, Lek pharmaceutical company, Maros company) or by public funds (Stožice sport park a green roof with size 8000 $m^2$).

## 12.6 Pilot Actions

The notion of spatial organisation as a method of urban management has been adopted in three typical areas of the City of Ljubljana: the city centre, suburban area and hilly hinterland. This division is merely schematic as it would be difficult to demark these areas where the contents are intertwined (Decree ... 1986).

In Ljubljana 21 streets in the centre of the city have been closed to traffic since 2006. The public cycling system Bicikelj has been established in 2011. In some areas in the city centre the allowed speed of traffic was reduced to 30 km/h, allowing safer cycling and pedestrian transport.

The 2010 Municipal spatial plan defines land use activities from urban land use, green areas, cycling routes and green areas points of view.

A number of drinking fountains were established in the city centre next to several fountains. In 2010 two new parks measuring 15 ha have been established: the Northern park and the park at the Šmartinska street. The city of Ljubljana is taking care of more than 72,000 trees in the city and its surroundings. More than one hundred trees were planted in the streets of the city centre in 2009 and more than 250 trees were planted in 2010 as a part of a project the Labyrnth of Art. About 1409 ha of forests on the nearby hills (Rožnik, Šišenski hrib, Grajski grič, Šmarna gora and Grmada) were given a status of forest of special importance, although almost 90 % of the forests are a property of private subjects. Next to a new stadium and sport hall about 90.000 $m^2$ of green areas (parks and sport areas including cycling and running tracks) next to 5000 $m^2$ of play grounds are going to be built. Also, the Ljubljanica river banks were reconstructed in order to allow public access to water in the city. Large areas of the city dump in the Ljubljansko barje area were transformed to golf links (http://www.ljubljanapametnomesto.si/). A list of buildings which are accessible for disabled people has been made public (http://www.disabledgo.com/sl/org/ljubljana).

The determination of pilot Urban Heat Island installations locations was a subject of numerous consultations and debates within the City administration and public. Some of them were realized in 2013, the others were prepared and installed this year.

### 12.6.1 Ambient Urban Intervention "Boats in the Fountain" on the Plateau in Front of Slovenian Ethnographic Museum (SEM), June, 2013

The architectural concept of the plateau in front of the SEM in the very city centre is a part of a collage of various surfaces and objects. Some of the surfaces are less used that expected, as is the case of the 260 m2 plateau, that was planned as a water surface, but was not realized for years. In the summer period, water is one of the most important elements of public space – it cools the ambient and represents an ideal playground for children. Water also reminds us on holidays, sea, boats, …, and is especially attractive for children. For those reasons we have decided to restore the project and to realize the long-time ago planned water surface in an innovative way that will be attractive for public.

In June 2013, the ambient intervention "Boats in the fountain" was realized, by:

– re-establishment of water surface,
– establishment of spaces where it is possible to play, work or rest,
– rearranging the area to bring it closer to the children and adults.

For the installation old boats were used, after the detailed renewal, each of them had its own name and own story. It became an ideal meeting place for all generations. Throughout the whole summer period numerous visitors enjoyed the vicinity of water surface, especially children. The project was concluded by the end of September (Figs. 12.9 and 12.10).

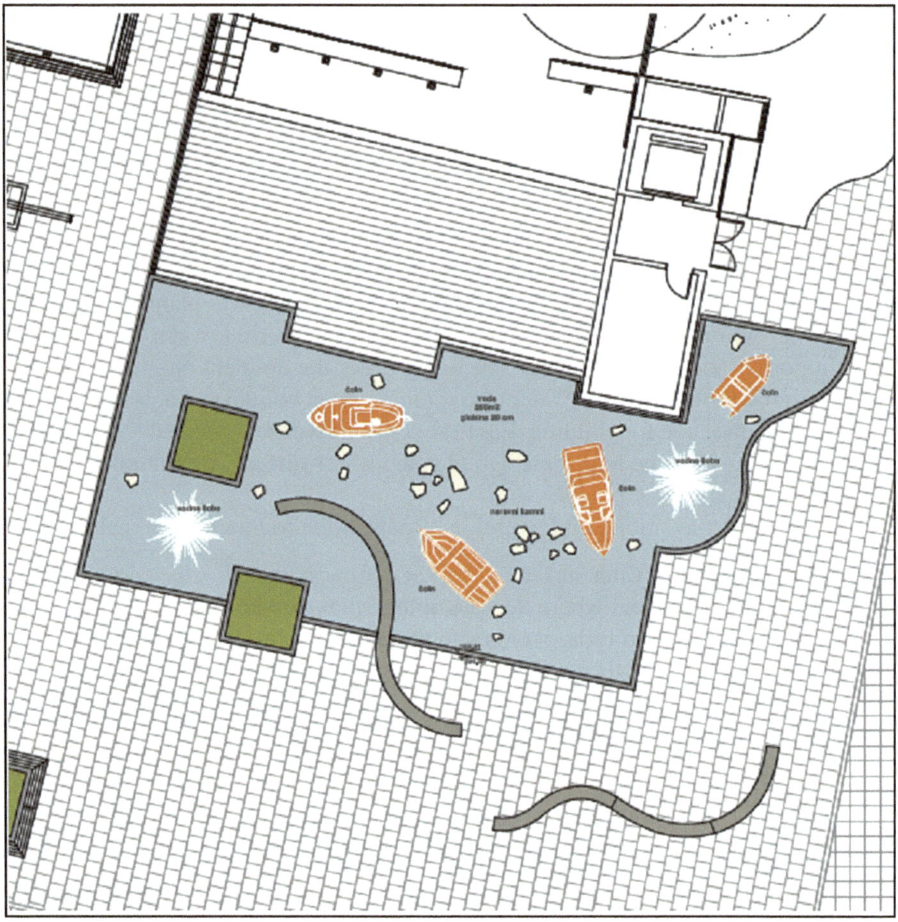

**Fig. 12.9** Project baseline – Slovenian ethnographic museum (SEM)

**Fig. 12.10** SEM – realized water surface

### 12.6.2   Temporary Installation on the Central Section of the Slovenska Street after its Closure to Traffic (September 2013)

Since September 22, 2014, the central part of Slovenska street, passing through the very city centre was closed to motorized traffic, with the exception of public transport. With this measure the transit of motorized vehicles is being reduced in the city centre and the priority of using it is given to pedestrians, cyclists and public transport. On the occasion of closing the street a temporary arrangement was designed and realized to show the citizens and tourists what they can gain from the street closure. The project of temporary arrangement was planned in a sustainable way according to the principle "Less is more" (or "More with less"). All elements may be moved from one location to another, whenever needed. Besides street markings no element will be used for the Slovenska street only. Temporary arrangement comprised of:

– information point (Info point) about new arrangement of the Slovenska street – citizens were invited to give additional proposals for additional installations or improvement of the existing,
– potted groups of trees (wooden pots),
– smaller permanent green plants in pots – pocket parks,
– Tables and chairs,

An info point-container in which the exhibition of the projects concerning the new look of the street was prepared together with the exhibition showing the history of the main street. Numerous citizens and other visitors gave their proposals for improvements and compliments. It is important that the citizens were satisfied with new arrangements, using the newly gained space daily. With additional plants (green areas) an effort was taken to improve local environmental conditions and to establish more attractive street ambient. The black-carbon and nitrogen dioxide measurements were performed on-site as well (Fig. 12.11).

### 12.6.3   Green Gym – Gymnastic House with Green Roof (April 2014)

The prototype of small green gymnastic house with green roof was designed and built. It was designed for adults, to be placed on public playground for children to serve for parents and other adults while their children are playing. The time, usually spent waiting for children on the bench, has changed into the active/healthy recreation and can serve as an example of a good practice for children, as well. The gym-house is placed on recently renewed playing ground in the city centre.

The gym-house was named "Always young" has a base of 6 m$^2$ and is about 3 m high. It consists of the metal construction that serves as well as a gymnastic object,

**Fig. 12.11** Temporary installation on the main street (Slovenska ulica) soon after its closure to personal motorized traffic

solid roof planted with extensive growing grass on wooden base reinforced by metal sub-construction. As the installation is placed outside, the rain will be enough for the watering of the roof. In the house a simple construction consisting of three basic gym equipment suiTab. for both genders and ages are available. The gym house is funded on concrete foundation. Within the area of 1.8 m around house, material that buffers falls is spread. On the walls there is a Table with instructions how to exercise. As the gym-house is a part of the playground for children it was planned and built in such a way that it meets the standards (the gym-house is certified) (Fig. 12.12).

**Fig. 12.12** Children and parents playground – the gym-installation for parents with the green roof was added to the classical playground for children, with the invitation letter (*right*)

# Vabilo

*Ob Dnevu zemlje vas vljudno vabimo*
*na otvoritev prenovljenega otroškega igrišča*
*na Prulah, na Prijateljevi ulici,*
*v torek, 22. aprila 2014, ob 16.00 uri.*

Tam nas bo pričakalo presenečenje - telovadna hiška z zeleno streho
za telovadbo staršev, babic in dedkov, pa še koga, ki se bo
najmlajšim pridružil pri igri. Pridružite se nam!
Z veseljem bomo delili dobro voljo z vami in z najmlajšimi iz Vrtca Pod Gradom.

V Ljubljani se vsak dan znova zavedamo pomembnosti varovanja našega planeta, kar z drobnimi in večjimi dejanji tudi
uresničujemo. Telovadna hiška z zeleno streho je nastala v okviru evropskega projekta UHI, katerega namen je priprava in
uporaba orodja za modeliranje pojava toplotnega otoka v mestu, ki bo služil predvsem načrtovalcem kot izhodišče za
umeščanje dodatnih zelenih in vodnih površin v mesto, s ciljem zmanjševanja pregrevanja mestnega središča.

### 12.6.4   Pocket-Parks in the Streets with Parking Places (April 2014)

Pocket parks are representing the long-term cultivation of public-spaces. With these installations we are trying to develop a method for revitalization and remediation of degraded spaces in the built part of the city centre. The interventions are not only artistic installations but also a serious search of new urban possibiliries for better use of urban spaces. The main objective is to pay special attention and to reveal selected city areas.

Pocket-parks are a kind of containers with the dimensions of a single parking place (car) that can play a certain role, with its planted part, to mitigate the UHI phenomenon on the micro-scale within the densely built area in the city centre.

Since May 2014 it is possible to enjoy on one of the pocket-parks, called PARKplac (PARKplace) on four streets and one square. They are equipped according to the reuse concept. Local inhabitants and local merchants will take care about them. With these installations we are addressing the citizens to the new approach to the public space and spreading the 3R idea (Reduce, Reuse, Recycle). According to the 3R approach all materials that are used were taken from past installa

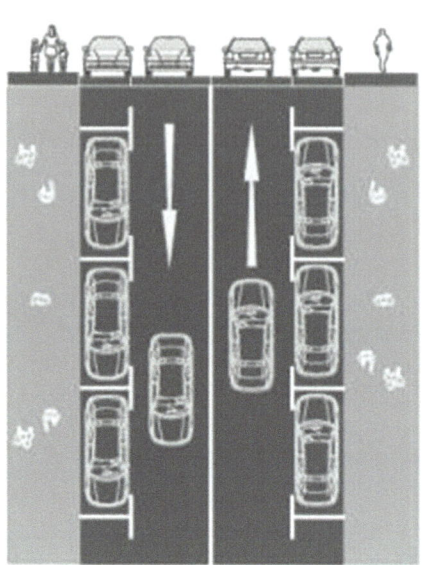 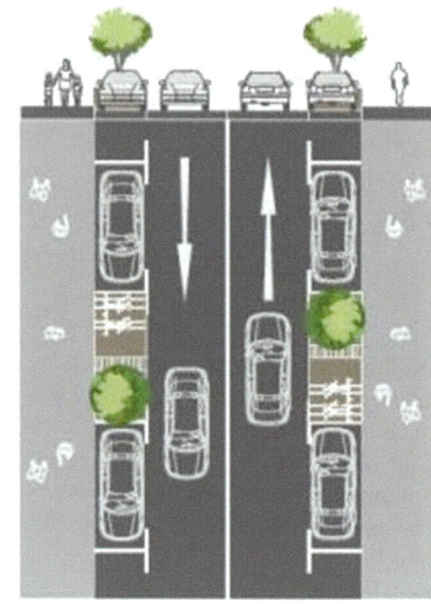

### 12.6.5   Planting Trees (April 2014)

By planting trees we are trying to increase the quantity of green areas in the city. Trees have been planted on three locations according to the standard SIST DIN 18916.

Near the Nove Fužine block-settlement the new park "Art Labyrinth" was planted. The Art labyrinth is bringing together the park as a green public area with the labyrinth art tradition, artistic approach and literature (book reading). The park is a kind of growing and development feature as it grows with its visitors. The more care that it is being given and the more trees that are planted, the bigger is the place where we can rest in peace and read a good book.

## 12.6.6   Promotion of the Project

On different occasions the UHI project was promoted (opening the pilot installations). One of them was also the project "For the cleaner city" that was taking place through the whole city area from 22th March to 22th April, 2014.

The project was promoted by the lectures about UHI phenomenon, public polls considering the pilot actions and general opinion on climate change, delivering the UHI gadgets and leaflets, see figures below.

## 12.7   Conclusions

The urban heat island project gave us a comprehensive scientific insight in the phenomenon, useful tools and instructions for adaptation and mitigation of urban heat island effects. At the same time it was proved that our efforts in the past were successful preserving the important green areas (wedges) and fresh air corridors.

We are satisfied with a big success of the project that can be recognized by the number of people that were involved and mobilized by its activities.

## References

Bechtel, B., Zakšek, K., Hoshyaripour, G. (2012). Downscaling land surface temperature in an urban area: A case study for Hamburg, Germany. *Remote Sensing, 4*.
Decree on the proclamation of the medieval city center – Old town and the Castle hill of cultural and historical monument and natural phenomenon, Official gazzette 5/1986-297.
Ciglič, R., Komac, B. (2015). Central-European urban heat island atlas. Ljubljana. URL: http://gismo.zrc-sazu.si/flexviewers/UHIAtlas/
Colombi, A., Pepe, M., Rampini, A. (2007). Estimation of daily mean air temperature from MODIS LST in Alpine areas. New Developments and Challenges in Remote Sensing. Rotterdam.
Internet    1:    http://meteo.arso.gov.si/uploads/probase/www/climate/text/sl/stations/ljubljana-bezigrad.pdf
Internet 2: https://urbanizem.ljubljana.si/index3/OPN_MOL_SD.htm
Internet 3: https://urbanizem.ljubljana.si/index3/files/OPN_MOL_SD_07_zelene.jpg
Jamnik, B., Smrekar, A., & Vrščaj, B. (2009). *Plot gardening in Ljubljana*. Geografija Slovenije 21. Ljubljana. http://giam2.zrc-sazu.si/sites/default/files/9789612541507.pdf
Komac, B., Ciglič, R. (2014). Urban heat island atlas: A web tool for the determination and mitigation of urban heat island effects. *Geographia Polonica* 87-1. Warszawa.
Ravbar, M., Bole, D., Nared, J. (2005). A creative milieu and the role of geography in studying the competitiveness of cities: The case of Ljubljana. *Acta geographica Slovenica* 45-2. Ljubljana.
Rejec Brancelj, I., Smrekar, A., & Kladnik, D. (Eds.). (2005). *Podtalnica Ljubljanskega polja*. Geografija Slovenije 10. Ljubljana. http://giam2.zrc-sazu.si/sites/default/files/9616500686.pdf
Smrekar, A., Šmid Hribar, M., Erhartič, B. (2016). Stakeholder conflicts in the Tivoli, Rožnik Hill, and Šiška Hill Protected Landscape Area. *Acta geographica Slovenica* 56-2. Ljubljana. doi: http://dx.doi.org/10.3986/AGS.895

Tiran, J. (2016). Measuring urban quality of life: Case study of Ljubljana. Acta geographica Slovenica 56-1. Ljubljana.

Zakšek, K., Oštir, K. (2012). Downscaling land surface temperature for urban heat island diurnal cycle analysis. *Remote Sensing of Environment* 117.

Žiberna, I. (2006). *Trendi temperatur zraka v mariboru kot posledica razvoja mestnega toplotnega otoka*. Revija za geografijo 1–1. Maribor. 81–98.

# Chapter 13
# Pilot Action in Budapest

**Györgyi Baranka, R. Ongjerth, F. Szkordikisz, and O. Kocsis**

**Abstract** Pilot area chosen for evaluation is one of the biggest green investments of the latest 30 years: the area of this public park is 3.5 ha cost 500 million HUF. 2000 m² water surface is 1.2 m deep. Rehabilitation of Millenáris Park including reconstructions of main building and establishment of a public park costs 15 billions HUF. This is one of the intervention areas of local government (District II) in Budapest. Municipality has determined boarders of pilot area. Pilot area contains brownfield area, street canyon, public park, which were rehabilitated in the past and there is a big building, which will be destroyed in the future. Surface of the area is 0.48 km². Local meteorological measurements are continuously available to characterize changing in microclimate of pilot action. Urban planners, experts of green roof planning were involved in choosing pilot area for the UHI assessment.

**Keywords** Pilot action • ENVI-MET simulation • Human comfort • Mitigation and adaptation strategies • Green roofs • Green facades • Single row of street trees • Double row of street trees • Planters • Heat waves alert system

## 13.1 Planning Framework

### 13.1.1 Legal Foundations of Urban Planning

In urban planning even the smallest municipalities (local government) have wide discretionary powers. Their planning decisions may be annulled only in cases of breaking the law (central state act or a government statute). Legal control of local plans is performed by the State Government Offices in each of the 19 Counties. The control in specific professional fields is exercised by 9 State Chief Architects with regional competencies. In cases of disputes the final authority is the Constitutional Court. Therefore, Hungarian urban planning is deeply embedded in codified law.

G. Baranka (✉)
Hungarian Meteorological Service, Kitaibel Pál u. 1,
Budapest, H-1024, Hungary
e-mail: baranka.gy@met.hu

R. Ongjerth • F. Szkordikisz • O. Kocsis
Hungarian Urban Knowledge Centre Non-Profit Ltd, Budapest, Hungary

© The Author(s) 2016    345
F. Musco (ed.), *Counteracting Urban Heat Island Effects in a Global Climate Change Scenario*, DOI 10.1007/978-3-319-10425-6_13

The base of urban planning law (Act on the Formation and Protection of the Built Environment) was adopted by the Hungarian Parliament in 1997, after a 7 year period of preparatory work. It was modelled on the German law (Baugesetzbuch). The Act was supplemented in the same year by a central government statute, the National Building Code (OTÉK). Besides the National Building Code some other governmental edicts give orders regarding the detailed contents of the particular plans. The Code is binding on all local planning decisions, but municipalities are permitted to render its maximum/minimum standards more "rigorous". The 1997 Act introduced four planning tools, namely:

- Urban-planning development strategy
- Urban Development Concept
- Structure Plan (preparatory land use plan)
- Regulatory Plan (binding land use plan)
- Local Building Code.
- the Action plan is though not part of the edict, but it is used quite often, if needed. (The action plan is a mide-term operational plan concerning to a particular part of a given territory.

The first two tools are adopted by the local authorities through a local government decision (e.g. they are "only" binding on the local government and it's organisations), the other two are adopted through a local government statute (i.e. they are binding for all concerned – e.g. property owners and developers).

Planning decisions are enforced by the Building Authorities functioning as departments of local government offices in cities and bigger villages. For some building affairs (i.e. heritage buildings, heritage areas, Natura 2000 districts) other state agencies function as first level building authorities and their full consent is needed for the issuing a building permit. As in the case of planning legislation, none of these authorities and agencies has discretionary powers, but operate in terms of the platform of law administered and enforced by them. The second level building authority – the place for appeals against decisions – operates within the State Government Offices in each County.

## 13.1.2  Growing Environmental Complexity of Urban Planning in Hungary

While the permitted regulatory content of urban physical building plans in Hungary is rather limited, another important trend facilitates their complexity. Most state agencies, representing specific professional fields look at local physical plans as "omnipotent" tools for the assertion of their interests. That is why a great – and growing – number of so called "supporting studies" should be worked out as part of a local plan.

These include:

- the protection of local historic and architectural heritage (including archaeology),
- environmental protection and control

- landscape and nature protection (including Natura 2000 areas)
- generation and management of local traffic
- development of public utilities
- rain water management.

These studies should be part of the non-binding Structure Plan, thus enhancing the complexity, specifically the environmental foundations of local planning. However, the 27 State agencies find it rather problematic to formulate clear-cut and legally sound regulations in these fields in Regulatory plans and in local building codes. It is also noteworthy that no social or housing studies are prescribed by law as "supporting studies". Social planning has never been strong in Hungary while environmentalists are gaining ground here as everywhere in the world.

## 13.2   The UHI Project and the Planning of the Pilot Area

Climate-conscious urban planning, especially that of public spaces, has little history in Hungary. Although Hungary's Environmental Law requires that each settlement prepare a program for the protection of the environment, these documents tend to be either overly theoretical studies or summaries of the initiatives of local non-governmental organisations. In urban planning documents, climate consciousness manifests itself mostly in the parroting of well-known slogans, without any concrete practical suggestions.

This situation – found not only in Hungary – is what the UHI Project wished to remedy: Relying on several years' worth of research identifying and evaluating factors that influence climate, and inviting the contribution of external partners, we tested the effects that climate-related factors of urban development had on a pilot area. Background support for the experiment was provided by a computer program called ENVI-MET, which, based on knowledge of the existing situation, is able to use several dozen climatic variables to calculate changes that would occur if the plans were realised.

## 13.3   Description of the Pilot Area

The Budapest pilot area lies on the Buda side of the city: an area of approximately 50 ha, bordered by Margit körút – Retek utca – Fillér utca – Garas utca – Alvinci út – Kapor utca – Felvinci út – Ribáry utca – Bimbó út – Keleti Károly utca. The pilot area fits the nature of the experiment rather well, from several points of view, since it is a rather multifaceted area. In terms of neighbourhood character, it is located at the meeting point of traditional high-density urban cores characteristic of the end of the nineteenth century; rather dense, quasi-urban areas along Margit körút, built in the 1930s and 1940s, with larger, green yards; and a high-prestige green belt with villas. The area is centred around a brownfield regeneration project, which used to be the site of an earlier turbine factory but which in the early 2000s gave way, with a complete change of its functions, to a recreational and cultural centre, retaining and utilising some of the earlier industrial structures and incorporating them into a public park. Another characteristic example of green space in the area is a park called Mechwart liget, which was renewed during the last few years, keeping intact its proportion of green space. The area also comprises several streets, such as Retek utca and Kis Rókus utca, which, along with Keleti Károly utca, provide typical examples of city canyons, with practically no trees.

Thus climate-conscious replanning of the area will offer, on the one hand, an opportunity for computer modelling that can project climatic conditions for a relatively diverse set of spaces, and, on the other hand, a potential starting point for the examination and analysis of many other modelling regions.

In addition, both the method of analysing data from the modelling regions and the content of the action plan can help to provide a framework for elected representatives in the local government of the targeted Budapest district to conduct informed discussions on the subject. The material may also provide ideas for the climate-conscious construction of projects for the new EU planning period.

## 13.4  Methodology – The Tools

The methods applied within the UHI Project are not new to public space planning: it is well known both in and outside professional circles that vegetation, for example, cools the environment through evapotranspiration; and these methods represent the primary tools employed in the redevelopment of outdoor public spaces in general. The novelty of the project – which is hopefully of revolutionary significance – lies in its ability to predict and calculate reliably the climatic effects of the tools employed, as well as the possibility to distribute these tools widely. Unfortunately, in Hungary – as in several other European countries – with the increase of solid paved surfaces, the currently fashionable trends in planning open public spaces often not only fail to improve but actually contribute to the deterioration of climatic conditions in the redeveloped areas, yet this effect is difficult to estimate in practice without reliable quantifiable methods. The methods tested within the UHI Project, designed to be made readily available for wide circulation, are expected to help specialists in organisations responsible for the regeneration of public spaces in accepting only commissioned plans that improve, rather than deteriorate, climatic conditions; otherwise visitors using public spaces will have a less comforTab. experience in the summer, and even though there might be temporary stop-gap measures taken, such as the installation of mist cooling gates, the new public spaces will be used less often than the old ones were.

Within the pilot area, the planners of the Budapest modelling regions used the following tools provided by the climate specialists of UHI:

- Single Row of Street Trees
- Double Row of Street Trees

- Planters
- Green Spaces
- Permeable Pavement
- Green Walls – Vertical Gardens
- Green Roofs

## 13.5  Suggestions for Application of the Tools

### 13.5.1  Open Space Characteristics of the Pilot Area

Rows of street trees appear only sporadically within the area (for example, in Lövőház utca, Bimbó út, Marczibányi tér, Fillér utca, Felvinci út, Kapor utca, Kitaibel Pál utca, Tizedes utca, Ribáry utca), and public green spaces or vegetation are present only to a minimal extent (Fillér utca).

The rows often have trees missing; there are single and double rows of street trees in the area (featuring Fraxinus, Koelreuteria, Acer, Robinia ssp.); and there are two areas of significant dimensions which also include water surfaces: Millenáris Park (3.5 ha) and Mechwart liget (1.8 ha).

Examples of permeable surfaces in the area are negligible; apart from isolated examples (such as Lövőház utca, Káplár utca, or Keleti Károly utca), the dominant surface is asphalt, on both the roads and the sidewalks.

### 13.5.2  Action Plan and the Tools Employed

Rows of Street Trees in Public Spaces

Main effects: shading; reducing the air and surface temperature; evapotranspiration; windbreaks; protection from UV rays; and reducing air pollution and greenhouse gas emission.

Street trees may be planted depending on the types of buildings, the forms of facades and streets, and the cross-section of the street; in many cases the dimensions of the sidewalk and parking lane will not allow for any green spaces other than rows of trees. Single or double rows may be planned (Figs. 13.1 and 13.2).

Street trees recommended for the area: a single row of Tilia x euchlora (15–20 m) in Keleti Károly utca; a single row of Fraxinus excelsior 'Jaspidea' (10–15 m, yellow-leaf variety) in Mechwart tér and Fillér utca and a double row in Fény utca (Fig. 13.3); a single row of Gleditsia triacanthos 'Sunburst' (10–15 m, yellow-leaf variety) in Buday László utca; a single row of Fraxinus excelsior 'Westhof's Glorie' (20–25 m) in Garas utca; a single row of Koelreuteria paniculata (5–8 m) in Ezredes utca and Pengő utca (Fig. 13.4); and a single row of Acer platanoides 'Emerald Queen' (10–15 m, yellow-leaf variety) in Retek utca.

**Fig. 13.1** Single row of trees on the pilot area

**Fig. 13.2** Double row of trees

On occasion, rows of street trees will be planted in combination with parking, at an appropriate distance from the building facades, taking into account the relatively narrow cross-section of the street – 10–11 m between facades – and the expected growth of the canopy. Based on these considerations, we recommend planting individual trees every 11.5 m, replacing every other parking space.

**Fig. 13.3** Double row of trees in Fény utca

**Fig. 13.4** Single row of trees in Pengő utca with permeable pavement and tree protection

To protect each individual tree, a Corten Steel tree hole is provided, with a base of at least 1.5 m×1.5 m (when combined with parking, the base will be of 1.5 m×2.3 m) and with a fully accessible finished surface. The species and types recommended are drought-tolerant, pollution-tolerant and at least relatively quick-growth. For planting in streets with more shade, the high-tolerance species of Koelreuteria paniculata is recommended.

**Fig. 13.5** Planters combined with rows of street trees

Planters
Planters may bring green spaces to streets that cannot accommodate trees.

In combination with rows of street trees, planters produce multi-zone vegetation (Fig. 13.5), but this solution is recommended only if the width of the sidewalk is at least 1.5–2 m – green surfaces that are too narrow can cause maintenance problems – and if parking allows enough space (that is, the parking lane is at least 2.5 m wide).

It is necessary to raise the level of the planters 50–60 cm from the surface of the sidewalk, because of their use within the city – litter, snow shovelling in the winter, salt, pedestrian traffic, dog walking, etc. – in order to ensure a longer lifetime for the vegetation.

Recommended locations: in the middle section of Fény utca, on both sides, with widths of 2.3 m and 2 m, respectively; in the southern section of Fillér utca, on one side, with a width of 1 m; and in the middle section of Bimbó út, on one side, with a width of 1.5 m (Fig. 13.6).

In terms of species and types to plant, low shrubs, perennials and grasses are preferable.

Species and types recommended for planting in full/partial sun: Helictotrichon ssp., Panicum ssp., Carex ssp., Calemagrostis ssp., Yucca ssp., Lavandula ssp., Rosmarinus ssp. For containers interspersed with rows of street trees, shade tolerant species and types are recommended: Carex ssp. (C. morrowii, C. sylvatica, C. oshimensis etc.), Deschampsia ssp., Hakonechloa ssp., Phalaris ssp., Vinca ssp.

Contiguous green spaces greatly contribute to decreased storm water runoff within inner city areas, and, due to the significant presence of ligneous plants, have advantages similar to those of planting rows of street trees. The plan recommends

**Fig. 13.6**  Planters in Bimbó street

the installation of new green spaces: a public park in the area bordered by Margit-körút – Kis Rókus utca – Fény utca, adjacent to Millenáris, which is to be called "Széllkapu Park" – a pun on the neighbouring public transport hub called Széll Kálmán tér and the Hungarian words szél ('wind') and kapu ('gate'); at the corner of Fillér utca and Garas utca (see Fig. 13.7); and in the inside courtyards of buildings Fillér utca 9–11 and Lövőház utca 24, 22, 20 and 16b.

These planned green spaces are characterised by multi-zone vegetation; the species and types recommended for planting in the Széllkapu area are similar to those planted in Millenáris (Fig. 13.8). Within green spaces, permeable pavement may be applied in areas paved for foot and service traffic: gravel, concrete tiles and light natural stone, such as limestone.

Permeable pavement is applied to decrease storm water runoff and desiccated soil in the city, thereby increasing evaporation even on paved surfaces.

Recommended locations: replacing older pavement on sidewalks (except for Lövőház utca, Káplár utca and Keleti Károly utca), and in parking lanes. As for the choice of materials, an important consideration is the use of light colours for a higher albedo, instead of dark, waterproof asphalt. For sidewalks, mostly used for foot traffic, the use of concrete tiles is recommended, while for areas of heavy use (surfaces in new green spaces) natural stone – limestone – is suggested.

For the installation, permeable pointing and foundation work must be applied.

Green Walls – Vertical Gardens

The use of green walls or vertical gardens contributes to the shading of walls, thereby lowering the surface temperature of wall facades, reducing heat gain and lowering energy use inside buildings, as well as cooling through evapotranspiration.

Green walls or vertical gardens are best used in locations where the width of the street does not allow for the placement of any other vegetation. Recommended locations: Kis Rókus utca 35, 18 (Fig. 13.9); Fény utca 21 (on the facade of the Melegpörgető building). Species and types of plants recommended: Parthenocissus ssp., Vitis ssp., Hedera ssp., Reynourtia ssp.

**Fig. 13.7** Green spaces in Széllkapu Park

**Fig. 13.8** View of Széllkapu Park

Using green roofs (Fig. 13.10) helps in retaining water and shading roofs; via evaporation, in reducing the surface temperature of roofs; in reducing air temperatures and cooling the city; in reducing heat gain and lowering energy use inside buildings; and in storing storm water within the substrate as well as evaporating it to the atmosphere.

Extensive green roofs – with a 5–10 cm soil bed – do not require watering and are low maintenance in general.

In the effort to encourage the installation of extensive green roofs, local authorities have a range of possibilities to intervene, depending on the ownership of the property: for buildings belonging to the local government, the local authorities themselves have full control over modifications, while in the case of buildings

**Fig. 13.9**  Green wall in Kis Rókus utca

**Fig. 13.10**  Green roof within the pilot area

owned by other legal entities – such as the state government or private companies – the local authorities may initiate the process or, in the case of buildings owned by private citizens (condominiums, for example), may offer financial support as part of the preferred solution. Recommended locations:

Green roofs on buildings owned by local government: Marczibányi tér 5a, 13, etc.;
Green roofs on buildings owned by other legal entities: Tulipán utca 24, Marczibányi tér 3, Kis Rókus utca 18, 16, 14, 12, 2, 4, 6, Lövőház utca 1–6, 12, 14, Fényes Elek utca 7–13, 14–18, etc.;

Green roofs on buildings with multiple private owners: Kis Rókus utca 33–31
   (Fig. 13.10), 1-1a-3-5-7, etc.
   Species and types of plants recommended: Sedum ssp., Euphorbia ssp.,
Delosperma ssp., Thymus ssp.


## 13.6   Effects of Interventions on the Local Climate

### 13.6.1   Characteristics of the Local Climate

The microclimate of the pilot area is influenced by several factors. On the one hand,
the prevailing north by north-westerly winds arrive from the direction of Hűvösvölgy,
while on the other hand the terrain plays a significant role in the formation of the
microclimate of the area, which lies on the south-western slopes of Rózsadomb,
reaching down to Margit körút. In addition, larger buildings have a significant
impact, especially in narrow streets with high facades, such as Lövőház utca, which,
moreover, runs parallel to the prevailing winds, making it the most significant city
canyon of the area. It is also important to mention that, as is true for District II as a
whole, the pilot area is also quite well endowed with urban trees. Yet the state of the
trees and, on occasion, run-down conditions in general offer sufficient justification
for the regeneration of the area.


### 13.6.2   Description of the ENVI-MET model

In order to examine the effects on the microclimate created by the most important
interventions planned in the area, a microclimate-modelling project was undertaken
using ENVI-MET modelling software. Since the entire area would have been too
extensive to model, three representative regions were selected, as shown by the map
in Fig. 13.11.
   For modelling microclimates, 4 m×4 m cells were used throughout. The direction of the wind was defined as north-westerly; wind speed at the elevation of 10 m
was defined as 3 m/s. Simulations were started on a typical summer day at 9 p.m.;
the period of simulation was 24 h. Initial air temperature was set to 23 °C; relative
humidity at the elevation of 2 m was set to 70 %. For each selected region the model
was projected in two versions: the first one represented the original situation, while
the second one represented the planned situation.

**Used Tools**

- alley
- green roof 1 – on buildings of other owners
- green roof 2 – on buildings of government
- green roof 3 – on dwellings
- green spaces
- planting
- green walls

**Fig. 13.11** Layout of modelling regions within the pilot area

## 13.6.3 Evaluation of the Simulation Results

Below, the effects of the intervention tools described in the previous section will be summarised briefly.

Rows of Street Trees: Planting single and double rows of street trees results in decreases local to each tree, both in terms of mean radiant temperature (MRT) and in terms of predicted mean vote (PMV). With expansion of the canopy of the trees, or by planting trees closer together, the distribution of these decreases tends to become more linear.

As is shown by Figs. 13.12, 13.13, 13.14, and 13.15, rows of street trees typically reduce predicted mean vote (PMV) value by 2, occasionally by 3. In our case this means that while in the initial situation 80 % of pedestrians on the sidewalk feel uncomforTab., after trees are planted this ratio is reduced to 10–30 %.

*Planters* In isolation, planters have little effect on the microclimate; however, they exercise an undeniably positive effect on the streetscape and on the psychological well-being of the population. In Bimbó út, planters appear in combination with rows of street trees; here their effects cannot be clearly distinguished from the microcli-

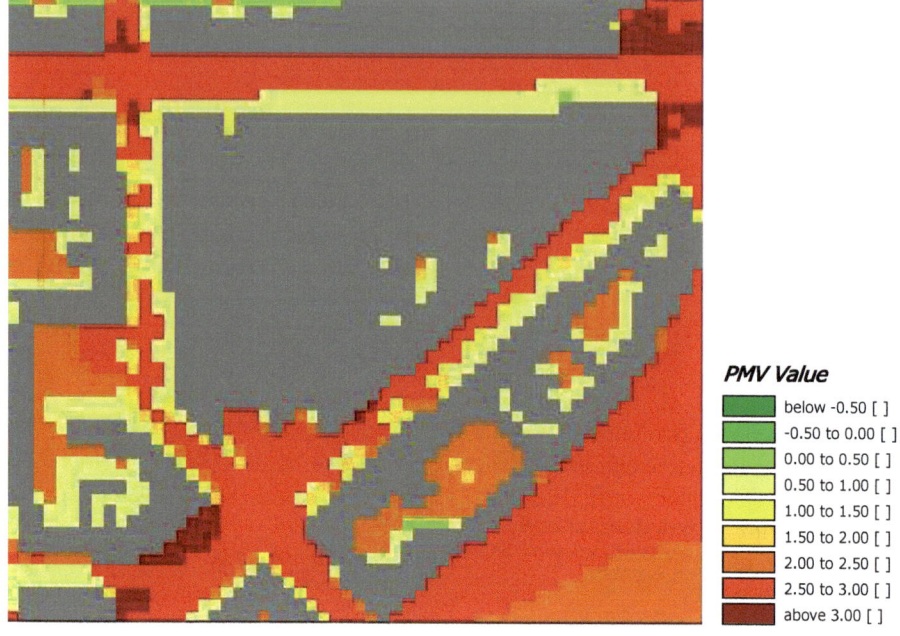

**Fig. 13.12–13.13** Effects of single and double rows of street trees on predicted mean vote (PMV) in Fény utca and Retek utca (summer status, 12:00 p.m., 1.6 m)

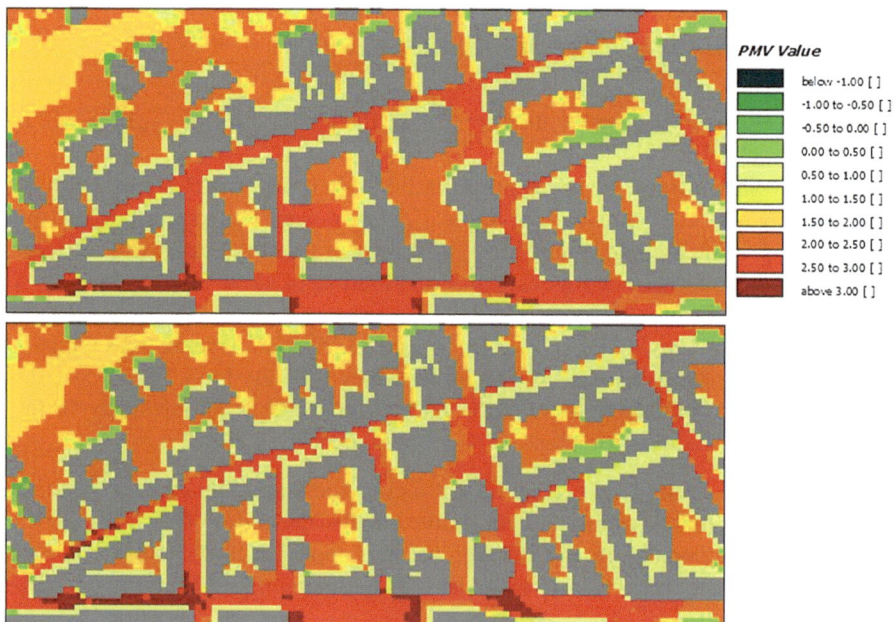

**Fig. 13.14–13.15** Effects of rows of street trees on predicted mean vote (PMV) in Keleti Károly utca (summer status, 12:00 p.m., 1.6 m)

matic effects of the trees, and, therefore, far-reaching conclusions cannot be drawn concerning their effectiveness.

*Permeable Pavement* Permeable pavement cannot be modelled using the ENVI-MET program, thus its effects can only be estimated based on descriptive studies. It is well known, however, that, depending on the base layer, permeable pavement is typically able to retain 35-60 % of the water. This has several advantages. Due to its porosity and absorbance, it warms more slowly and cools more quickly than traditional pavement surfaces and, therefore, has a positive effect on the microclimate. It decreases storm water runoff, thus allowing water to reach the trees along the street, and helps replenish the water Tab.. Using permeable pavement therefore addresses two important problems of urban heat islands: it improves the radiation and water balances.

*Green Walls – Vertical Gardens* For the scale of the modelling regions, the planned green walls and vertical gardens could not be modelled; earlier research shows, however, that green walls – vertical gardens that are nearly parallel to the direction of the prevailing winds significantly decrease mean radiant temperature (MRT) and predicted mean vote (PMV) values in their immediate surroundings.

*Green Roofs* The effects of the planned green roofs are rather complex, and, therefore, difficult to model. Green roofs not only decrease the intensity of urban heat islands but also decrease storm water runoff, thus also reducing the amount of greywater to be treated. They also play a significant role in improving the quality of life for people working or living in the buildings by offering a natural area for relaxation and recreation.

*Green Spaces* Within the pilot area, there are two significant public green spaces, Millenáris Park and Mechwart liget, where no changes were recommended. The model did, however, examine the effect of a new green space: the new park, to be created at the site of a soon-to-be-demolished ministry building, which will significantly increase the green spaces of Millenáris Park. The disappearance of the ministry block and the creation of the new green space will decrease air temperatures by 1.5–2.5 °C, according to the microclimate modelling results (Figs. 13.16 and 13.17).

In summary, it can be stated that within the modelled regions the microclimate – following the localised nature of the intervention – improves in discrete areas due to the proportionate increase of green spaces: cross-ventilation improves, relative humidity increases, mean radiant temperature (MRT) significantly decreases and, in cases of drastic intervention, air temperatures also show significant decreases.

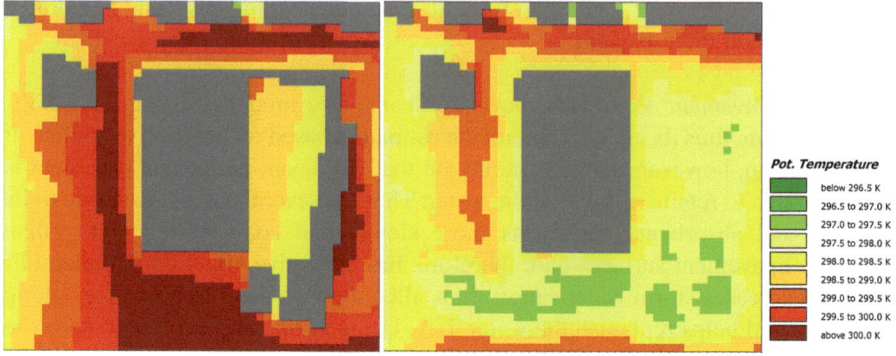

**Fig. 13.16–13.17** Area of Széllkapu Park: status before (*left*) and after (*right*) the demolition of the ministry building and the installation of the planned park (air temperature, summer status, 12:00 p.m., 1.6 m)

## 13.6.4  Suggestions Concerning Organisation and Logistics

For the realisation of climate-conscious development of the area, the suggested solutions are similar to traditional solutions, undertaken in such a way that the role of the local government is in harmony with existing tools on the one hand and with the ownership relations of the properties to be developed or regenerated on the other, with special attention to properties owned by the local government.

Due to the diversity in ownership, interests and abilities, it cannot be hoped that the suggested plans could be realised in their entirety in one, well organised initiative, in a short period of time – yet most elements of the plan could indeed be carried out in such fashion, while a smaller proportion of the plans could be realised as – preferably coordinated – individual actions carried out by the other parties but initiated, and on occasion supported, by the local authorities.

For the preparation and organisation of tasks, especially those to be carried out by the local authorities, the relevant local government body responsible for urban development and planning should be appointed as coordinator – who will, then, invite the contributions of other parties, such as planners and constructors, as necessary.

As for the timing of individual tasks, priority should be given to properties owned by the local authorities, to tasks that can be carried out within the jurisdiction of the local authorities, primarily to public spaces that could serve as direct examples, such as green roofs on buildings of institutions controlled by the local authorities. An equally important recommendation is the simultaneous commencement of negotiations with the property managers of national institutions, to invite them to participate in the program and to help realise climate-conscious redevelopment for elements of their buildings. The installation of a green wall or vertical garden on the Kis Rókus utca facade of the buildings in Millenáris Park may have a special position in this respect, as the soil in which the vegetation is planted is in a public space owned by the local authorities, while the vegetation climbs on the walls of a building under national ownership. Here a special agreement will be necessary between parties concerning maintenance, the recommended solution being that the local government take responsibility for planting and watering the plants.

The inclusion of local businesses and condominiums in the program requires yet another set of solutions. In both cases the recommended procedure would be for the local government to invite owners to participate in the program, and to explain in the course of preparatory negotiations how advantages and savings will accrue as a result. In addition, for businesses, PR-based incentives might be introduced (such as the establishment of the title 'Climate-Conscious Business of the Year'), while for condominiums a system of financial support might be successful.

### 13.6.5  Cost Estimate and Suggestions for Financing

Costs and sources of funds for realising the climate-conscious urban development action plan are estimated below. The sums presented are to be viewed as an order-of-magnitude estimate and relate primarily to the establishment of cost ratios, intended to help make decisions about subsequent steps and actions.

Individual trees

Keleti Károly utca: Tilia x euchlora (15–20 m) single row
45 @ 50,000 HUF, total: 2,250,000 HUF
Mechwart tér: Fraxinus excelsior 'Jaspidea' (10–15 m, yellow-leaf variety) single row
22 @ 25,000 HUF, total: 550,000 HUF
Fillér utca: Fraxinus excelsior 'Jaspidea' (10–15 m, yellow-leaf variety) single row
38 @ 25,000 HUF, total: 950,000 HUF
Fény utca: Fraxinus excelsior 'Jaspidea' (10–15 m, yellow-leaf variety) double row
28 @ 25,000 HUF, total: 700,000 HUF
Buday László utca: Gleditsia triacanthos 'Sunburst' (10–15 m, yellow-leaf variety) single row
10 @ 20,000 HUF, total: 200,000 HUF
Garas utca: Fraxinus excelsior 'Westhof's Glorie' (20–25 m) single row
17 @ 50,000 HUF, total: 850,000 HUF
Ezredes utca: Koelreuteria paniculata (5–8 m) single row
14 @ 20,000 HUF, total: 280,000 HUF
Pengő utca: Koelreuteria paniculata (5–8 m) single row

14 @ 20,000 HUF, total: 280,000 HUF
Retek utca: Acer platanoides 'Emerald Queen' (10–15 m, yellow-leaf variety) single row
14 @ 20,000 HUF, total: 280,000 HUF

Trees total: 6,340,000 HUF + 27 % VAT = 8,000,000 HUF

Tree hole grilles
202 @ 200,000 HUF, total: 40,400,000 HUF

Rows of street trees, total: 48,400,000 HUF

**Planters**
Fény utca: 2 m wide, on both sides
200 m² @ 20,000 HUF, total: 4,000,000 HUF
Fillér utca: 1 m wide, on one side
70 m² @ 20,000 HUF, total: 1,400,000 HUF
Bimbó út, 1.5 m wide, on one side
270 m² @ 20,000 HUF, total: 5,400,000 HUF
Plants, planters, total: 10,800,000 HUF
**Green Spaces**
SzéllKapu area
12,000 m² @ 20,000 HUF, total: 240,000,000 HUF
Green spaces, total: 240,000,000 HUF
**Permeable Pavement**
Total of 5 km pavement, on average 3.5 m wide
5000 lm @ 50,000 HUF, total: 250,000,000 HUF
Permeable pavement, total: 250,000,000 HUF
**Green Walls – Vertical Gardens**
Trellises
1100 m² @ 5,000,000 HUF, total: 5,500,000 HUF
Plants
110 lm @ 5000 HUF, total: 550,000 HUF
Green walls – vertical gardens, total: 6,050,000 HUF
**Green Roofs**
On buildings owned by local government
3900 m² @ 15,000 HUF, total: 58,500,000 HUF
On buildings owned by other legal entities
26,000 m² @ 15,000 HUF, total: 390,000,000 HUF
On buildings with multiple private owners
21,000 m² @ 15,000 HUF, total: 315,000,000 HUF
Green roofs, total: 763,500,000 HUF
**Reconstruction of Public Utilities**
In streets with rows of trees, organised in tandem with the introduction of district heating, in cooperation with FŐTÁV, the Budapest district heating agency
5 km @ 400,000,000 HUF, total: 2,000,000,000 HUF
Complete reconstruction of public utilities in affected streets, total: 2,000,000,000 HUF

Climate-conscious regeneration of complete area, sum total: 3,318,750,000 HUF
Costs to be paid directly by local government: 613,750,000 HUF
Costs to be paid by local government without the relatively low-benefit yet high-cost permeable pavement: 363,750,000 HUF

Costs of climatically most efficient and also most spectacular investment (street trees and planters), which are within the jurisdiction of local authorities: 59,200,000 HUF

A fundamental prerequisite for planting rows of street trees, however, is coming to a satisfactory arrangement with public utilities. Planting may be coordinated with complete upgrades carried out by the utility companies, or with the installation of a significant new type of utility – such as district heating – which necessitates the relocation of underground utility installations. It is therefore recommended that preliminary negotiations be conducted with utility companies, primarily with the representatives of FŐTÁV Zrt, the Budapest district heating agency, concerning the submission of an EU pilot project, such as Horizont 2020.

## 13.7   Action Plan for Local Governments in Heatwave Alert Situations

The following legal regulations apply, and should be considered, before raising a heat alert:

Act LXXII of 2012 on the Amendment of Act CXXVIII of 2011 Concerning Disaster Management and Amending Certain Related Acts;

Decree of the Minister of the Interior No. 61/2012. (XII. 11.) on the Disaster Management Categorisation of Municipalities, and on the Amendment of the Decree of the Minister of the Interior No. 62/2011. (XII. 29.) on Certain Rules of Disaster Management.

Act LXXXII of 1995 on the Promulgation of the UN Framework Convention on Climate Change;

Act CXXIX of 2000 on the Amendment of Act LIII of 1995 on the General Rules of Environmental Protection;

Act XV of 2005 on Greenhouse Gas Emission Allowance Trading; and 213/2006 (X. 27.) Governmental Decree Implementing Act XV of 2005 on Greenhouse Gas Emission Allowance Trading;

Act LX of 2007 on the implementation framework of the UN Framework Convention on Climate Change and the Kyoto Protocol thereof; and The National Climate Change Strategy (NCCS), which was based on the act on the implementation framework of the Kyoto Protocol;

"Adapting to Climate Change in Europe" European Commission Green Paper, June 2007;

Act XXXVII of 1996 on Civil Protection;

Act LXXIV of 1999 on Disaster Management (direction and structure of protection
   against disasters and the protection against major accidents involving hazardous
   materials);
Within the system of civil protection planning, the 'Decree of the Minister of the
   Interior No. 20/1998. (IV. 10.) on the System and Requirements of Civil
   Protection Planning' classifies the types of hazards that necessitate planning;

   Due to climate change caused by global warming, extreme weather incidents
have occurred in Hungary with increasing frequency; summer heatwaves, lasting
for several weeks, are particularly stressful for the population. In Hungary, disaster
management is conducted within a strict legal framework; the tasks of local authorities are specified in Act CXXVIII of 2011 on Disaster Management and Amending
Certain Related Acts.
   The steps for preparing for longer heatwaves, along with special points for consideration, will be detailed below. The European Commission Green Paper, June
2007, draws attention to, among other things, the dangers posed by heat to human
health; thus, it is extremely important to prepare an action plan to protect the population during heat alerts.
   There are several groups within the population of Hungary that are especially at
risk. The age distribution of the population categorizes Hungary as an ageing society. A significant proportion of people performing manual labour are outdoors in the
summer during heatwaves, for extended periods of time or all of the time. Summer
is also the period for open air events, with cultural and sports events attracting large
numbers of participants. Special attention must be paid to minimizing health risks
for participants in these programs. Employers must seek to protect the health of
employees working outdoors; during heatwaves, they must provide for their
employees ample drinking water, periods of rest in shady areas, appropriate working clothes and protective gear.
   From the point of view of the environment and of environmental health, special
consideration must be given to those utilities and public services that influence positively or negatively the quality of life of the population, as well as the quality of the
environment, during heatwaves.
   Special protection must be provided for the drinking water infrastructure, particularly strategic reservoirs and water mains, which are maintained by water utilities to provide high-quality drinking water for the population.
   In the event of disruptions to electricity service during heatwaves, all customers
can be provided with electricity with blackouts averaging 2–3 h. During longer
heatwaves, energy demands may increase, which might require the imposition of
limitations on consumption.
   Since Hungary is a country along major transit routes, dangers affecting branches
of transportation during heatwaves must also be taken into account. A relatively
minor disruption in railway services might cause serious delays. In extensive heat,
rail lines can be deformed; overhead wires and pylons can be damaged. During a
temporary suspension of railway services, railway stations must offer shelter for
railway passengers until such time as they can continue their journeys. While these
passengers remain stranded, their basic needs must be addressed.

Settlements are often situated near motorways and major transit routes. During periods of continuously high daily average temperatures, the number of accidents may increase in the affected stretches. When congestion builds up following an accident, people sitting in cars will need to be supplied with liquids.

Transit lines run by public transportation companies work continuously on days of high average temperatures as well. Newer vehicles have been equipped with air conditioning, which must be used. On older vehicles, cross ventilation and the use of ventilators must be applied continuously to offer heat protection for the passengers.

During periods of high daily average temperatures, communal waste must be collected more frequently in order to prevent epidemics.

### 13.7.1   Communicating with the Public

Heatwaves are usually possible to predict, and, therefore, the public must be informed about preventive measures they can take to protect their health. If the local authorities are to issue a warning, attention must be paid to the following considerations: Communication may be effected in writing or as a personal announcement. It may take the form of a press release, a public announcement, a briefing or interactive communication, during which attention must be called to the negative effects of heatwaves on health; the necessity of remaining hydrated; the need to seek shelter in shady places; and the dangers of leaving one's place of residence.

The public must be informed that they should take along some drinking water if they leave their homes. Warnings must be issued about the need to be extra careful and circumspect in traffic. The information issued must always be authenticated; creating a sense of panic must be avoided.

*The tasks of the preparatory phase are as follows:*

The forecasts and news bulletins of the Hungarian Meteorological Service and the National Public Health and Medical Officer Service (NPHMOS) must be followed.

The public must be informed.

People with severe medical conditions and others at special risk must be identified.

Reserves of drinking water must be prepared.

Vehicles delivering drinking water must be arranged.

Any available street watering trucks must be pressed into service.

Local institutions with air conditioned buildings that may accommodate large numbers of people must be identified.

*The tasks of the protection phase are as follows:*

Local authorities must remain in contact with the local offices of NPHMOS and with disaster management authorities.

Communication with the public must be continuous.

Access to public buildings with air conditioning must be arranged.

Shelters must be opened, if necessary.

Drinking water distribution in the busiest centres of the community must be organised.

Major routes and public spaces must be watered down several times a day.

Logistical support, with special attention to the availability of equipment and manpower, must be established and operated continuously.

**Stages of heatwave alerts, action items, and the raising of alerts**

### Level One (Heatwave Advisory)

Criteria: according to forecasts, the daily median temperatures exceed 25 °C for at least 1 day

Actions required: Within its own internal system, NPHMOS provides information to its regional and micro-regional institutions. Local authorities will decide whether to inform the public through local media.

### Level Two (Stage 1 Alert)

Criteria: according to forecasts, the daily median temperatures exceed 25 °C for at least 3 days

Actions required: The Chief Medical Officer will use the institutional network of NPHMOS to inform public health institutions, emergency services, doctors' offices and nurses' stations within the primary healthcare network and local authorities concerning the duration and degree of the heat alert. The task of the local authorities is to inform their own institutions (primarily those providing social services) to warn members of the public to take preparatory measures to protect their health.

### Level Three (Stage 2 Alert)

Criteria: according to forecasts, the daily median temperatures exceed 27 °C for at least 3 days

Actions required: The responsible authorities must verify that actions taken at Level Two for the Stage 1 Alert are maintained. The public must be informed of the Stage 2 Alert through the media and via the local authorities. Protective measures must be initiated by health care institutions, nursing homes, chariTab. institutions, crèches, nursery schools, day care centres, and summer camps.

**Rapid alert system for heatwave alerts**

The first step is to set up a list of officials authorised to raise, and call off, a heatwave or UV alert. The appropriate level of heatwave or UV alert will be raised, and called off, by the Mayor, based on the alert issued by NPHMOS, the UV index values published by the Red Cross, or both. Based on the weather forecasts made by the Hungarian Meteorological Service, the Chief Medical Officer and NPHMOS will issue a warning. They will send a letter to inform the doctors' offices and nurses' stations of the primary healthcare network and the micro-regional local authorities concerning the duration and stage of the heatwave alert, asking for their cooperation in abiding with the orders issued. The Mayor's Office will issue a warning to the institutions of the rapid alert system, as well as the general public, to take preventive action to protect their health (Table 13.1). The Mayor's Office must notify without delay the following institutions concerning the situation that has developed:

Civil Protection Office
Municipal Police Headquarters, Department of Public Order and Safety
Hospitals
National Ambulance Service
Social care institutions and sanatoriums
Local water works
Local energy providers
The Hungarian Labour Inspectorate
Local crèches
Preschools in operation at the time
Day camps and summer camps
The affected population, institutions and organisations, through the local media

Advice for various age demographic groups in the event of heatwave alerts is presented below in Tables 13.2, 13.3, and 13.4 in an easily understandable form, which can be converted to flyers. The colourful, printed flyers with illustrations can help draw attention to the information and reinforce the message (Tables 13.2–13.4).

Useful advice during heatwaves for the prevention and treatment of heat strokes for the elderly (Table 13.2).

**Table 13.1** Advice for the general public

| Suggestions and important messages for the general public in the event of heatwave alerts | |
|---|---|
| How to avoid heat | Important comments |
| Cool your home | Monitor room temperatures |
| Keep the windows closed during the day; use curtains, shutters, or other means of keeping the room dark. Open the windows to air the room at night, if possible. Switch off non-essential electric appliances (even including lights). If you have air conditioning, keep windows and doors closed | During heatwaves, when external temperatures reach 35–39 °C, the ideal indoor temperature is about 28 °C. Very cold settings for the air conditioner should be avoided. Electric fans should be used only for short periods of time |
| If these measures cannot be taken, spend at least 2–3 h in air conditioned places | People can be directed to the list of air conditioned places open to the public, as arranged during the protection phase |
| Avoid heavy physical work and stay in the shade during the hottest hours | |
| For the next summer, consider cooling your home ("cool" paint, humidifier, green plants) | |
| Keep your body temperature low and drink plenty of liquids to prevent dehydration | |
| Take frequent showers or baths in lukewarm water | Showers may increase the risk of falling for the elderly |

(continued)

**Table 13.1** (continued)

| Suggestions and important messages for the general public in the event of heatwave alerts | |
|---|---|
| Use wet bandages and cool your feet in lukewarm water | |
| Wear loose garments of light colour and natural materials. If you go out in the sun, wear a large-brimmed hat and sunglasses | |
| Drink liquids regularly. Do not drink beverages with alcohol or high sugar content | During hydration, it is important to replenish lost salt and to avoid water intoxication. Caffeine acts as a diuretic |
| If you take medications regularly, ask your doctor about the effect of your medications on your internal fluid balance | People with elevated body temperatures require special attention |
| Monitor your body temperature | It is important to realise that body temperatures above 38 °C are detrimental to one's health. Heat strokes can occur at body temperatures exceeding 39 °C. Body temperatures above 40 °C present a life-threatening situation |
| Keep your medications at the appropriate temperature | If room temperature exceeds 25 °C, medications should be kept refrigerated, even if their boxes do not say so |
| Contact your doctor if you suffer from a chronic illness or if you take several different medications. If you experience unusual symptoms, contact your doctor immediately | |
| Inform yourself about the forms of assistance available | |

**Table 13.2** Brief, easily understandable advice for the elderly

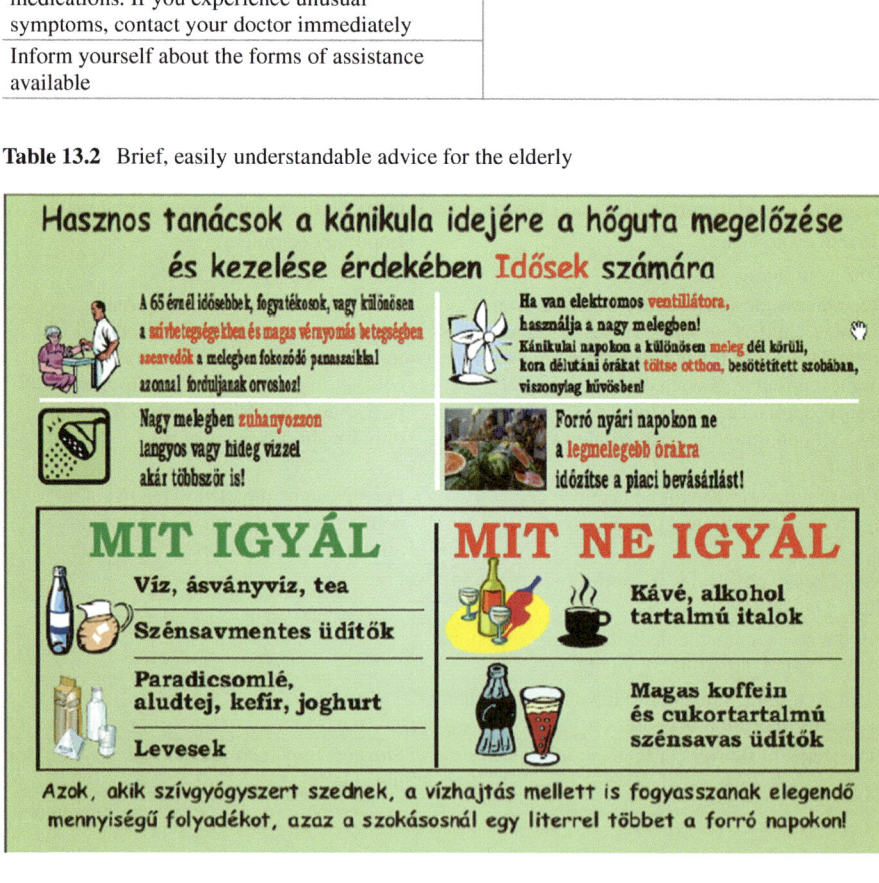

Useful advice during heatwaves for the prevention and treatment of heat strokes for young mothers and small children (Table 13.3).

| | |
|---|---|
| Those above 68, the disabled, or, especially, those with heart problems or high blood pressure should seek medical help immediately if their symptoms become worse in the heat | If you have an electric fan, use it in hot weather. During heatwaves, stay home in a darkened room, in a relatively cool place, for the hottest hours of the day, especially around noon or in the early afternoon |
| In very hot weather, take frequent showers or baths in lukewarm water | In hot summer days, do not go to the market during the hottest hours |
| WHAT TO DRINK | WHAT NOT TO DRINK |
| Water, mineral water, tea | Coffee, alcoholic beverages |
| Non-carbonated soft drinks | |
| Tomato juice, curd, kefir, yoghurt | Carbonated beverages with high caffeine and sugar content |
| Soups | |
| If you are taking medication for a heart condition, consume sufficient amounts of liquid while taking diuretics, that is, 1 l per day more on hot days than the usual amount. | |

**Table 13.3**  Brief, easily understandable advice for young mothers and small children

Useful advice during heatwaves for the young (Table 13.4)

| Babies and small children should be taken out for air only in shady places. Do not take babies for walks in extreme heat | If you have an electric fan, use it in hot weather. If possible, leave windows open at night for ventilation |
| --- | --- |
| Never leave children or animals (dogs) in locked, unventilated, parked cars | Protect yourself and your child by wearing wide-brimmed hats and sunglasses |
| WHAT TO DRINK | WHAT NOT TO DRINK |
| Water, mineral water, tea | Coffee, alcoholic beverages |
| Non-carbonated soft drinks | |
| Tomato juice, curd, kefir, yoghurt | Carbonated beverages with high caffeine and sugar content |
| Soups | |

Babies in particular need a lot more liquid, even in addition to breast milk. Always offer them water or some tea with lemon and a little salt after breastfeeding.

**Table 13.4** Brief, easily understandable advice for the young

| Protect yourself from the heat of the sun by wearing wide-brimmed hats and sunglasses. Wear light, loose cotton garments of light colour on hot days. | Apply sun protection lotion appropriate for your skin type several times a day. If you have very light skin and blue eyes, use sun lotion with a sun protection factor (SPF) above 10. |
|---|---|
| In very hot weather, take frequent showers or baths in lukewarm or cold water. Spend 1–2 h in air conditioned spaces. | If you are doing sports outdoors during heatwaves, cool yourself frequently and drink at least 4 l of liquids. It is also important to replenish lost salt. |
| WHAT TO DRINK | WHAT NOT TO DRINK |
| Water, mineral water, tea | Coffee, alcoholic beverages |
| Non-carbonated soft drinks | |
| Tomato juice, curd, kefir, yoghurt | Carbonated beverages with high caffeine and sugar content |
| Soups | |

## 13.7.2 The Latest Tools for Communicating with the Public

Along with the public media (television, radio), web applications can also help in communicating with the public. In 2013, the National Directorate General for Disaster Management (NDGDM) developed and launched a disaster alert information application that can be downloaded on smart phones and Tab.ts free of charge. The mobile application makes it possible for people to be notified immediately about local weather anomalies or other dangers relevant to their homes or current locations. Through these devices, users with an Internet connection can conFig. the application to receive alert information relevant to the whole country or to a specific county, depending on their personal preferences. If the device has a GPS, that is, a unit able to define their position using a space-based satellite system, users can receive information for actual locations, for a planned route, or even for great distances.

In conjunction with its weather forecast and alert system, the Hungarian Meteorological Service (OMSZ) developed an application called Meteora, which can be downloaded from http://meteora.met.hu/meteora.html. The Meteora application is a clock that runs on mobile devices and also provides weather information. The program uses the positioning and Internet data access capabilities of the mobile device. Based on cell information, Meteora defines the position of the user and, using the data connection, automatically downloads local alerts and forecasts from the computers of the Hungarian Meteorological Service. Meteora is a widget developed for Android devices; after downloading, the version designed for the device screen size must be installed (Fig. 13.18). Installing the clock automatically activates the alert function.

Updates calling off alerts are passed on via the same rapid alert system.

*The tasks for resuming normal life are as follows:*
Establish responsibility.
Draw conclusions.

**Fig. 13.18** Alert system application developed by the Hungarian Meteorological Service, displaying a clock on the mobile device

Evaluate efficacy and efficiency.
Provide relief as long as needed.
Resume the operation of institutions.
Reconcile costs incurred during disaster management.

Earlier experience indicates that the negative effects of extreme heat can affect extended areas and large segments of the population. It is therefore of paramount importance to be prepared for extended heatwaves and to utilize the protective powers of local governments to their full extent. Special attention must always be given to the protection of human life and property, as well as to the execution of any reconstruction work immediately required. Local authorities have a significant role in disaster management; they must devote special attention to the careful preparation of protective measures against the detrimental effects of extreme weather phenomena, because these preparations are crucial for developing an effective adaptation strategy.

# Chapter 14
# Pilot Actions in European Cities – Prague

**Michal Žák, Pavel Zahradníček, Petr Skalák, Tomáš Halenka, Dominik Aleš, Vladimír Fuka, Mária Kazmuková, Ondřej Zemánek, Jan Flegl, Kristina Kiesel, Radek Jareš, Jaroslav Ressler, and Peter Huszár**

**Abstract** This chapter describes results of pilot actions in Prague. Two different pilot areas were selected (Legerova street and Bubny-Holesovice quarter) with different modelling approach. Finally, the Green belt around Prague is studied as well. Different scenarios are tested and their results discussed. The matter of air quality is also analysed.

**Keywords** Urban heat island of Prague • Potential equivalent temperature • Mitigation of urban heat island • Effects

M. Žák (✉)
Department of Climatology, Czech Hydrometeorological Institute,
14306 Praha 4, Prague, Czech Republic

Faculty of Mathematics and Physics, Department of Atmospheric Physics,
Charles University Prague (CUNI), Prague, Czech Republic
e-mail: michal.zak@chmi.cz

P. Zahradníček • P. Skalák
Department of Climatology, Czech Hydrometeorological Institute,
14306 Praha 4, Czech Republic

T. Halenka • V. Fuka • P. Huszár
Faculty of Mathematics and Physics, Department of Atmospheric Physics,
Charles University Prague (CUNI), Prague, Czech Republic

D. Aleš • M. Kazmuková • O. Zemánek • J. Flegl • R. Jareš
Prague Institute of Planning and Development, Prague, Czech Republic

K. Kiesel
Department of Building Physics and Building Ecology,
Vienna University of Technology, Vienna, Austria

J. Ressler
Institute of Computer Science, The Czech Academy of Sciences, Prague, Czech Republic

© The Author(s) 2016    373
F. Musco (ed.), *Counteracting Urban Heat Island Effects in a Global Climate Change Scenario*, DOI 10.1007/978-3-319-10425-6_14

## 14.1 Urban and Environmental Framework

### 14.1.1 General Remarks on Prague

Prague is the capital of the Czech Republic. As the largest city in the country by its area (496 km$^2$) and by population (1.2 million inhabitans) faces the same environmental challenges, as the other large cities in the world. The city strives for the substantial reduction of the environmental burden as to become clean, healthy, and harmonic place for living. Prague is situated on the banks of the Vltava river and its tributaries. The complicated morphology creates limitation for a good ventilation of the area. The lowest point is in 149 m and the surrounding hills at the southwestern part of Prague are almost 400 m above sea level.

### 14.1.2 Prague Urban Heat Island Analysis

Prague and especially its city centre belongs to the warmest regions of the Czech Republic with annual average air temperature around 10 °C in the city centre (see Fig. 14.1). This is partly caused by the urban heat island (UHI) of Prague.

The intensity of UHI is about 1.6 °C when we use the average daily temperatures. The highest intensity occurs during June, while the lowest intensity in September. It has to be noted, that the UHI intensity of Prague is considerably higher when looking at minimum temperatures (annual average is approximately 3 °C) but smaller for maximum temperatures (annual average approx. 1 °C). The intensity of Prague's UHI has increased during the last 50 years. This increasing is caused due to the city enlargement and transport intensification.

The magnitude of this intensification is 0.15 °C/10 years for minimum temperatures, 0.07 °C/10 years for average daily temperatures and 0.02 °C/10 years for maximum temperatures. This intensification of UHI is documented on Fig. 14.2 where differences of daily air minimum temperatures between period 2001–2010 and 1961–1970 are demonstrated. While the temperature in the whole Prague area increases (due to the climate change in the Central Europe), the largest increment of the temperature can be seen in city centre, close to the Vltava river, in the densely built up part of the city.

Another point of view of the intensity of UHI can be obtained by using physiological equivalent temperature PET (Matzarakis et al. 2010). This temperature describes the temperature really felt by the human being standing outside and includes not only influence of air temperature and wind, but also humidity and radiation including the radiation coming from buildings in the streets. The average annual and daily course of PET for the Praha–Karlov stations is given in Fig. 14.3. It can be seen, that the highest values occur during summer months, July and the first half of August.

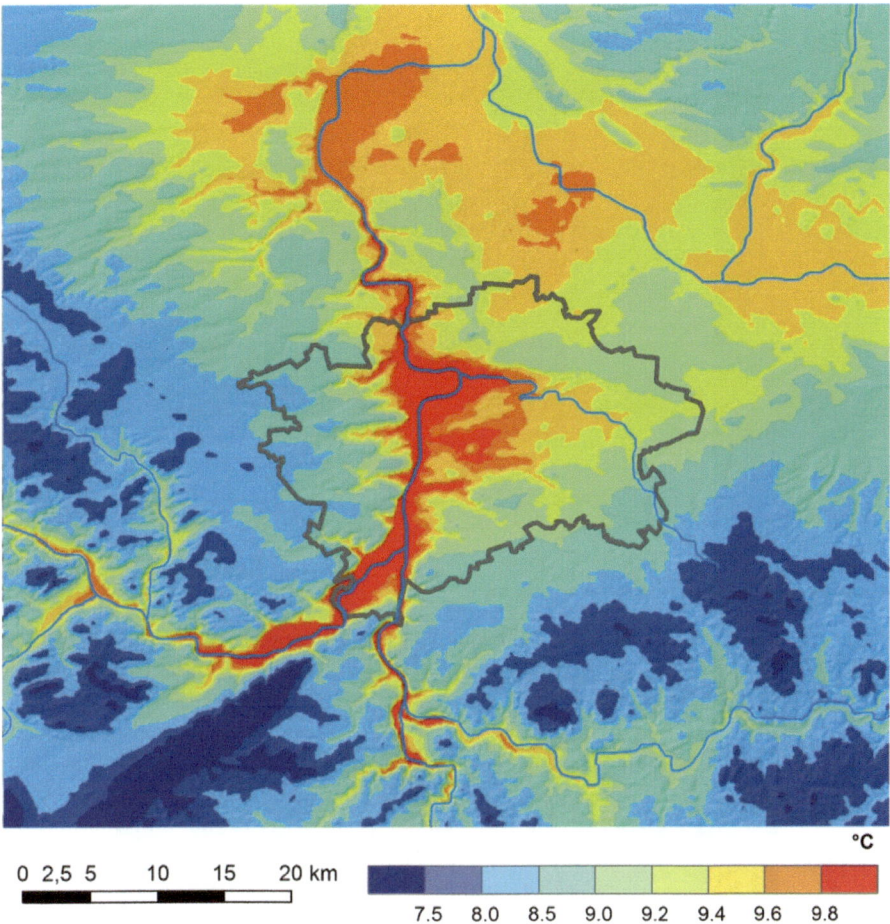

0 2,5 5    10    15    20 km

°C

7.5   8.0   8.5   9.0   9.2   9.4   9.6   9.8

**Fig. 14.1** Annual average air temperature for Prague and surrounding, period 1961–2010

   The differences of PET values between Praha–Karlov station located in the city centre and Praha–Ruzyně station situated on the periphery of the city (Fig. 14.4) show the highest values in the summer half of the year starting after the sunset and vanishing during the morning hours. Few hours after the sunrise this differences can be even negative indicating lower PET values in the city centre. Regarding long term changes, there is positive trend in PET values in Prague during spring and summer indicating greater human stress during the warm summer half-year. It should be noted that PET was computed for the location with ideal horizon without obstacles – this is also the case of the following case study of a hot day.
   The largest UHI negative effects (in the sense of bioclimatic discomfort) usually occur during the summer months. For demonstration, case study of hot day, the day of 28th July 2013 has been chosen. Maximum air temperature on that day reached

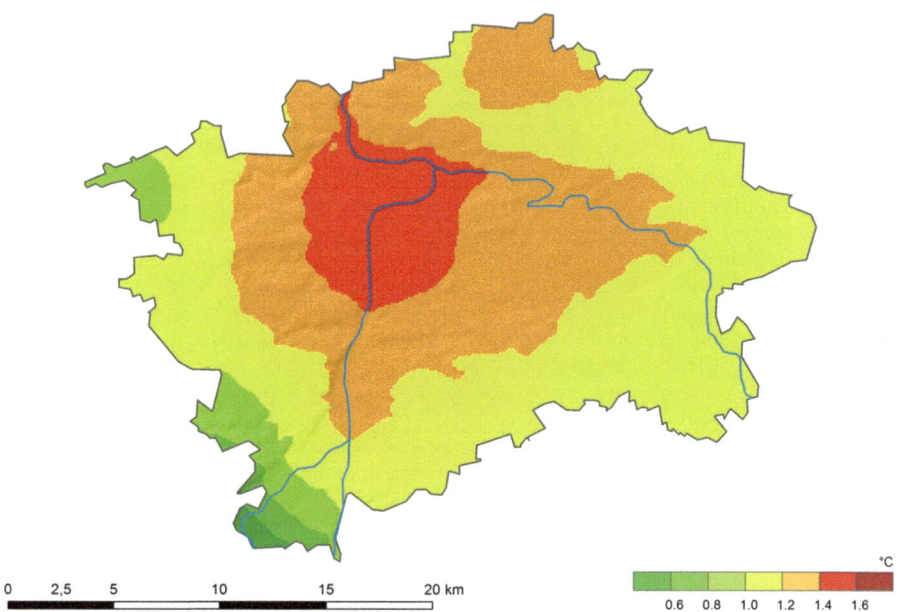

**Fig. 14.2** Difference of daily air minimum temperature between period 2001–2010 and 1961–1970

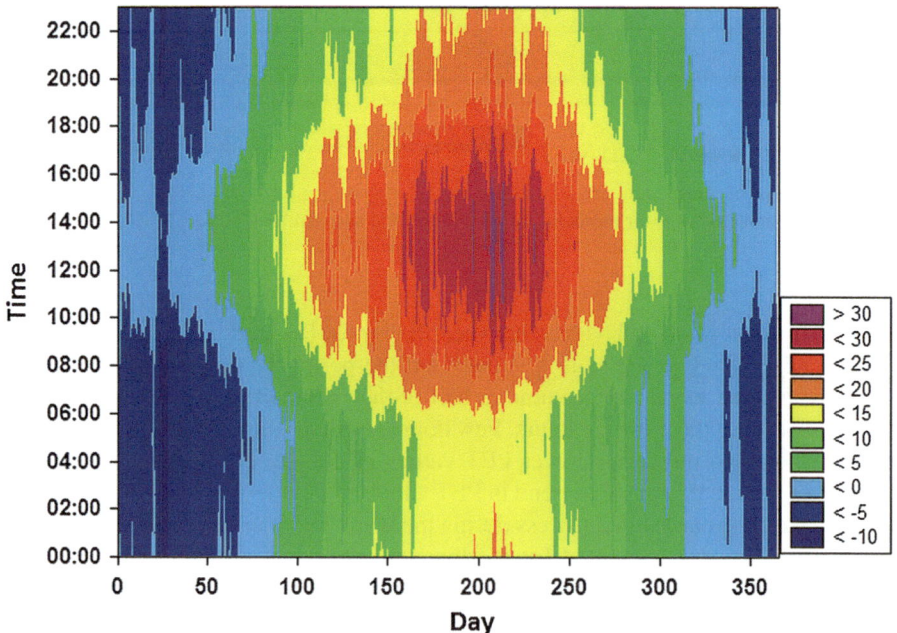

**Fig. 14.3** Annual course of PET for Praha-Karlov station, period 2005–2013

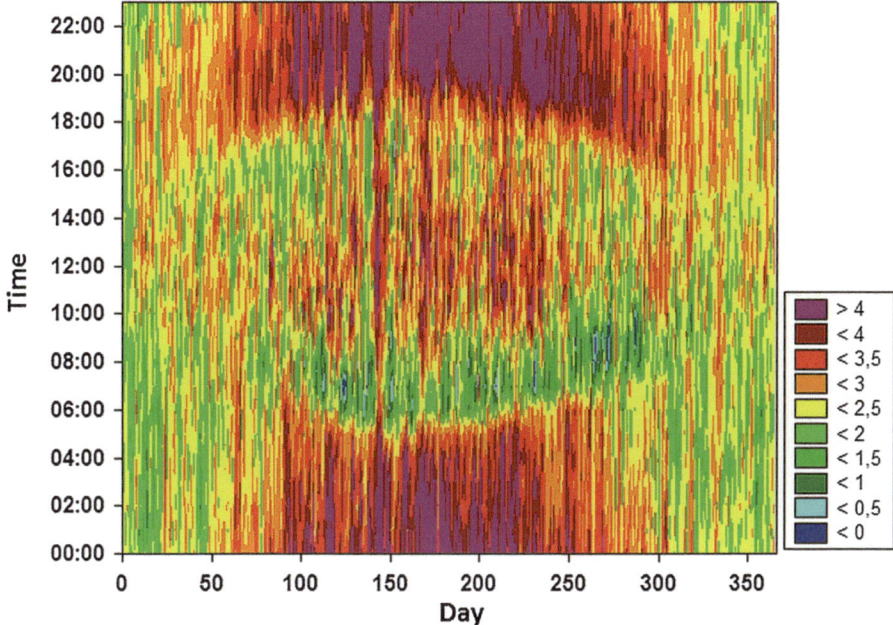

**Fig. 14.4** Annual course of PET of differences between stations Praha-Karlov and Praha-Ruzyně, period 2005–2013 (positive values mean Karlov is warmer than Ruzyne)

values around 37 °C (Fig. 14.5), with differences among different parts of the city being maximal about 2 °C. PET values (Fig. 14.6) on that day reached over 46 °C in the city centre, with difference through the city being a bit higher compared to the differences in the air temperatures. Values of PET over 40 °C started already around 9 in the morning and continued to 5 in the afternoon. Especially in the evening, the differences between city centre and periphery exceeded 8 °C.

## 14.1.3   Air-Quality Issues

Mária Kazmuková
Prague Institute of Planning and Development, Prague, Czech Republic

Peter Huszár and Tomáš Halenka
Faculty of Mathematics and Physics, Department of Atmospheric Physics, Charles University Prague (CUNI), Prague, Czech Republic

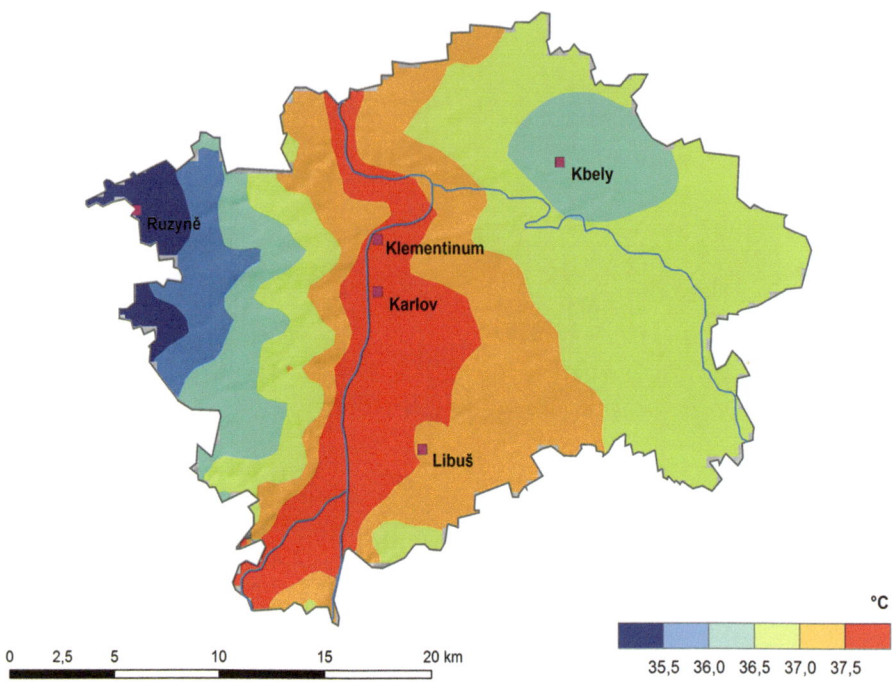

**Fig. 14.5** Maximum air temperature in Prague on 28th July 2013

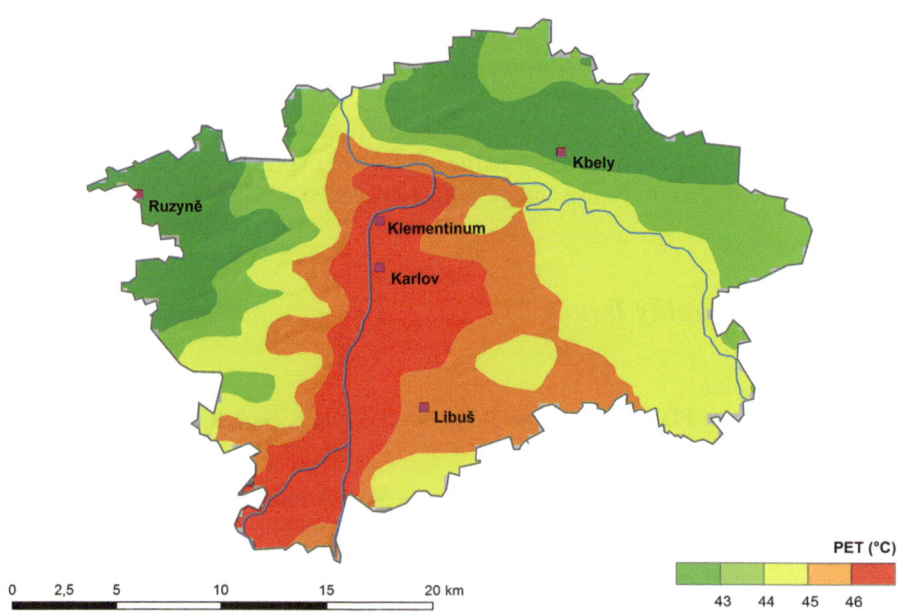

**Fig. 14.6** Maximum PET in Prague on 28th July 2013

Despite the substantial reduction of the emissions from industrial sources in the past years, air quality is still influenced by the emissions from automotive traffic, as a main source of air pollution. In the suburban residential areas air quality is influenced by the emissions from local heating burning solid fuels.

In the agglomeration of Prague the limits for air quality are exceeded, especially for particular matter PM10, NO2, O3 and benzo(a)pyren. The majority of exceedances is connected with the high traffic loads in Prague, but also with the domestic heating in family houses in the residential areas in Prague.

The share of mobile sources on the total of PM emissions is more than 85 %, on the total of $NO_x$ ca 75 %. The contribution of the household heating to PM emissions is almost of 16 %.

In the last years the ambient air quality has been improved. The annual limit concentration $NO_2$ (40 $\mu g.m^{-3}$) has been exceeded only on two traffic monitoring stations in Prague, however it can be supposed that the exceeding could occur in other areas with a similar traffic volume.

Also the $PM_{10}$ concentrations dropped significantly, nevertheless the average 24 h $PM_{10}$ concentrations are exceeding the limit at the 13 monitoring stations.

The concentrations of benzo(a)pyren measured at two monitoring stations in Prague were exceeded only at one of theme, the results of monitoring fluctuates around the limit of 1 $ng.m^{-3}$.

The concentrations of $O_3$ are regularly exceeded only at the one background station in the suburb over years.

The rest of the air quality limits are usually met in the area of Prague (Figs. 14.7 and 14.8).

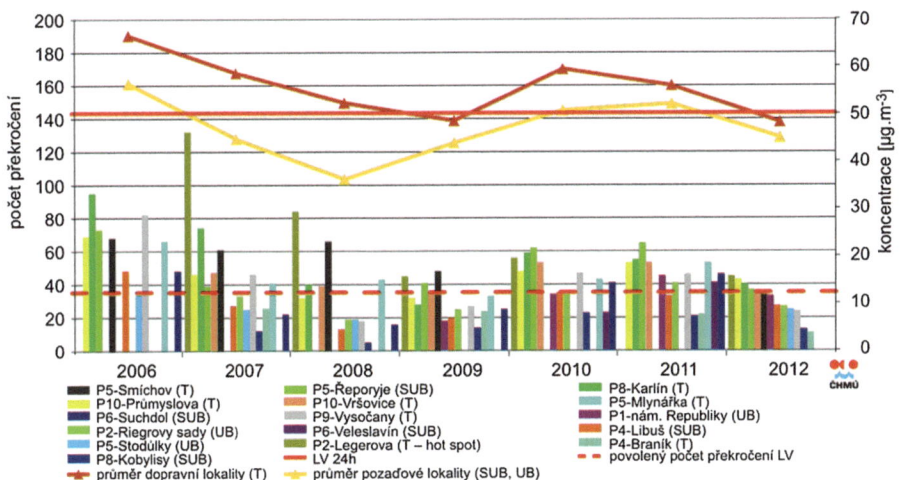

**Fig. 14.7** Trends in yearly characteristics of the fraction $PM_{10}$ and the 36th highest 24-h $PM_{10}$ concentration in selected monitoring stations in Prague

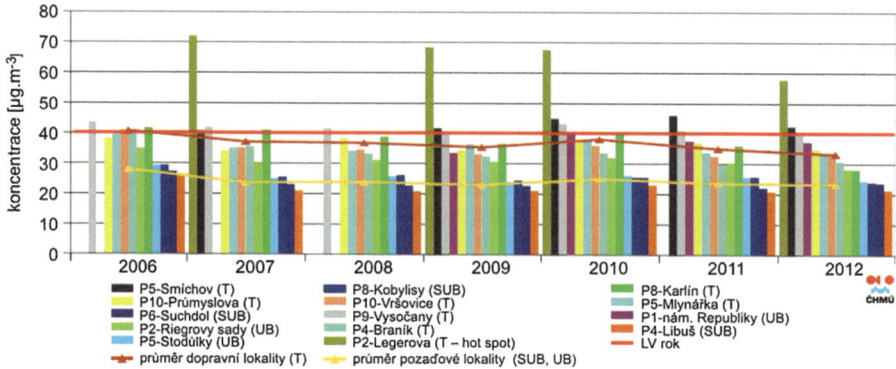

**Fig. 14.8** Trends in yearly characteristics of NO$_2$ in selected monitoring stations in Prague

## 14.2   Pilot Areas Identification Methodology

The pilot areas in Prague were selected with the aim to enable the simulation of UHI mitigation strategies in different scales.

For the scale of a street canyon the Legerova street was selected, representing one of the streets with a very high traffic volume crossing the residential area.

For the Pilot Area 2 was chosen the brownfield Bubny – Holešovice, an abandoned railway area, which aspires to be a new city quarter. Microclimatic simulations were performed for the central part measuring 500 m by 500 m.

The Pilot Area 3 as the whole territory of Prague has been chosen to enable simulations of the mitigation effects as a green belt around Prague or traffic emission reduction in all Prague agglomeration.

## 14.3   UHI Phenomena in the Pilot Area and Connection with Specific Aspects of Urban Form and Built Environment

### 14.3.1   Pilot Area 1 Legerova street

Dominik Aleš and Mária Kazmuková
Prague Institute of Planning and Development, Prague, Czech Republic

Vladimír Fuka
Faculty of Mathematics and Physics, Department of Atmospheric Physics, Charles University Prague (CUNI), Prague, Czech Republic

Michal Žák and Pavel Zahradnicek
Department of Climatology, Czech Hydrometeorological Institute, Prague, Czech Republic

**Fig. 14.9** Prague

Legerova Street represents a corridor with the width of 25 m, surrounded by 21 m high buildings. The traffic density is approximately 45 000 cars per day in 4 lines. The street leads through a residential area in north-south direction. During summer months it is fully open to sunshine and the incoming solar energy is largely absorbed by asphalt and concrete as well as by facades of the buildings. There are only a few sparse parts of grass beds and no availability for shade (Figs. 14.9 and 14.10).

Implementation of tree alleys was assessed as a mitigation measure. Three different scenarios varying the form and position of the alleys were tested in cooperation of Prague Institute of Planning and Development, Czech Hydrometeorological Institute, and the Department of Meteorology and Environment at Faculty of Mathematics and Physics, Charles University in Prague.

Besides the thermal comfort, also the air pollution concentrations were taken in mind. The initial $NO_x$ concentrations in Legerova Street could reach ca. 33 µg/m³ on the east windward sidewalk and even around 160 µg/m³ on the west leeward sidewalk due to the prevailing west wind direction which causes this unbalanced air pollution dispersion.

The thermal comfort was simulated by using of the microclimate model RayMan (Matzarakis et al. 2010). The ventilation conditions for air pollution were simulated by a model developed at the Department of Meteorology and Environment, CUNI Prague.

**Fig. 14.10**  Aerial view of Legerova street in Prague

A mild summer day of 21st June 2013 was chosen for the simulation of all proposed scenarios. Temperature effect was also modelled on a tropical day of 18th
June (Fig. 14.11).

The simulation results for all scenarios in the mild summer day ($T_{A,max} = 26$ °C)
show a possible effect of PET reduction of 2.3° in shade. During the tropical day
($T_{A,max} = 37$ °C) the reduction can reach 3.5°. However, all scenarios show also more
or less negative impact on the ventilation conditions for air pollutants.

In the scenario with small trees positioned densely along the sidewalks, there is
quite short time period of shade provided to one assessed point. On the other hand,
the street canyon ventilation in not worsened noticeably (Fig. 14.12).

The effect of PET reduction in the scenario with large trees along the sidewalks
lasts for a longer afternoon period due to a larger shade. However, the large crowns
significantly impact the air flow and cause a serious concentrations increase on the
windward sidewalk (Fig. 14.13).

The simulation of the scenario with an axial position of small trees in one row in
the centre of the street shows no impact on PET, providing no shade on the sidewalks. At the same time, this arrangement constitutes an obstacle to the vortex and
thus causes additional increase of already worse high concentrations on the leeward
sidewalk (Fig. 14.14).

High trees in the street bring more shade with a positive effect on PET, but also
create less favourable ventilation conditions. The scenario with the small trees
planted densely along the sidewalks seems to be the optimal solution for UHI
mitigation for Legerova Street. This scenario does not have such a negative effect
on ventilation conditions and provides shade and a positive effect on PET.

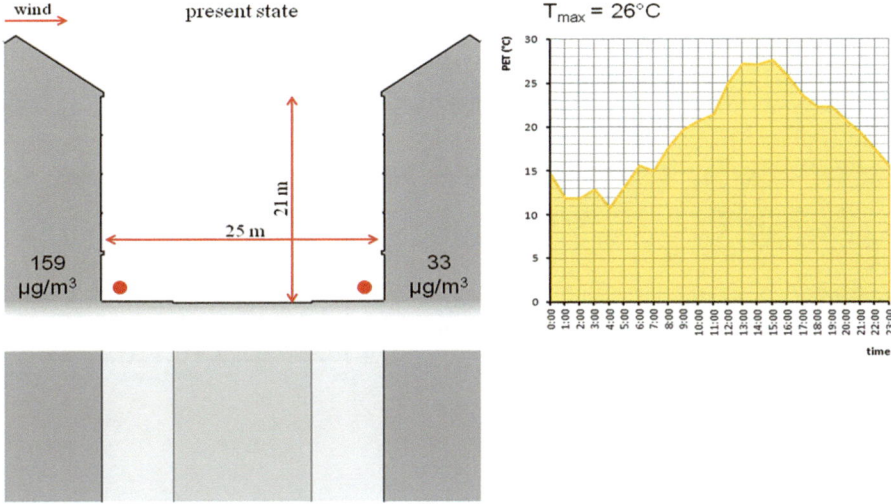

**Fig. 14.11** Mild summer day scenario

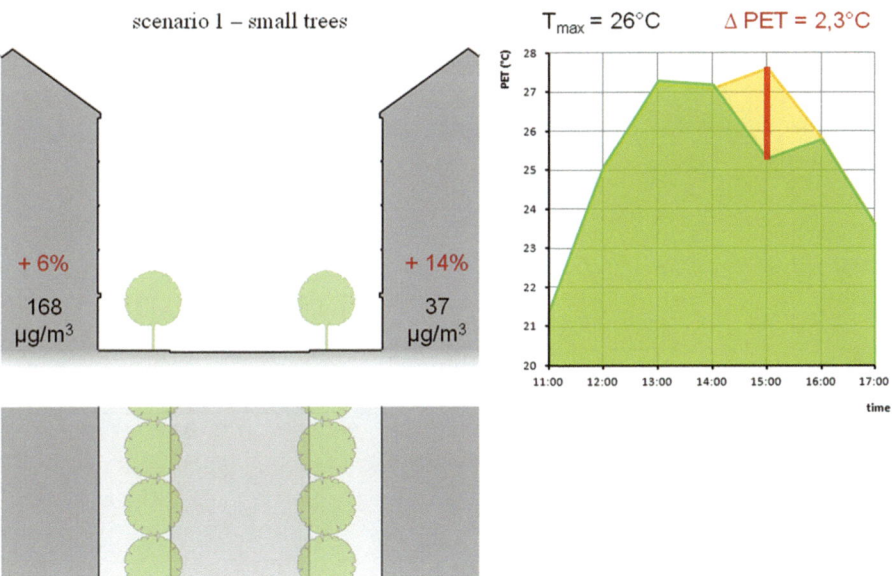

**Fig. 14.12** Effect of PET reduction

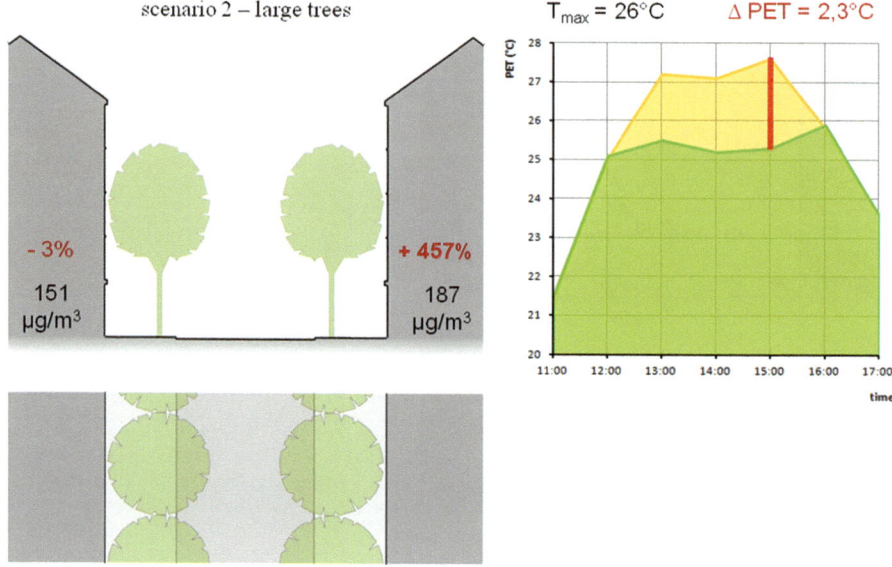

**Fig. 14.13** Scenario 2

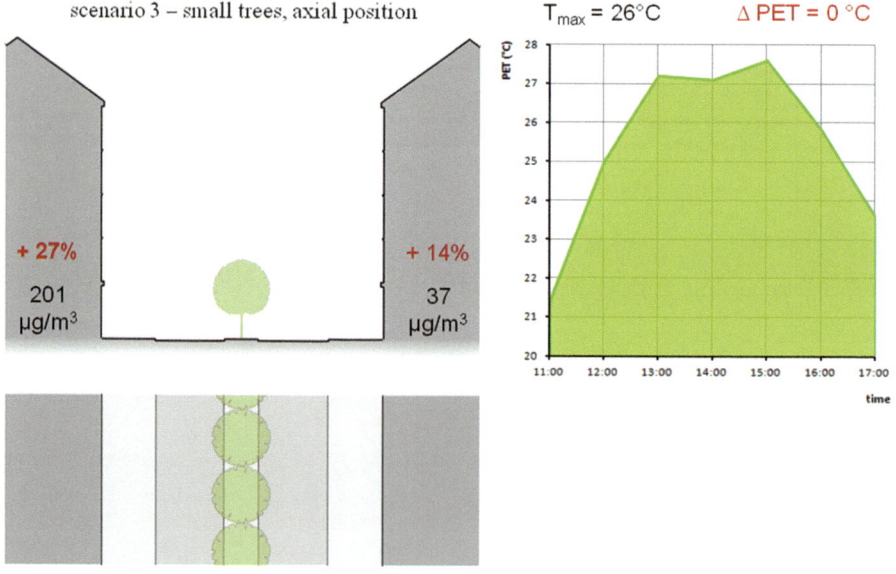

**Fig. 14.14** Scenario with the small trees planted densely along the sidewalks

## 14.3.2   Pilot Area 2 Holešovice – Bubny

Mária Kazmukova, Ondrej Zemánek, Jan Flegl, and Radek Jareš
Prague Institute of Planning and Development, Prague, Czech Republic

Michal Žák
Department of Climatology, Czech Hydrometeorological Institute,
Prague, Czech Republic

Faculty of Mathematics and Physics, Department of Meteorology
and Environment Protection, Charles University Prague (CUNI),
Prague, Czech Republic

Pavel Zahradníček
Department of Climatology, Czech Hydrometeorological Institute,
Prague, Czech Republic

Kristina Kiesel
Department of Building Physics and Building Ecology,
Vienna University of Technology, Vienna, Austria

Lokalita Holešovice – Bubny

Holešovice – Bubny was chosen to be one of the Prague's pilot areas due to its strong development potential and proximity to the city centre. Once used as a freight station the site is nowadays a brownfield that aspires becoming a living city quarter. A significant part of the area is occupied by the transport infrastructure, the rest is scattered with isolated buildings and fragments of block structure. Existing greenery is not properly maintained. In the east and west the site is adjoining various urban structures and the Vltava River in the north and south. The selected pilot area is about 82.5 ha large (Fig. 14.15).

The site is considered to be the future residential and commercial district. According to the current urban study the area shall be converted into a block structure, accompanied by small-scale parks and alley-like streets. Mean building height (between 25 and 26 m) shall be pierced with several landmarks (height from 50 up to 70 m). These adjoin to park areas as well as to the northern river bank. Existing railway tracks shall be reduced and elevated to enable streets to pass beneath (Fig. 14.16, 14.17 and 14.18).

The aim of the research was to examine the benefits of the current study (scenario 1) and to compare it with alternative urban studies proposing different urban structures and larger park areas (scenarios 3 to 6) (Fig. 14.19, 14.20 and 14.21).

In terms of land use, greenery types and building characteristics the following GIS models were performed:

Scenarios 3 and 4 propose a massive east-west oriented park strip located in the middle of the pilot area. A loose urban structure with high buildings of small footprints adjoin to the park in the north (Fig. 14.22).

This arrangement should leave more space for greenery and enable better ventilation of the area.

**Fig. 14.15** Future residential and commercial district

**Fig. 14.16** Series of small-scale parks shall by hooked up through alley-like streets

**Fig. 14.17** Inside yards of housing blocks shall be used for greenery and be walkthrough

**Fig. 14.18** Scenario 1

**Fig. 14.19** Scenario 2

Similarly to the current study scenarios 5 and 6 propose a block structure com-bining it with another arrangement of the central park: (Figs. 14.23 and 14.24)

In collaboration with the Meteorological Institute of the University of Freiburg and ATEM Prague following microclimate models were carried out with use of the ENVI-met software. The models simulate life conditions 1.5 m above the ground level during the day with the strongest insolation on June 20th at 3.00 PM.

We are conscious that the following conclusions are badly one-sided. For acquir-ing more realistic climatic conditions it would be necessary to perform a higher number of simulations.

Street canyons shaded by buildings have cooling effect (depending on the aspect ratio): (Figs. 14.25 and 14.26)

Wide streets not surrounded by buildings and not shaded by trees have a desic-cating effect:

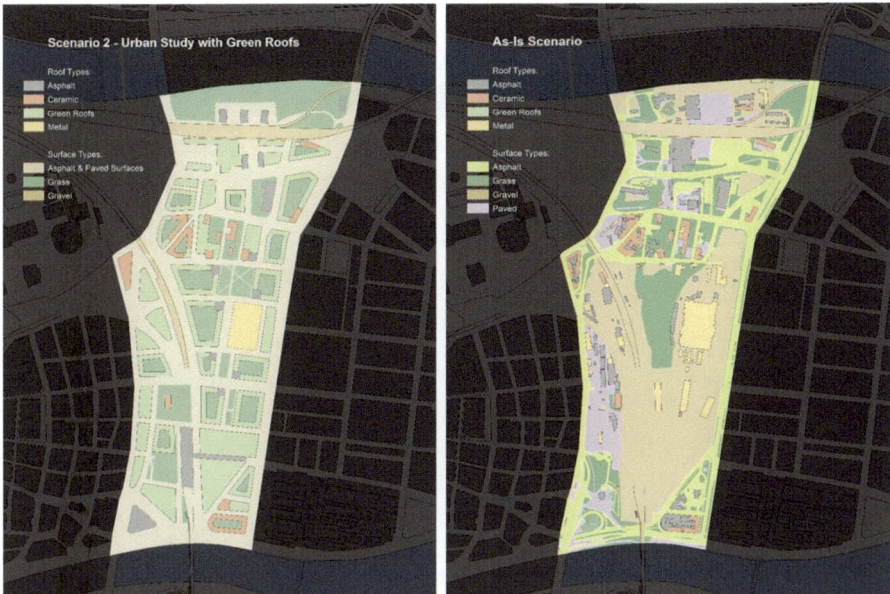

**Fig. 14.20–14.21**  (Scenario 3 and Scenario 4)

Busy streets have a warming effect due to the heat output from motor transport
(see the bottom Fig.). Still water basins have no cooling effect: (Figs. 14.27, 14.28,
14.29, 14.30 and 14.31)

As mentioned above the block structure offers better day time conditions than the
loose urban structure. However further research should explore the cooling effect of
the parks during the night time.

## 14.3.3   Green Belt

Mária Kazmukova and Radek Jareš
Prague Institute of Planning and Development, Prague, Czech Republic

Tomáš Halenka
Faculty of Mathematics and Physics, Department of Meteorology
and Environment Protection, Charles University Prague (CUNI),
Prague, Czech Republic

Jaroslav Ressler
Institute of Computer Science, The Czech Academy of Sciences,
Prague, Czech Republic

**Fig. 14.22** Better ventilation scenario

The assessment was focused on the issue of modelling of meteorological fields and air quality in conditions of conurbation with regard to presence of urban heat island phenomenon. Within the project framework, modelling tools for air quality evaluation were tested while meteorological parameters and chemistry of the atmosphere were taken into account. Based on the acquired findings from the base state, the horizon of fulfilment of land use plan in its present form and the variants of urban and traffic concept were assessed.

The project assessed the following scenarios: baseline state, fulfilment of the land use plan, low-emission zone and implementation of a green belt. In addition, sensitivity to the expected climate change was studied.

In terms of UHI, the most important was evaluation of green belt scenario, i.e. state when the transport concept and vehicle fleet composition corresponds to the year 2020 and fulfilment of the land use plan is presumed with the exception of areas defined as green belt whose land use is assumed to be changed into forest area or forest park.

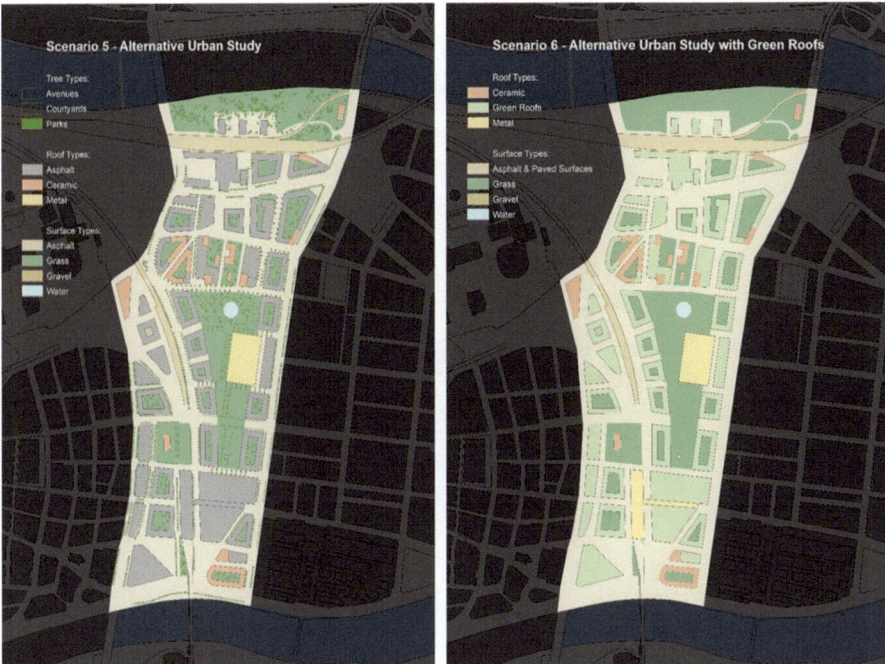

**Fig. 14.23–14.24**  Scenario 5 and Scenario 6

The method used for modelling of transport of chemical substances required to include a large territory in which boundary conditions were modelled, influencing meteorological quantities and concentrations of pollutants within the area of interest. The entire modelled area covered Europe, i.e. area measuring 4 $644 \times 3$ $294$ km with its centre being located in Prague. Assessment with the finest resolution was carried out for Prague and its surroundings where grids of 1 km and 333 m were used.

Input data for the project were prepared in such detail that has not yet been realized. For this purpose, data from regularly updated study Evaluation of Air Quality in the Territory of the Capital City of Prague Based on Mathematical Modelling as well as data about the area of interest provided by the IPR institute and available databases from other sources were utilized.

The project involved both modelling of meteorological fields using the WRF model and modelling of air pollutant dispersion using the CMAQ model. The meteorological model was con Fig.d with an urban surface impact model; the emission flux model contained an anthropogenic emission model, a biogenic emission model, a chemical transport model and modules of data post processing and statistical processing of the outputs. For long term experiments, urbanized RegCM was used with 10 km resolution, allowing SUBBATS (Pal, J. S., F. Giorgi, X. Bi, N. Elguindi, F. Solomon, X. Gao, R. Francisco, A. Zakey, J. Winter, M. Ashfaq, F. Syed, J. L.

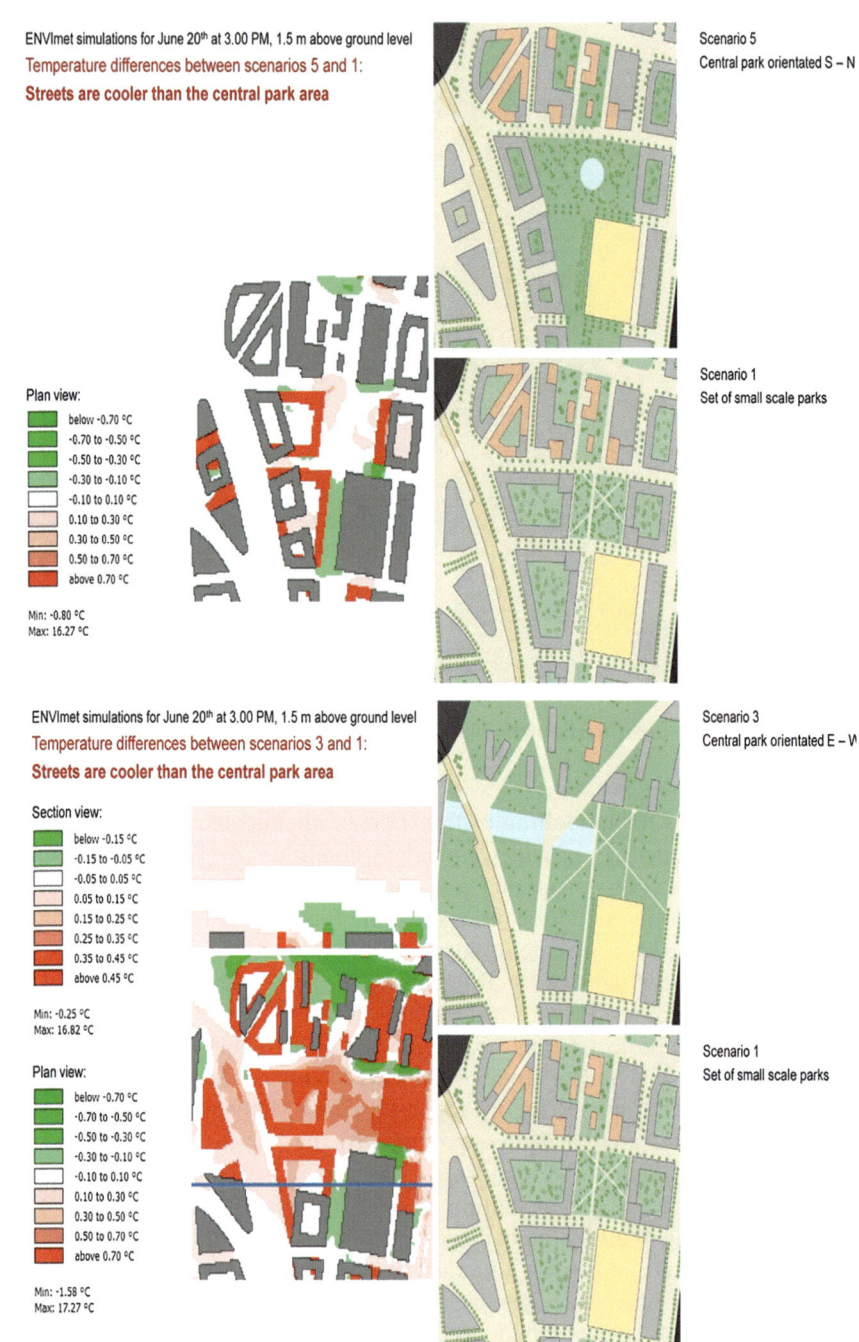

ENVImet simulations for June 20th at 3.00 PM, 1.5 m above ground level

Temperature differences between scenarios 5 and 1:

**Streets are cooler than the central park area**

Scenario 5
Central park orientated S – N

Plan view:

| | |
|---|---|
| | below -0.70 °C |
| | -0.70 to -0.50 °C |
| | -0.50 to -0.30 °C |
| | -0.30 to -0.10 °C |
| | -0.10 to 0.10 °C |
| | 0.10 to 0.30 °C |
| | 0.30 to 0.50 °C |
| | 0.50 to 0.70 °C |
| | above 0.70 °C |

Min: -0.80 °C
Max: 16.27 °C

Scenario 1
Set of small scale parks

ENVImet simulations for June 20th at 3.00 PM, 1.5 m above ground level

Temperature differences between scenarios 3 and 1:

**Streets are cooler than the central park area**

Section view:

| | |
|---|---|
| | below -0.15 °C |
| | -0.15 to -0.05 °C |
| | -0.05 to 0.05 °C |
| | 0.05 to 0.15 °C |
| | 0.15 to 0.25 °C |
| | 0.25 to 0.35 °C |
| | 0.35 to 0.45 °C |
| | above 0.45 °C |

Min: -0.25 °C
Max: 16.82 °C

Plan view:

| | |
|---|---|
| | below -0.70 °C |
| | -0.70 to -0.50 °C |
| | -0.50 to -0.30 °C |
| | -0.30 to -0.10 °C |
| | -0.10 to 0.10 °C |
| | 0.10 to 0.30 °C |
| | 0.30 to 0.50 °C |
| | 0.50 to 0.70 °C |
| | above 0.70 °C |

Min: -1.58 °C
Max: 17.27 °C

Scenario 3
Central park orientated E – W

Scenario 1
Set of small scale parks

**Fig. 14.25–14.26** Scenari

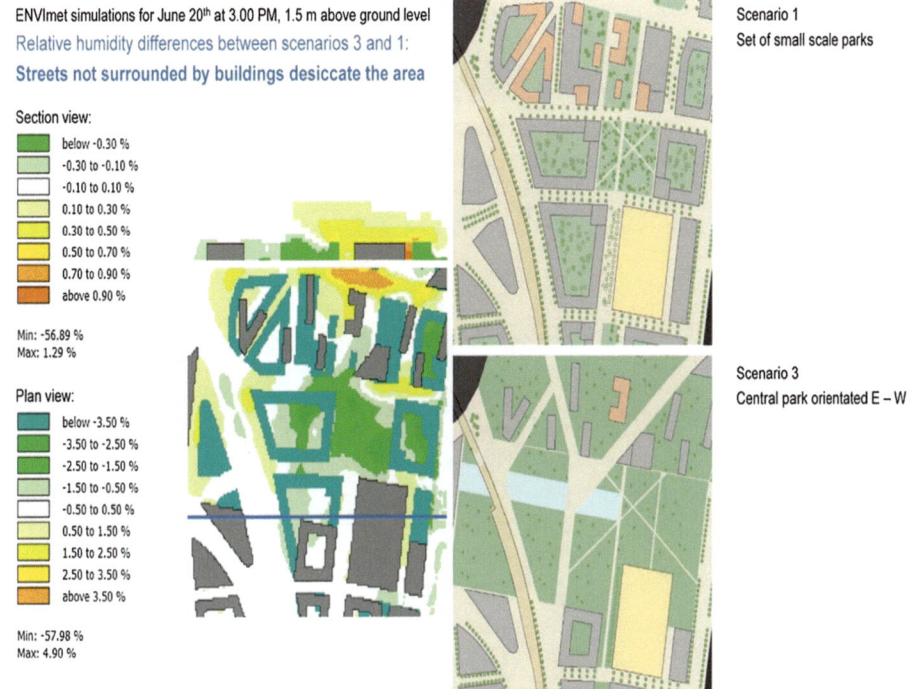

ENVImet simulations for June 20th at 3.00 PM, 1.5 m above ground level
Relative humidity differences between scenarios 3 and 1:
Streets not surrounded by buildings desiccate the area

Section view:

| | below -0.30 % |
| | -0.30 to -0.10 % |
| | -0.10 to 0.10 % |
| | 0.10 to 0.30 % |
| | 0.30 to 0.50 % |
| | 0.50 to 0.70 % |
| | 0.70 to 0.90 % |
| | above 0.90 % |

Min: -56.89 %
Max: 1.29 %

Plan view:

| | below -3.50 % |
| | -3.50 to -2.50 % |
| | -2.50 to -1.50 % |
| | -1.50 to -0.50 % |
| | -0.50 to 0.50 % |
| | 0.50 to 1.50 % |
| | 1.50 to 2.50 % |
| | 2.50 to 3.50 % |
| | above 3.50 % |

Min: -57.98 %
Max: 4.90 %

Scenario 1
Set of small scale parks

Scenario 3
Central park orientated E – W

**Fig. 14.27**  Scenari 1/3

Bell, N. S. Diffenbaugh, J. Karmacharya, A. Konare, D. Martinez, R. P. da Rocha, L. C. Sloan and A. Steiner, 2007: The ICTP RegCM3 and RegCNET: Regional Climate Modeling for the Developing World, B. Am. Meterol. Soc., 88, 1395–1409) in 2 km.

Both meteorological parameters (temperature, humidity, wind speed) and concentration of pollutants were modelled.

Outcomes of the assessment allow evaluating not only long-term indicators (annual average) but also characteristics that have not yet been assessed sufficiently accurately, such as exceedance period of an air pollution limit and nth highest values of short-term average values in accordance with legislation. Also ozone concentrations can be evaluated. The results are available both for current state, for future scenario of land use plan and other scenarios being assessed.

The evaluation points out the following:

By 2020 or by the horizon of land use plan fulfilment, respectively, reduction of concentrations of pollutants in Prague can be expected with the exception of vicinity of large transport structures where the impact of newly introduced car traffic outweighs.

Humidity differences between Scenario 1
and As-Is Scenario: **Increase in the central
part due to the unshaded park area?**

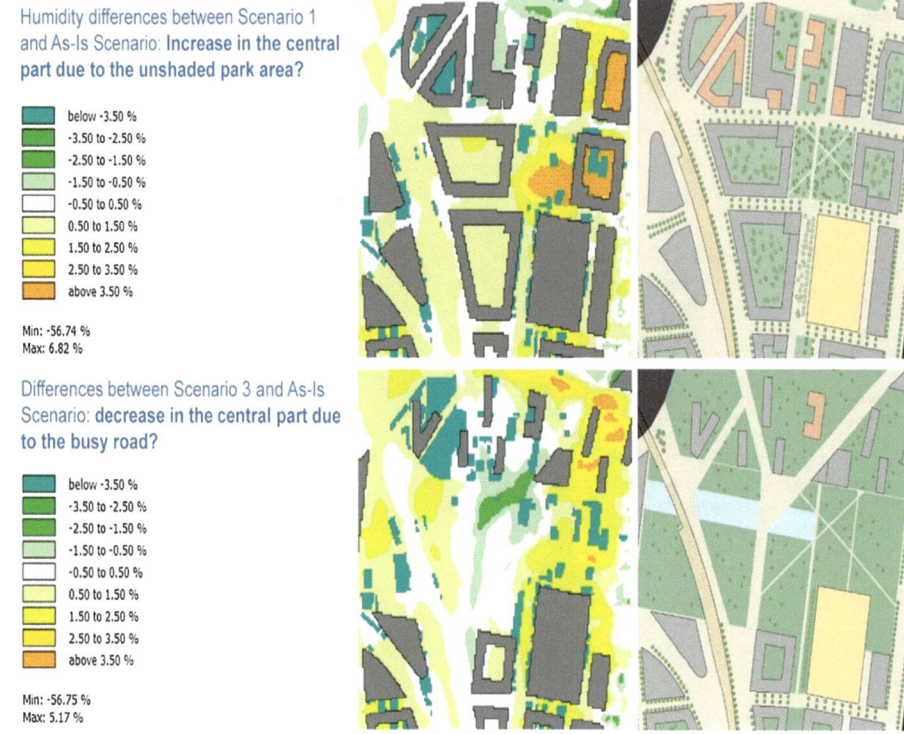

Min: -56.74 %
Max: 6.82 %

Differences between Scenario 3 and As-Is
Scenario: **decrease in the central part due
to the busy road?**

Min: -56.75 %
Max: 5.17 %

**Fig. 14.28–14.29** Humidity differencies

Improvement in vehicle fleet composition has a positive effect on $NO_2$ and suspended particulate matter concentrations and also on benzene (but less). In the case of ozone, the peak concentrations (which are limited in terms of health) decrease and average annual concentrations increase (higher concentrations at night time).

The evaluation showed that secondary aerosols have a relatively high contribution to air pollution load by suspended particulate matter. This issue requires further specification because higher concentrations of $PM_{10}$ are one of the major problems of air quality protection in the capital city of Prague.

The influence of the green belt is rather small; it affects only few sites by small decrease in temperature and consequent change in concentrations of pollutants

Temperature differences between Scenario
1 and As-Is Scenario: **Decrease in the part
due to the unshaded park area?**

- below -0.70 °C
- -0.70 to -0.50 °C
- -0.50 to -0.30 °C
- -0.30 to -0.10 °C
- -0.10 to 0.10 °C
- 0.10 to 0.30 °C
- 0.30 to 0.50 °C
- 0.50 to 0.70 °C
- above 0.70 °C

Min: -2.05 °C
Max: 16.60 °C

Differences between Scenario 3 and As-Is
Scenario: **increase in the central part due
to the busy road?**

- below -0.70 °C
- -0.70 to -0.50 °C
- -0.50 to -0.30 °C
- -0.30 to -0.10 °C
- -0.10 to 0.10 °C
- 0.10 to 0.30 °C
- 0.30 to 0.50 °C
- 0.50 to 0.70 °C
- above 0.70 °C

Min: -1.63 °C
Max: 16.81 °C

**Fig. 14.30**   The cooling effect of green roofs is negligible

caused by a change in biogenic emissions production and by a change in chemical
reactions in the atmosphere. Certain changes can be seen in long term simulation,
with small temperature decrease, especially in summer night. Remote effects are
rather climatic noise. Changes in the future go with the overall temperature change,
but they are again rather small.

The presented project practically verified the potential use of chemical transport
models for air quality and UHI assessment in a small scale. The fine resolution that
reaches up to 333 m for the innermost domain allows assessing air quality and UHI
effect in cities in detail.

Proposed green belt around Prague (Figs. 14.32, 14.33, and 14.34).

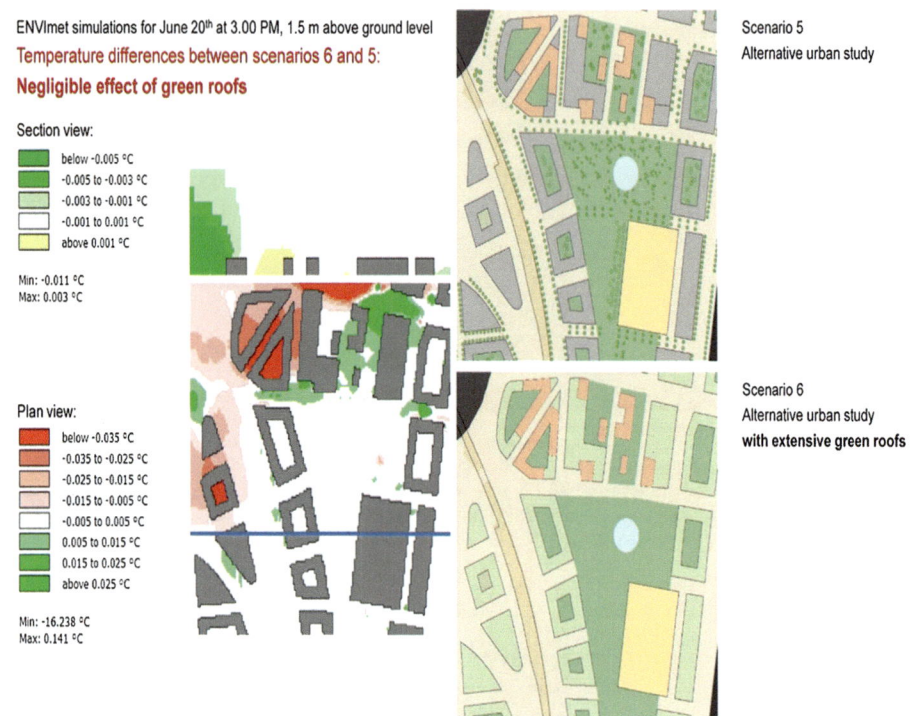

**Fig. 14.31** The block structure offers better day time conditions

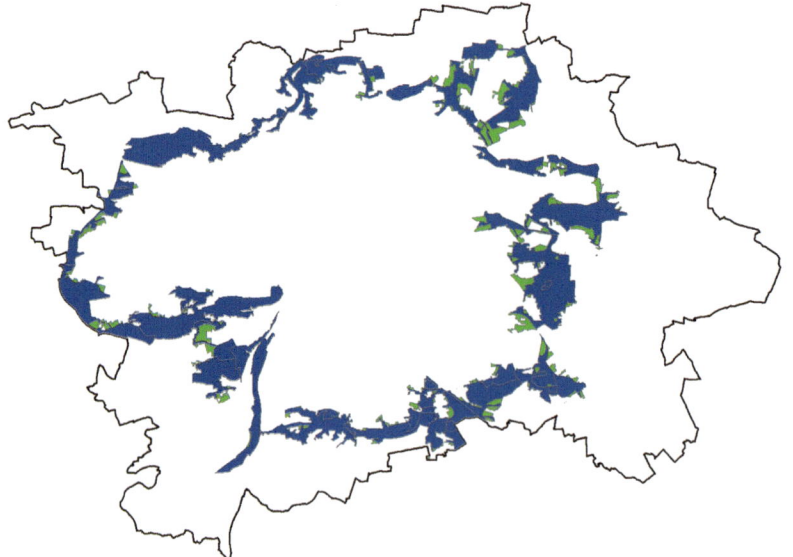

**Fig. 14.32** Temperature shift caused by green belt in a hot July day

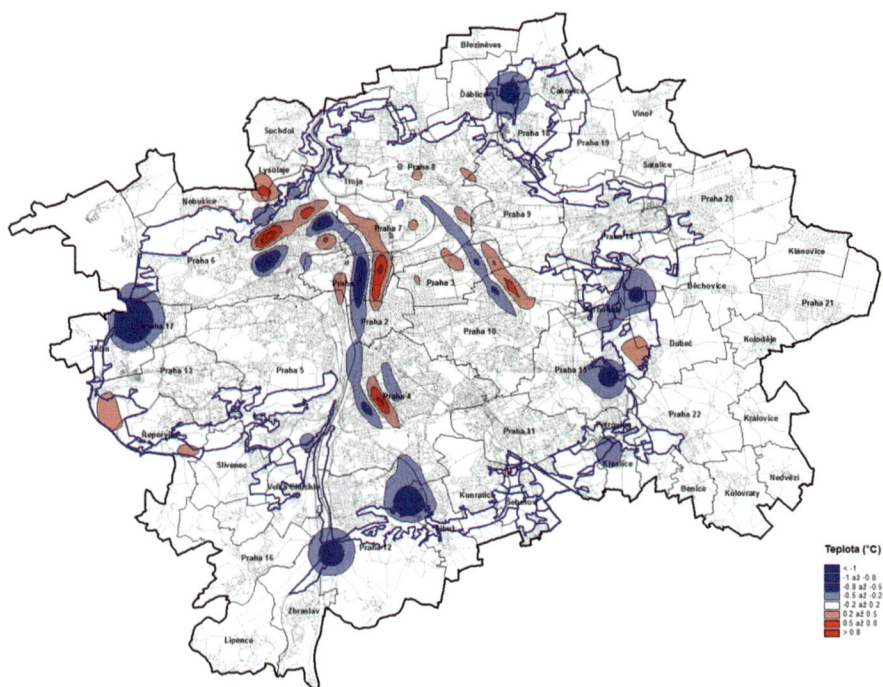

**Fig. 14.33–14.34** Temperature changes in long term simulation of 2001–2010 (JJA, night) caused by green belt

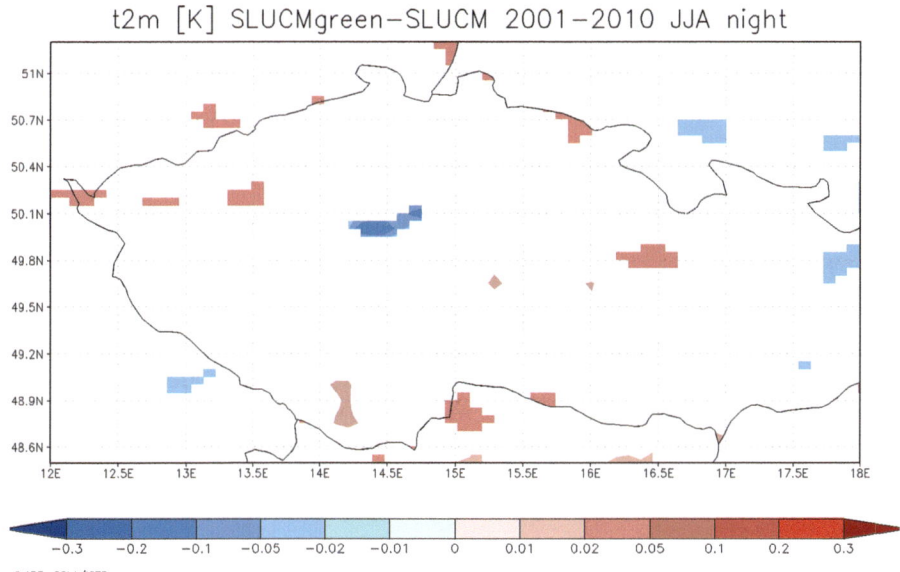

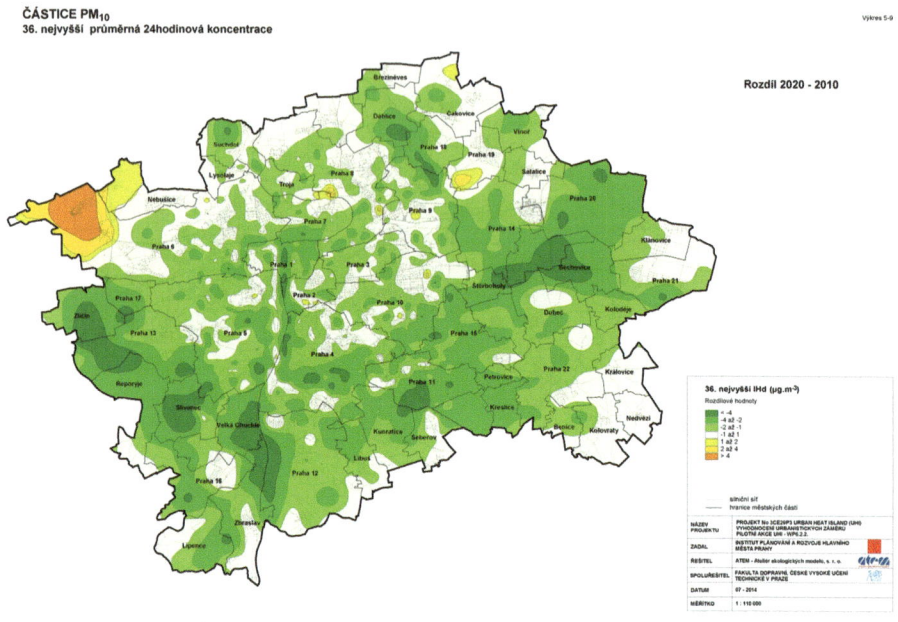

**Fig. 14.35** Differences in 24-h-averaged PM$_{10}$ concentrations

Differences in 24-h-averaged PM$_{10}$ concentrations between years 2010 and 2020 (Fig. 14.35).

## 14.4 General Strategic Vision to Mitigate UHI Effects-Counteracting Measures

### 14.4.1 Street Canyons

As a result of the simulations, the scenario with the small trees planted densely along the sidewalks seems to be the optimal solution for UHI mitigation for Legerova Street and other similar street canyons with a heavy traffic volume. This scenario does not have such a negative effect on ventilation conditions and provides shade and a positive effect on PET.

## *14.4.2   Development Areas*

The aim of the research was to examine the benefits of the different scenarios compared with alternative urban studies proposing different urban structures and larger park areas.

The simulations have shown that the block structure offers better day time conditions than the loose urban structure. However further research should explore the cooling effect of the parks during the night time.

## *14.4.3   Green Belt*

The assessment of the effect of a proposed green belt as a scenario for UHI mitigation in Prague was focused on the issue of modelling of meteorological fields and air quality in conditions of conurbation with regard to presence of urban heat island phenomenon.

Within the project framework, modelling tools for air quality evaluation were tested while meteorological parameters and chemistry of the atmosphere were taken into account.

In terms of UHI, the most important was evaluation of green belt scenario, i.e. state when the transport concept and vehicle fleet composition corresponds to the year 2020 and areas defined as green belt are assumed to be changed into forest area or forest park.

The method used for modelling of transport of chemical substances required to include a large territory in which boundary conditions were modelled, influencing meteorological quantities and concentrations of pollutants within the area of interest

The project assessed the following scenarios: baseline state, fulfilment of the land use plan, low-emission zone and implementation of a green belt.

The presented project practically verified the potential use of chemical transport models for air quality and UHI assessment in a small scale. The fine resolution that reaches up to 333 m for the innermost domain allows assessing air quality and UHI effect in cities in detail.

As a result of the recent modelling, the influence of the green belt showed to be rather small; it affects only few sites by small decrease in temperature and consequent change in concentrations of pollutants caused by a change in biogenic emissions production and by a change in chemical reactions in the atmosphere. Further simulations with traffic modifications are needed. It should be pointed out, that the effect of climate change for the year 2020 is negligible compared to the effects by changes in land-use and transport concepts with vehicle fleet changes.

Adaptation strategies to contrast bioclimatic emergencies were included in a proposition for the Prevention plan in cooperation with the State Institute of Health. The proposition of the Prevention Plan includes the instructions for people, especially for sensitive groups how to react and what measures to take in extreme hot periods in cities.

The proposed HEAT Warning System will help to coordinate adaptative strategies and the reaction of City Authoritities to the extreme weather phenomena as to protect citizens against the harmful effects of heat and the UHI.

# Reference

Matzarakis, A., Rutz, F., & Mayer, H. et al. (2010). Modelling radiation fluxes in simple and complex environments – Basics of the RayMan model. *International Journal of Biometeorology, 54*, 131–139.

**Erratum to**

## Chapter 2
# Urban Heat Island Gold Standard and Urban Heat Island Atlas

Gold Standard for UHI Measurements and Introduction of The Central-European Urban Heat Island Atlas

Györgyi Baranka, L. Bozó, Rok Ciglič, and Blaž Komac

© The Author(s) 2016
F. Musco (ed.), *Counteracting Urban Heat Island Effects in a Global Climate Change Scenario*, DOI 10.1007/978-3-319-10425-6

---

**DOI 10.1007/978-3-319-10425-6_15**

Chapter 2 in the original version of this book was inadvertently published without its original publication being cited throughout. The original publication and Publisher, Stewart and Oke (2012), has been cited at various instances as well as added in the *References* list in the correct version of this chapter, as given below:

Typically, a "three-step process" is suggested by Stewart and Oke (2012) "to users when classifying field sites into LCZs" (p. 1889):

"Step 1: Collect site metadata. Users must collect appropriate site metadata to quantify the surface properties of the source area (as defined in Step 2) for a temperature sensor. This is best done by a visit to the field sites in person to survey and assess the local horizon, building geometry, land cover, surface wetness, surface relief, traffic flow, and population density [...]. If a field visit is not possible, secondary sources of site metadata include aerial photographs, land cover/land use maps, satellite images (e.g., Google Earth©), and published tables of property values (e.g., Davenport terrain roughness lengths)" (Stewart and Oke, 2012, pp. 1889–1890).

---

The updated original online version for this chapter can be found at DOI
http://dx.doi.org/10.1007/978-3-319-10425-6_2

---

© The Author(s) 2017                                                                                          E1
F. Musco (ed.), *Counteracting Urban Heat Island Effects in a Global Climate Change Scenario*, DOI 10.1007/978-3-319-10425-6_15

"Step 2: Define the thermal source area. The thermal source area for a temperature measurement is the total surface area «seen» by the sensor […]." (Stewart and Oke, 2012, p. 1890) "Sources will include upwind buildings, the walls and floor of an upwind street, and perhaps a branching network of more distant street canyons" (p. 1890).

"Quantifying the surface properties for field sites and source areas located on or near the border of two (or more) zones is problematic. If the location of the sensor can be moved, it should be placed where it samples from a single LCZ. […]. If the location of the sensor cannot be moved, temperature data retrieved from that site should be stratified first according to wind direction, then to LCZ. […] A site with a split classification is less ideal for heat island studies because changes in airflow and stability conditions interfere confuse the relation between surface form/cover and air temperature. It is recommended that transitional areas be avoided when siting meteorological instruments" (Stewart and Oke, 2012, p. 1891).

"Step 3: Select the local climate zone. Metadata collected in Step 1 should lead users to the best, not necessarily exact, match of their field sites with LCZ classes. Metadata are unlikely to match perfectly with the surface property values of one LCZ class. If the measured or estimated values align poorly with those in the LCZ datasheets, the process of selecting a best-fit class becomes one of interpolation rather than straight matching. Users should first look to the surface cover fractions of the site to guide this process. If a suitable match still cannot be found, users should acknowledge this fact and highlight the main difference(s) between their site and its nearest equivalent LCZ" (Stewart and Oke, 2012, p. 1891).

Stewart and Oke (2012) note that "updating LCZ designations is crucial for all sites, particularly those used in long-term temperature studies" (p. 1893). They add that "sites located on the edges of cities where urban growth and environmental change are rapid, or in the cores of cities where land redevelopment and large-scale greening projects are taking place, should be surveyed and classified annually. For sites used in mobile or short-term stationary surveys, the frequency of updates is dictated largely by day-to-day variations in weather and soil moisture […]" (p. 1893).

# Reference

Stewart, I. D., & Oke, T. (2012). "Local Climate Zones" for urban temperature studies. *Bulletin of the American Meteorological Society, 93*, 1879–1900.

GPSR Compliance

*The European Union's (EU) General Product Safety Regulation (GPSR) is a set of rules that requires consumer products to be safe and our obligations to ensure this.*

*If you have any concerns about our products, you can contact us on ProductSafety@springernature.com*

In case Publisher is established outside the EU, the EU authorized representative is:

Springer Nature Customer Service Center GmbH
Europaplatz 3
69115 Heidelberg, Germany

**Batch number: 09478213**

Printed by Printforce, the Netherlands